Cladistic Analysis of North American Platynini and Revision of the *Agonum extensicolle* Species Group (Coleoptera: Carabidae)

Cladistic Analysis of North American Platynini and Revision of the *Agonum extensicolle* Species Group (Coleoptera: Carabidae)

by James Kenneth Liebherr
(Current address: Department of Entomology,
Cornell University, Ithaca, New York 14853)

University of California Press
Berkeley • Los Angeles • London

UNIVERSITY OF CALIFORNIA PUBLICATIONS IN ENTOMOLOGY

Volume 106
Issue Date: May 1986

UNIVERSITY OF CALIFORNIA PRESS
BERKELEY AND LOS ANGELES, CALIFORNIA

UNIVERSITY OF CALIFORNIA PRESS, LTD.
LONDON, ENGLAND

ISBN 0-520-09958-3
LIBRARY OF CONGRESS CATALOG CARD NUMBER: 85-29034

PRINTED IN THE UNITED STATES OF AMERICA

Library of Congress Cataloging in Publication Data

Liebherr, James Kenneth.
Cladistic analysis of North American Platynini and revision of the Agonum extensicolle species group (Coleoptera, Carabidae).

(University of California publications in entomology ; v. 106)
Bibliography; p.
1. Agonum—Classification. 2. Agonum extensicolle—Classification. 3. Insects—Classification. 4. Insects—North America—Classification. I. Title. II. Title: Platynini. III. Series.
QL596.C2L45 1987 595.76′2 85-29034
ISBN 0-520-09958-3 (alk. paper)

CONTENTS

ACKNOWLEDGMENTS AND ABBREVIATIONS

I thank the following curators and institutions for loan of material: G. E. Ball and D. Shpeley, Univ. of Alberta (UASM); L. H. Herman, American Museum of Natural History (AMNH); F. G. Werner, Univ. of Arizona (UAZ); R. T. Allen, Univ. of Arkansas (UARK); J. A. van Reenen, Univ. of British Columbia (UBC); N. E. Stork, British Museum of Natural History (BMNH); D. H. Kavanaugh, California Academy of Sciences (CAS); F. G. Andrews, California Dept. of Food and Agriculture (CDFA); J. A. Chemsak, California Insect Survey, University of California (CISC); R. O. Schuster, University of California, Davis (UCD); S. I. Frommer, University of California, Riverside (UCR); A. Smetana, Canadian National Collection of Insects, Arachnids and Nematodes (CNC); L. L. Pechuman, Cornell University (CUIC); E. H. Smith, Field Museum of Natural History (FMNH); R.E. Woodruff and P.M. Choate, Florida State Collection of Arthropods (FDPI); A. F. Newton, Jr., Museum of Comparative Zoology, Harvard University (MCZ); W. F. Barr, University of Idaho (UIDA); C. L. Hogue, Natural History Museum of Los Angeles County (LACM); T. E. Moore and B. O'Connor, Museum of Zoology, University of Michigan (MICH); R. L. Fischer, Michigan State University (MSUC); G. R. Noonan, Milwaukee Public Museum (MPM); R. L. Brown, Mississippi State University (MSSU); N. B. Nikitsky, Moscow University Zoological Museum (ZMM); C. A. Triplehorn, Ohio State University (OSUC); R. L. Westcott, Oregon Department of Agriculture (ODA); H. Perrin and A. Villiers, Museum National D'Histoire Naturelle, Paris (MNHP); K. C. Kim, Pennsylvania State University (PSUC); H. R. Burke and S. J. Merritt, Texas A. & M. University (TAMU); T. L. Erwin, U.S. National Museum of Natural History (NMNH); R. T. Bell, University of Vermont (RTBe); W. J. Turner, Washington State University (WSUC).

Fellow coleopterists allowed me to borrow specimens and/or extract data from their collections. The use of the following personal collections is gratefully acknowledged: P. M. Choate (PMCh); R. L. Davidson (RLDa); D. H. Kavanaugh (DHKa); L. L. Lampert (LLLa); S. McCleve (SMcC); R. D. Ward (RDWa); D. K. Young (DKYo).

This project could not have been accomplished without extramural funding. I acknowledge the following granting institutions: Theodore Roosevelt Memorial Fund of the American Museum of Natural History; and the Snodgrass Memorial Fund of the Entomological Society of America, administered through the Society of Sigma Xi. Travel funds for museum-based research were provided in part by the CanaColl Foundation of the Canadian National

Collection, and the Department of Entomology, Museum of Comparative Zoology, Harvard University. Publication costs were paid in part by a grant from the Grace Griswold Fund, Department of Entomology, Cornell University.

I thank P. G. Sanchez, Death Valley National Monument for permission to collect at Saratoga Springs; D. Giuliani, Big Pine, and E. P. Pister, California Department of Fish and Game, Bishop for advice on aquatic habitats in eastern California. Yves Bousquet, Canadian National Collection, provided inspired discussion of pterostichine affinities. R. B. Madge and N. E. Stork, British Museum of Natural History, graciously provided me a photocopy of Bonelli's "Tabula Synoptica." H. V. Daly allowed me to use his digital plotter, facilitating portions of the biometric studies.

My dissertation committee has provided me with guidance and inspiration as varied as their backgrounds. I thank John Doyen for his efforts on my behalf throughout the degree program. He and Jerry Powell provided my introduction to the California insect fauna. Tom Duncan helped sort out the vagaries of systematic philosophy and methodology, and Jim Patton helped me understand the limits of genetic data.

The electrophoretic analysis would not have been possible without the cooperation and support of the staff of the Museum of Vertebrate Zoology. I thank M. M. Frelow, R. D. Sage, and D. B. Wake for support, training, and advice during the electrophoretic phase of this research.

Live specimens of *Agonum* beetles were obtained from E. R. Hoebeke, Cornell University, and J. A. and R. T. Bell, R. L. Davidson, and B. Farrell of the University of Vermont.

Many colleagues aided my research through discussions at any and odd times during the past several years. I would like to thank John Chemsak, Charles Griswold, Dave Kavanaugh, Dave Wagner, Jim Whitfield, and Bob Zink for their criticism, ideas, and inspiration. I thank Jamie Throgmorton of Community Systems Associates for expert composition of the manuscript, and Carolyn Mullinex-Tibbets for the habitus drawing of *Agonum extimum*. Cindie Frost of Cornell University Graphic Purchasing Services oversaw the preparation of camera-ready copy. Ann E. Hajek assisted me in proofreading of the manuscript.

This material is based upon research supported in part by the National Science Foundation under Grant No. DEB-7924298 and is published with support of National Science Foundation Grant No. BSR-8501435.

The Foundation provides awards for research and education in the sciences. The awardee is wholly responsible for the conduct of such research and preparation of the results for the publication. The Foundation, therefore, does not assume responsibility for such findings or their interpretation.

Any opinions, findings, conclusion, or recommendations expressed in this publication are those of the author and do not necessarily reflect the views of the National Science Foundation.

ABSTRACT

Cladistic analysis based on internal male and female reproductive characters and external characters is used to group exemplar taxa in the carabid tribe Platynini. A classification, key to genera in North America, and a key to species groups of *Agonum* in North America north of Mexico are presented.

The *Agonum extensicolle* species group comprises seven species: *A. cyanope* (Bates); *A. extimum* Liebherr, n. sp.; *A. parextimum* Liebherr n. sp.; *A. texanum* (LeConte); *A. extensicolle* (Say); *A. decorum* (Say); *A. elongatulum* (Dejean). Analyses of infraspecific geographic variation show: 1) *A. texanum* is biometrically uniform over the center of its range whereas individuals from outlying populations deviate in several measurements; 2) *A. extensicolle* is a variable species, with clinal changes in biometry and color occurring across its range; 3) *A. decorum* is polymorphic for color and setation, and clinally variable in biometric characters. Across the group, flight apparatus development is inversely correlated with the amount of genetic heterogeneity measured by starch-gel electrophoresis.

Electrophoretic, qualitative morphological, and biometric data are used to estimate phylogenetic relationships in the *A. extensicolle* group. The electrophoretic and morphological data produce compatible estimates of phylogeny. The biometric data are incompatible with the other data and are judged less useful for estimation of affinities. Distributional data are utilized in conjunction with the proposed phylogeny to investigate speciation events in the group. The principal mechanism is allopatric speciation brought about by vicariance across the lowlands of southeastern Arizona; the Cochise filter barrier. A second pattern involves a peripheral isolate of Antillean stock diverging on the Florida peninsula. A third speciation event involves a habitat shift in which a lowland desert form produced a species which now inhabits the pine-oak zone in the Sierra Madre Occidental. The area-taxon relationships are compared with those in other groups. Based on an electrophoretic clock calibrated using data from *Drosophila*, the timing of the initial speciation event in the group is estimated at 6-12 million years b.p. Other speciation events occurred throughout the Pliocene and Pleistocene, with the most recent divergence of *A. decorum* and *A. elongatulum* estimated at less than two million years b.p.

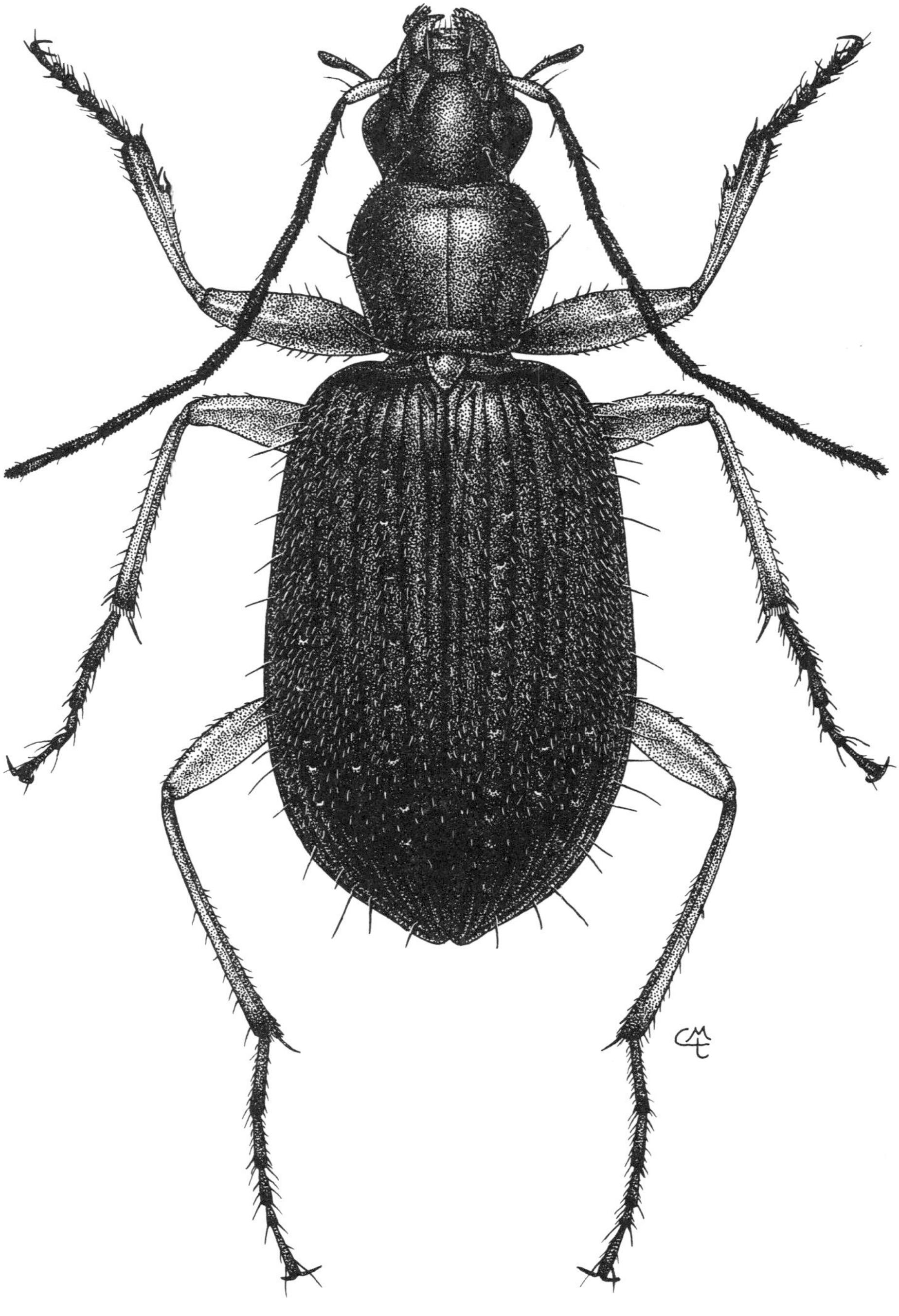

Figure 1. *Agonum extimum* Liebherr, n. sp., MEX: Coahuila, San Pedro, 31.2 km E., 900 m el., 31-V-1981, J. T. Doyen and J. K. Liebherr collectors, paratype male.

INTRODUCTION

John L. LeConte wrote of the *Agonum extensicolle* species group in 1854, saying that it is "a group of extreme complexity, containing winged species of metallic colors, bluish or greenish, with the base of the antennae, the feet, and occasionally the thorax rufous or pale piceous." LeConte recognized that the species within the *Agonum extensicolle* group are difficult to diagnose. Several of the species are widely distributed geographically, and the accompanying variability in biometric features has caused past authors to describe and name many invalid taxa. Six of the seven species are found in the American Southwest and in adjacent Mexico, one of the last areas in North America for which the carabid fauna remains poorly understood. This study circumscribes the species using qualitative morphological characters, electrophoretic data analyzed at the population level, and discriminant function analysis based on biometric measurements.

The *A. extensicolle* group is distinct within the genus, supporting recognition of it as a monophyletic lineage. Relative confidence that a taxon is monophyletic is a prerequisite for certain aspects of systematic study. This study explores the phylogenetic affinities of the species in the *A. extensicolle* group based on genetic data obtained from starch-gel electrophoresis and on internal and external morphological characters. The hypothesis of phylogenetic relationships is then analyzed in light of the present-day distributions of the species. This analysis provides insight into the speciation events and mechanisms that have brought about the present day diversity of the *A. extensicolle* group.

Although the main body of this study deals with the *Agonum extensicolle* species group, a more general analysis of the carabid tribe Platynini is first provided. The genus *Agonum* and allied taxa in the Platynini are analyzed cladistically using internal and external morphological characters. Characters of the internal reproductive tract of the females are used to group the lineages within *Agonum* and allied genera. Based on this cladistic analysis, a classification is presented that recognizes the various lineages comprising the Platynini.

CLADISTIC ANALYSIS OF THE PLATYNINI

History of Classification. Bonelli (1810) described a number of genera and proposed a higher classification for the Carabidae in his "Tabula Synoptica." In this table he supplied diagnostic combinations for the genera *Agonum, Anchomenus*, and *Platynus* among many others, placing the former two in the Anchomenii and *Platynus* in the Platynii. The "Tabula Synoptica" is not present in most copies of Bonelli's first volume of *Observationes Entomologiques*, leading Andrewes (1919) to hypothesize that the "Tabula" was not properly published, and was inserted separately after the date of publication. This interpretation has held sway until recent times, upsetting the stability of platynine generic and tribal level names. Habu (1973, 1978) considered the "Tabula" validly published, and recognized Platynini as the valid tribal name for *Platynus* and allied genera based on Audouin and Brullé's 1834 ranking of *Platynus* as senior synonym to *Agonum* and *Anchomenus*. Since *Platynus* is a large cosmopolitan genus, such action is logical.

Mroczkowski (1977) applied for approval of the "Tabula" as an accepted work in the belief that it was never properly published. Madge (1978) presented opposing evidence in support of Habu's interpretation that it was validly published: the "Tabula" was an integral part of *Observationes Entomologiques*, part 1, since the number of pages in the signature just preceding the "Tabula" has been reduced; and part 1 of *Observationes Entomologiques* was offered for sale. Madge recommended that part 1 of *Observationes Entomologiques* be placed on the Official List of Available Works for Zoological Nomenclature, and that its date of publication be fixed as 1810. By vote of the International Commission of Zoological Nomenclature, Madge's recommendations were accepted (Melville 1982).

LeConte (1854) provided a synopsis of *Platynus sensu lato* and allied genera, recognizing six genera in this lineage in North America: *Calathus, Pristodactyla* (*Synuchus*), *Anchus, Rhadine, Platynus* (including *Agonum*), and *Olisthopus*. LeConte's concept of the platynine lineage is used in this work.

Casey (1920) investigated the lineages allied to *Platynus*, proposing many new generic level names. Casey's work suffers from a plethora of invalid species level names. However, the lineages he recognized as genera, subgenera and sections are somewhat natural. He did not know the European literature, as evidenced by his improper choice of type species for the genera he recognized. His work leaves little lasting mark aside from supplying subgeneric names that might be used in a formal classification.

Lindroth (1956) revised *Synuchus* Gyllenhal and proposed a classification for it and related groups. On the basis of the male aedeagus and parameres, he recognized four groups in the

Platynini (= Agonini). One group, the Pristosiae, has the aedeagus lying on its left side in repose, and being everted toward the right side, inverse from the condition in other groups. This difference is dramatic. However, the inverse condition also occurs in *Calathus ovipennis* Putzeys, whereas most other *Calathus* have the aedeagus lying on the right side in repose. The Sphodri constitute the second subtribe, diagnosed by a long styloid right paramere (fig. 7a) that is much longer than the other paramere. In the Synuchi the right paramere lacks the long styloid apex and the shorter rounded apex is at an angle to the narrowed basal part (fig. 7b). The fourth group, the Agoni, possesses subequal parameres (figs. 7c-h; 8).

Habu (1978) revised the Platynini of Japan, using characters found on the female gonostyli to delineate three groups. His subtribe Platynina consists of species that have a fringe of setae along the apex of the penultimate segment of the gonostylus, external lateral setae, a dorsal seta, and preapical sensory setae set in a depression on the apical segment of the gonostylus (figs. 5g,h; 6). This group corresponds to Lindroth's (1956) Agoni. The second group in Habu's classification is the subtribe Sphodrina. Sphodrina lack the fringe of setae on the penultimate gonostylus segment, have the dorsal seta on the apical segment moved to the inner lateral margin (fig. 5e), and retain the preapical sensory setae on the apical segment of the gonostyli. Habu's Sphodrina corresponds to the union of Lindroth's Sphodri and Pristosiae. Habu's third subtribal group is the Dolichina. Like the Sphodrina, members of the Dolichina lack the apical fringe on the penultimate segment of the gonostyli and possess an internal lateral seta instead of a dorsal seta on the apical segment of the gonostyli. In addition, preapical sensory setae set in a depression on the apical segment of the gonostyli are lacking (fig. 5f). Habu's Dolichina corresponds to Lindroth's Synuchi.

Habu's classification is important for two reasons. The female gonostyli offer a new suite of characters for assessing phylogenetic affinities. In addition, Habu presents his classification in comparison to Lindroth's, permitting assessment of phylogenetic relationships in light of the suites of characters found in the male and female genitalia.

Rank of the Platynini. This study recognizes the platynine lineage as a tribe in Carabidae. This level of recognition has persisted in classifications from LeConte and Horn (1883) to Habu (1978). Alternatively, the platynine lineage has been recognized as a subtribe of Pterostichini (LeConte 1861, Erwin 1979b, among others). With the increased use of Hennigian classifications, the latter placement of platynine taxa has become commonplace (Kryzhanovsky 1976; Reichardt 1977). This interpretation assumes the platynine lineage to be a group of generalized carabids that gave rise to a number of lineages, but that possess no derived feature among themselves by which they can be distinguished from Pterostichini.

Carabids possess a well-developed pair of defensive glands that open onto the eighth abdominal tergum. Forsyth (1972) surveyed taxa from throughout Carabidae, presenting data on sectioned as well as whole glands. Nebriini, Notiophilini, Elaphrini, Loricerini, Trachypachini, Omophronini, Metriini, and Rhysodini have an ovoid storage reservoir, with the main collecting canal of the secretory gland entering near the junction of the reservoir and efferent duct. This type of gland assembly is also found in Scaritini, Cicindelinae and aquatic Caraboidea, and is considered the primitive condition. Platynini, Harpalini, Anthiini, Dryptini, and Licinini differ in possessing a secondary dorsal lobe attached to the reservoir. The main collecting canal from the secretory lobe empties into the vicinity of the main lobe and dorsal lobe junction. The shared derived character of the dorsal lobe unites these tribes. Pterostichini and Amarini differ in possession of a well-muscled, heart-shaped reservoir lacking a dorsal lobe.

Kanehisa and Murase (1977) investigated the defensive gland components and gland morphology. They found that methacrylic acid is the most common defensive secretion component throughout Carabinae, Bembidiini, Patrobini, Amarini, and Pterostichini (*sensu stricto*). Principal use of methacrylic acid is considered the primitive state of the defensive secretion. Harpalini, Licinini, Platynini, Lebiini, Dryptini and Zuphiini use formic acid as the principal component in the defensive secretion. This shared derived state unites the Platynini with higher Harpalinae, and refutes recognition of it as a subtribe of Pterostichini. Kanehisa and Murase (1977) corroborate Forsyth's less extensive but more intensive analysis in finding that Platynini possess the same gland configuration as Harpalini, Licinini, Lebiini, Dryptini, and Zuphiini.

Based on these studies, and the growing amount of supporting chemical evidence (Schildknecht, Maschwitz, and Winkler 1968; Moore and Wallbank 1968; Moore 1979), it is appropriate to recognize the Platynini at the tribal level, recognizing its affinities with Harpalini and Lebiini. The Pterostichini represents a distinct lineage, allied to Amarini. Additional shared derived character states are necessary to aid in resolution of the tribal affinities within the pterostichoid-platynoid-lebioid nexus of the Carabidae. Platynini may prove to be a generalized group ancestral to Harpalini or Lebiini. The phylogenetic hypothesis supported by the defensive gland system must now be tested using other characters.

MATERIALS AND METHODS

Taxa Included in Analysis. Because the relatively large number of lineages in the Platynini contain too many species to include in a study of this type, exemplars of taxa allied to *Agonum* have been chosen for inclusion in this analysis. The use of exemplars brings the associated risk that the ranges of character states within taxa are inadequately represented.

The 43 species included in the cladistic analysis are the first ones listed in Appendix 1. This list includes representatives of all taxa heretofore considered genera in North American Platynini, and subgenera or species groups within the genus *Agonum* in North America. In addition, type species of several generic level taxa are included. *Anchomenus dorsalis* (Pontippidan) is the type species of *Anchomenus* Bonelli. The inclusion of this European species permits estimation of its affinities with North American species. *Paranchus albipes* (Fabricius), introduced into the Maritime Provinces of Canada from Europe, was made the type of *Paranchus* (Lindroth 1974a). *Tanystoma striata* Dejean is the type of *Tanystoma* Motschulsky, a genus comprised of species found west of the Sierra Nevada and Cascade mountain ranges. *Agonum sexpunctatum* (L.), a European species of *Agonum sensu stricto* is also included. *Pristonychus complanatus* Dejean, a cosmopolitan species originally from the Mediterranean region, is included to increase the diversity of sphodrine genera analyzed. Other platynine taxa that were studied but not included in the analysis are noted in Appendix 2.

To conduct a cladistic analysis, one must specify the ancestral and derived states for the characters that are used in the analysis. The most reliable means of determination of such character state polarities is out-group comparison. The essence of out-group comparison resides in the character state distributions of a three taxon statement: two sister taxa and the taxon most closely related. If a character state is restricted to only one sister species and is absent in the out-group, it can be considered the derived state. Out-group analysis is easily understood in principle, yet very difficult in practice. The proper choice of an out-group is critical, and any uncertainty clouds the assignment of derived and primitive states of a character. Watrous and Wheeler (1981) discuss at length the problems inherent in the method.

Taxa outside the limits of the Platynini which were studied to ascertain the more generalized character states for characters used in this analysis are listed last in Appendix 1. The character state distributions of these taxa aid in determining the primitive and derived states of characters in the Platynini.

The female internal reproductive tract proved a rich source of characters. This system has proved very useful for higher classification in Tenebrionoidea (Doyen and Tschinkel 1982), for classification within the elaterid genus *Agriotes* (Becker 1956) and the carabid genus *Nebria* (D. H. Kavanaugh, pers. comm.), and within the carabid tribe Lachnophorini (Liebherr 1983b). Iuga (1966) describes the male and female reproductive systems for representatives of the carabid genera *Carabus, Cychrus, Abax, Pterostichus*, and *Harpalus*. Freitag (1979) includes drawings of *Cicindela* spermathecae in his revision of Australian species. The spermathecae of *Bembidion* are described by Schuler (1960) and Erwin and Kavanaugh (1981). Schuler (1962) reports on the spermathecae of several groups within the Chlaeniini. Use of this literature aids evaluation of the primitive and derived states associated with this system.

Descriptions and Polarities of Characters. Based on the examination of taxa outside the Platynini, and the survey of literature, the following characters were coded for primitive and derived states. To facilitate analysis, characters with more than two states were unit coded using the method of Farris, Kluge, and Eckhardt (1970). In the analysis, transformation of such characters was constrained so that the more primitive character state changes occur basad to the more derived state changes. Character states are listed in Appendix 1.

Twenty-three unit characters were used from the female external genitalia and internal reproductive tract, seven from the male external genitalia, and 26 from the external structure of the beetles. The descriptions of the characters and assessment of their primitive and derived states follow. In all characters, 0 codes for the primitive state and 1 codes for the derived state. Figures are listed only to illustrate the character states, and are not an inclusive listing of all taxa possessing that character state.

The female genitalia and reproductive tract characters (1-23) are:

1. Spermatheca entering reproductive tract near junction of bursa copulatrix and median oviduct (state 0; figs. 2b,d; 3b; 4d); spermatheca entering tract at point on median oviduct (state 1; figs. 2a,c).

2. Spermatheca entering reproductive tract near junction of bursa copulatrix and median oviduct (state 0; figs. 2b,d; 3b; 4d); spermatheca entering tract on bursa copulatrix (state 1; fig. 4c).

3. Spermathecal duct of uniform diameter basally at attachment point to reproductive tract (state 0; figs. 2g; 3b-h; 4); spermathecal duct basally widened at some point from attachment to one-third length of spermathecal assembly (state 1; figs. 2d,e,f,h; 3a). The taxa coded as derived for this character have a basal expansion of the spermatheca. This expanded area is from two to ten times the diameter of the more apical portions of the spermatheca. The function of the apical filament is unknown, but the expanded basal portion would appear to be the sperm reservoir.

4. Spermathecal assembly with median area of same diameter as basal third (state 0; figs. 2f; 4a-h); median third of spermatheca of larger diameter than basal or apical third (state 1; figs. 3b-h). All the taxa exhibiting a spermatheca with a medial expansion have a duct that is not extremely short, and an angle in the expanded portion of the spermatheca (state 1, character 8). Whether these character states occur independently in any other taxon is unknown.

5. Spermathecal duct with apical portion of same diameter or smaller than base or median third (state 0; figs. 2e,f; 3b,d); spermathecal duct with apical reservoir (state 1; figs. 4; 5a-d). In these taxa the sperm reservoir is located at the apex of the duct. The length of the spermathecal duct is irrelevant to coding of the derived state, and the derived state is probably not homologous across the taxa that possess it. *Pristonychus complanatus* has a symmetrical bursa copulatrix without a lateral pouch (state 1, character 6) whereas *Agonum* and *Platynus* have bursae that are asymmetrical (state 0, character 6). The apical expansion of the sperm reservoir has probably occurred independently in these two groups.

6. The bursa copulatrix possesses an asymmetrical lobe in the Patrobini, Carabini (Iuga 1966), *Agonum* and *Platynus* (figs. 2a, d-h; 3a; 4a). This is assumed the ancestral state. The sphodrine genera *Synuchus, Calathus*, and *Pristonychus* possess bursae that are more elongate (fig. 2b), and lack the asymmetrical lobe. The function of the lobe is assumed to be an egg "holder" as one gravid *A. ferruginosum* was found with an egg lodged transversely across the bursa, with one end in the lobe. The egg may be held in this position for fertilization, or held just prior to being ejected. The derived state, absence of the lobe (figs. 2b, c) occurs in a number of pterostichine lineages and in *Lebia grandis*.

7. Primitively the combined length of the spermatheca and any narrowed duct is on the order of one-half to two times the length of the bursa (figs. 2b; 3b-h; 4a-d). This is the configuration seen in *Cicindela* (Freitag 1979), *Carabus* (Iuga 1966), *Bembidion* (Erwin and Kavanaugh 1981), and *Pterostichus*. The spermathecal duct and spermatheca can be elongate in relation to the rest of the reproductive tract (state 1), being more than two times the length of the bursa. This occurs in taxa with a basally expanded spermatheca (figs. 2d,e,f,h) and in taxa with an apical expansion (figs. 4e-h).

8. The spermathecal assembly is usually lightly sclerotized, and may be straightened during dissection (state 0); the widest portion of the spermatheca may be angulate, and sclerotized to the degree that it regains this shape even if moved or distorted (state 1; figs. 2g; 3b-h; 5a,b).

9. The spermatheca is primitively fusiform, and more or less straight (with the exception of the corkscrew configuration of *Atranus*; see character 10). In *Platynus decentis*, the spermatheca assumes an S-shape (fig. 5a). This configuration is assumed to be a derived specialization of the derived state of character 8.

10. The spermathecal configuration of *Atranus* (fig. 2g) is also considered a derivation of the angulate spermatheca (state 1, character 8). A similar configuration is present independently in *Myas coracinus*.

11. The spermatheca is normally free, and connected to the reproductive tract through a duct (state 0); however, in *Platynus hypolithos* the body of the spermatheca is appressed to the bursa (state 1; fig. 5d).

12. The spermathecal gland usually enters the spermathecal assemblage at or adjacent to its widest point (state 0; figs. 2; 3; 4; 5c, d); in a limited number of cases the gland duct enters the spermathecal duct some distance from the sperm reservoir (state 1; figs. 5a,b; 38c).

13. The spermatheca is generally a fusiform structure, which may have very fine or irregular constrictions on its surface (state 0; figs. 2b; 3b,e; 4b,g); the spermatheca is moniliform due to regular, well-developed constrictions in several lineages (state 1; figs. 3c; 4a,c; 5a,b,c; 38c).

14. Generally the spermathecal duct is membranous and flexible, or lightly sclerotized so that it regains its position if distorted and then released (state 0; figs. 2b-h; 3; 4; 5a-d); in *Olisthopus*, the duct is short and cuplike, and is rather heavily sclerotized (state 1; fig. 2a). This state is also seen in genera of the carabid tribe Lachnophorini (Liebherr 1983b).

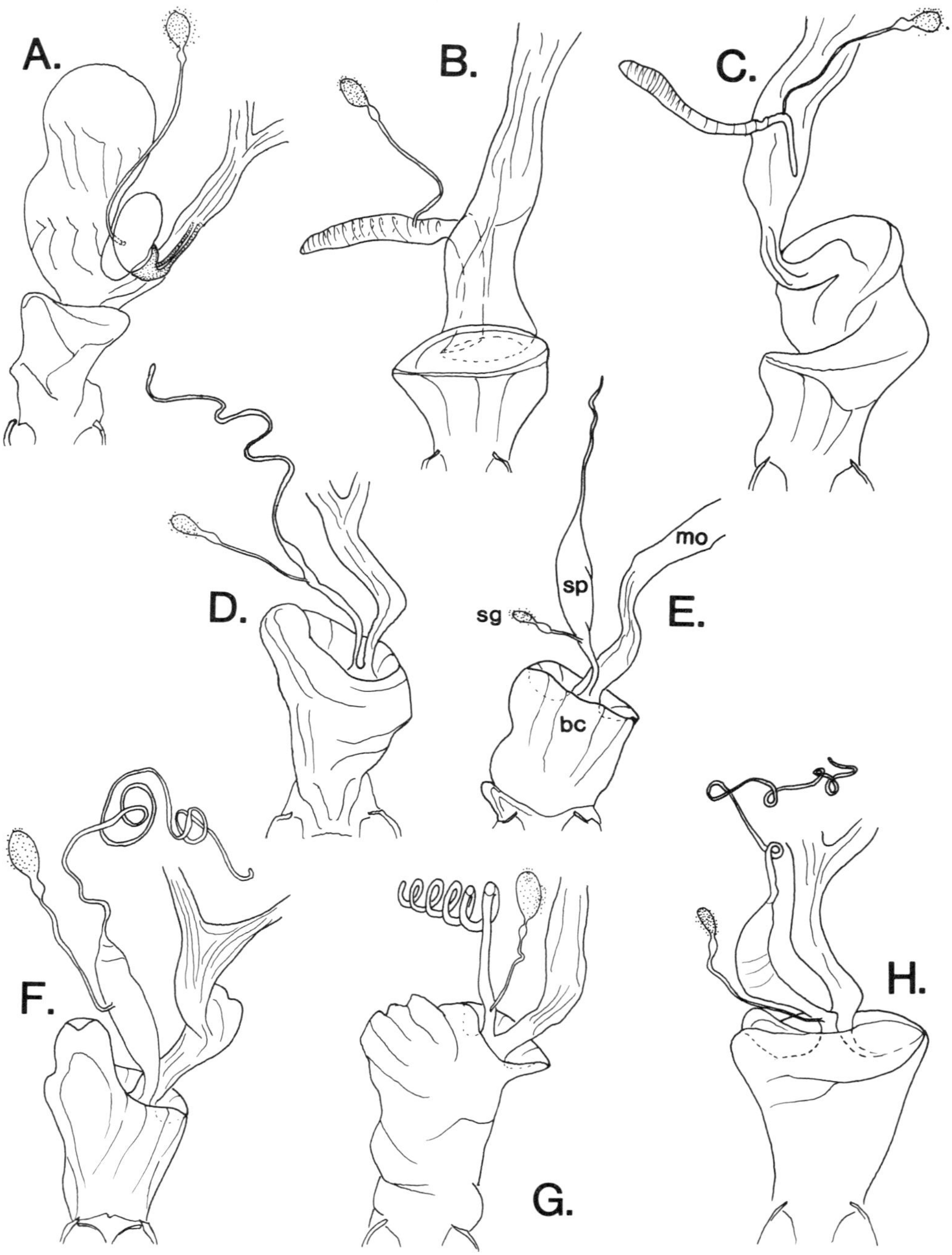

Figure 2. Bursa copulatrix (bc), median oviduct (mo), spermatheca (sp), and spermathecal gland (sg) from ventral aspect, penultimate segments of gonostyli at bottom of view. Spermathecal gland has spherical reservoir stippled. A. *Olisthopus parmatus* Say. B. *Calathus ruficollis* Dej. C. *Synuchus impunctatus* Say. D. *Elliptoleus* sp. E. *Sericoda obsoleta* Say. F. *Anchomenus dorsalis* Pont. G. *Atranus pubescens* Dej. H. *Anchomenus picticornis* Newman.

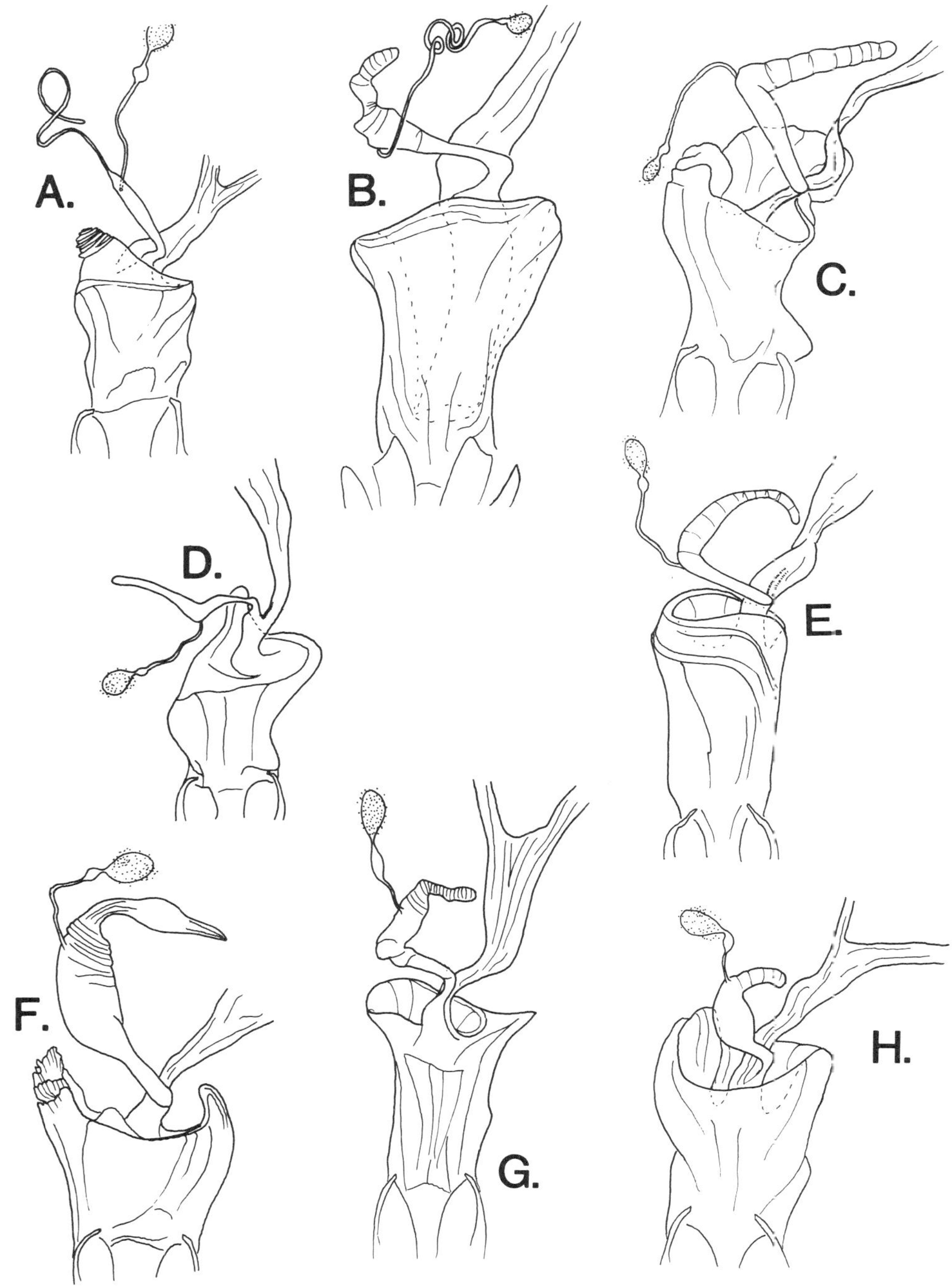

Figure 3. Bursa copulatrix, median oviduct, spermatheca,and spermathecal gland from ventral aspect, penultimate segments of gonostyli at bottom of view. Spermathecal gland has reservoir stippled. A. *Pseudanchus funebris* LeC. B. *Rhadine nivalis* Horn. C. *Tanystoma maculicolle* Dej. D. *Paranchus albipes* Fabr. E. *Tanystoma striata* Dej. F. *Anchus pusillus* LeC. G. *Agonum mutatum* Gemm. and Harold. H. *Agonum collare* Say.

15, 16. The penultimate segment of the gonostyli has an apical fringe of setae on its ventral surface in a broad array of carabid tribes (figs. 5g,h; 6c-j; 45b-d). This primitive state is coded 0,0 for characters 15 and 16. This fringe may be reduced to a few setae found on the ventro-lateral margin of the penultimate segment (figs. 6a,b). This condition is coded 1,0 for characters 15 and 16. The fringe may be completely reduced, leaving a glabrous penultimate segment of the gonostylus (figs. 5e,f). This complete reduction is assumed a derivation of the intermediate condition, and coded 1,1 for characters 15 and 16.
17. The apical segment of the gonostyli bears a seta on its dorsal surface in many tribes of Carabidae (state 0; figs. 5g; 6d,f,h-j). This seta is lost in several lineages in the Platynini (state 1; figs. 6b,e).
18. The dorsal setae of the apical gonostylus segment may be located on the inner lateral margin (state 1; figs. 5e,f,h; 6a,c,g). In those taxa in which the dorsal/marginal seta is absent, character 18 is scored no comparison (n.c.). The taxa coded n.c. are restricted to positions on the cladogram that are basal to the derived state of character 18.
19, 20. The apical segments of the gonostyli generally bear two lateral setae (figs. 5e,g,h;, 6a-d,f,g). This state is coded 0,0 for characters 19 and 20. If more than two lateral setae are observed consistently (figs. 6e,i,j), the state is coded 1,0. The lateral setae in *Onypterygia tricolor* are located some distance from the cutting edge of the style, and such a configuration recurs in several platynine taxa, among them *Atranus* (fig. 6b). The lateral setae on the ventral surface of the *Onypterygia* gonostylus are assumed homologous to the lateral setae seen elsewhere in the Platynini. If fewer than two lateral setae are consistently observed in a taxon (figs. 5f; 6h), the state is coded 0,1 for these characters. Often the number of lateral setae varies within a specimen or among specimens of a species. If any of the specimens possess two lateral setae the taxon is scored as 0,0.
21. The apical segment of the gonostyli usually bears only lateral setae and a dorsal or inner marginal seta (figs. 6a-i). The specialization of the *Onypterygia* gonostylus, in which setae cover the ventral surface, is considered a further derivation of character 19 (state 1, fig. 6j).
22. The apical segment of the gonostyli generally bears two fine setae set into a depression (state 0; figs. 5e,g,h); in *Synuchus* these apical setae are absent (state 1; fig. 5f).
23. The gonostylus is used to deposit eggs into a substrate, and the apical segment is generally long and bladelike (state 0; figs. 5e-h; 6); in *Pristonychus complanatus* this segment is short and blunt (state 1).

The male genitalic characters (24-30) are:

24, 25. The primitive carabid aedeagus probably was symmetrical and bore equally developed parameres. Such is the condition in *Trachypachus* and *Cicindela*. The aedeagus of Platynini is asymmetrical, as in most Carabidae, and the parameres are subequal. The ventral or right paramere is somewhat more twisted than the dorsal (left) paramere, but there are no drastic differences in size or shape. Examples of this condition can be seen in *Agonum* and *Platynus* (figs. 7h; 8b,c,d,f; 45e) and are coded 0,0 for characters 24 and 25. The genus *Synuchus*, and other taxa in what Lindroth (1956) designates the Synuchi, possess unequal parameres. The right paramere has a narrowed basal portion which forms an angle with the broad apical portion (fig. 7b). This is considered a derived state, and is coded 1,0 for character 24 and 25. In *Calathus,* and the taxa considered Sphodri in Lindroth (1956), the right paramere retains the angle between base and apex, but the apical portion is elongate (fig. 7a). This is termed the styloid condition. Lindroth (1956) did not discuss the derivation of this state, simply considering it distinct from the parameral condition seen in *Synuchus*. In this analysis, the styloid condition is considered to be derived from the less modified *Synuchus* condition, based on the shared attribute of an angle between basal and apical parts of the right paramere. The styloid condition seen in *Calathus* and *Pristonychus* is coded 1,1 for character 24 and 25.

26. The parameral apex bears setae in *Trachypachus* spp., *Haliplus* spp. (Haliplidae), *Gyrinus* spp. (Gyrinidae), *Dytiscus* spp. (Dytiscidae), several *Patrobus* spp., *Platidiolus vandykei* Kurnakow, *Trechus* spp. (Lindroth 1961; Sharp and Muir 1912). This generalized distribution implies that this is the ancestral state for this character (figs. 7c,d). The absence of setae on the parameral apex is considered the derived state (figs. 7a,b,e-h; 8).

27. The generalized condition of the aedeagal sac is unornamented (state 0; figs. 7a,b; 8a,b,c,f). Different lineages possess spines on the sac, which may or may not be homologous (state 1; figs. 7e-h; 8d,e). In this study set, there appear to be four different configurations of spinose aedeagal sacs exemplified by the *Rhadine-Tanystoma* lineage, *A. collare, A. languidum,* and *Onypterygia* sp. These are all coded as derived for this character. This coding introduces false synapomorphies into the analysis if saccal spines have arisen repeatedly. Uncertainty regarding homology of these spines is the basis for the present simplified coding scheme.

28. Throughout the Carabidae, the aedeagal shaft is generally smooth (state 0; figs. 7; 8a,c-e); within several lineages in *Agonum* and *Platynus* the basal area of the aedeagus bears wrinkles (state 1; figs. 8b,f; 45e).

29. The aedeagus is generally a moderately sclerotized structure, moderately transparent when prepared in cold 10% KOH (state 0; figs 8c,d; 49a-c). In relatively few species, mostly those that possess dark and metallic body surfaces, the aedeagus is melanic and heavily sclerotized (state 1; fig. 8b).

30. In most carabid lineages, the apex of the aedeagus is acuminate, without hooklike modifications (state 0; figs. 7; 8b-f); in *Atranus pubescens* the apex of the aedeagus is modified to form a hooklike protuberance (state 1; fig. 8a).

The external characters (31-56) and their primitive and derived states are:

31. The generalized carabid antenna has the basal three segments glabrous except for apical setae (state 0); the pubescence seen on segments 4-11 is present on the third segment in some taxa (state 1). The entire third segment may bear fine pubescence or the pubescence may be restricted to the apical half of the segment.

32. Generally, the apex of the scape has one large seta on its outer face, with several smaller setae distributed elsewhere on the apex of the segment. If the total number of setae is three or fewer, this character is considered the primitive state. If four or more setae are present on the apex of the scape, this character is coded as derived.

33. In taxa with less developed antennal pubescence, three or fewer setae are present at the apex of the pedicel (state 0). If more than four setae are present at the apex of antennomere 2, this character is coded as 1.

34. Throughout the Carabidae, the antennal segments are generally concolorous (state 0); in various lineages, the antennae possess segments that are differentially pigmented. In several taxa in this study set, the apical antennal segments are pale to white (state 1).

35, 36. Most carabids possess a mentum with a unidentate tooth. This state is coded 0,0 for characters 35 and 36. This tooth may be absent (state 1,0). Alternatively the mentum tooth may be bidentate due to a median invagination (state 0,1).

37. Generally, the submentum has two setae laterally: a larger inner seta and a smaller outer seta (state 0; fig. 29g). In a limited number of taxa the outer seta is absent (state 1).

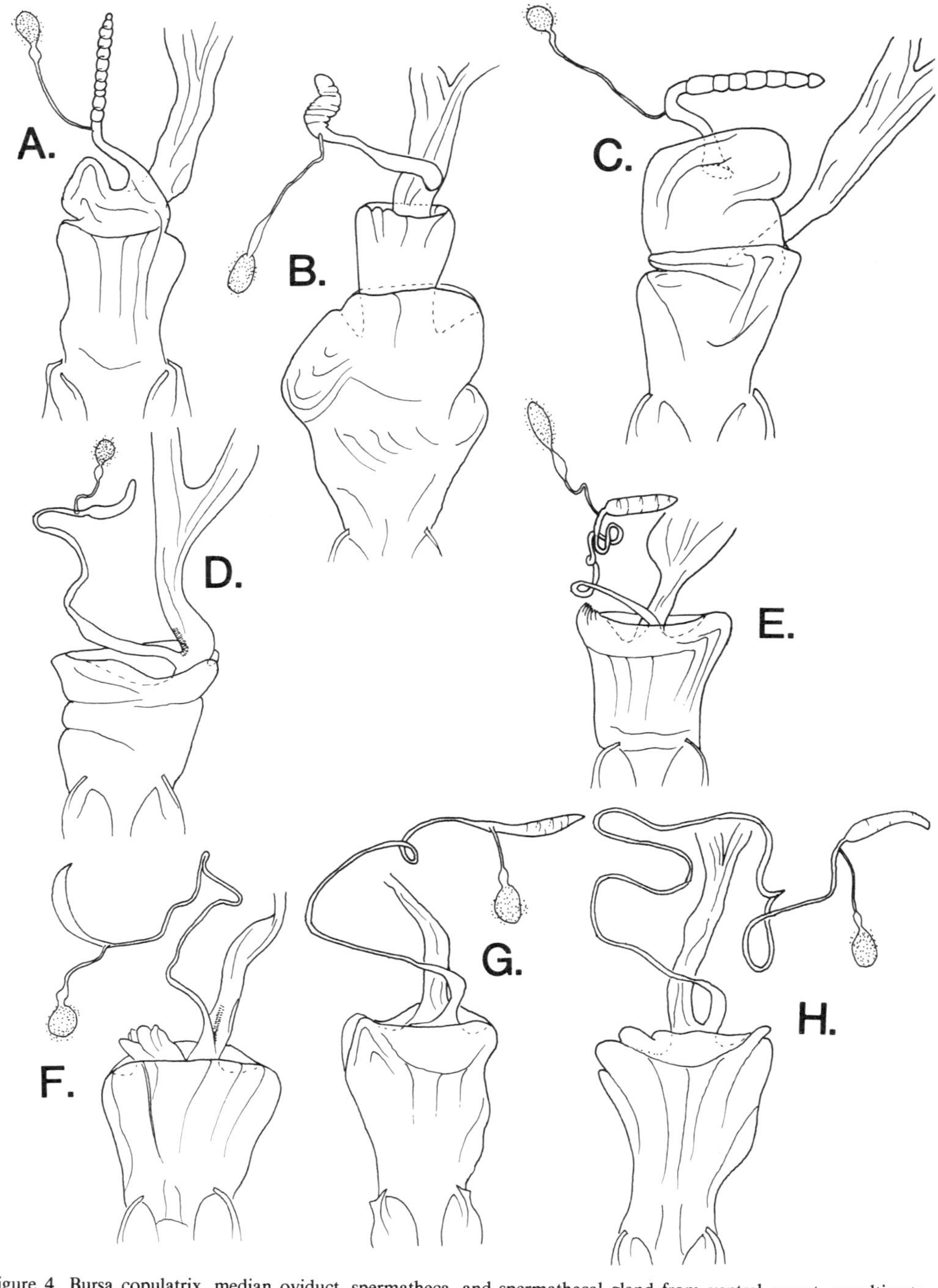

Figure 4. Bursa copulatrix, median oviduct, spermatheca, and spermathecal gland from ventral aspect, penultimate segments of gonostyli at bottom of view. Spermathecal gland has reservoir stippled. A. *Agonum ferreum* Hald. B. *Agonum languidum* Horn. C. *Agonum sulcipenne* Horn. D. *Agonum nigriceps* LeC. E. *Agonum punctiforme* Say. F. *Agonum anchomenoides* Rand. G. *Agonum sexpunctatum* L. H. *Agonum harrisi* LeC.

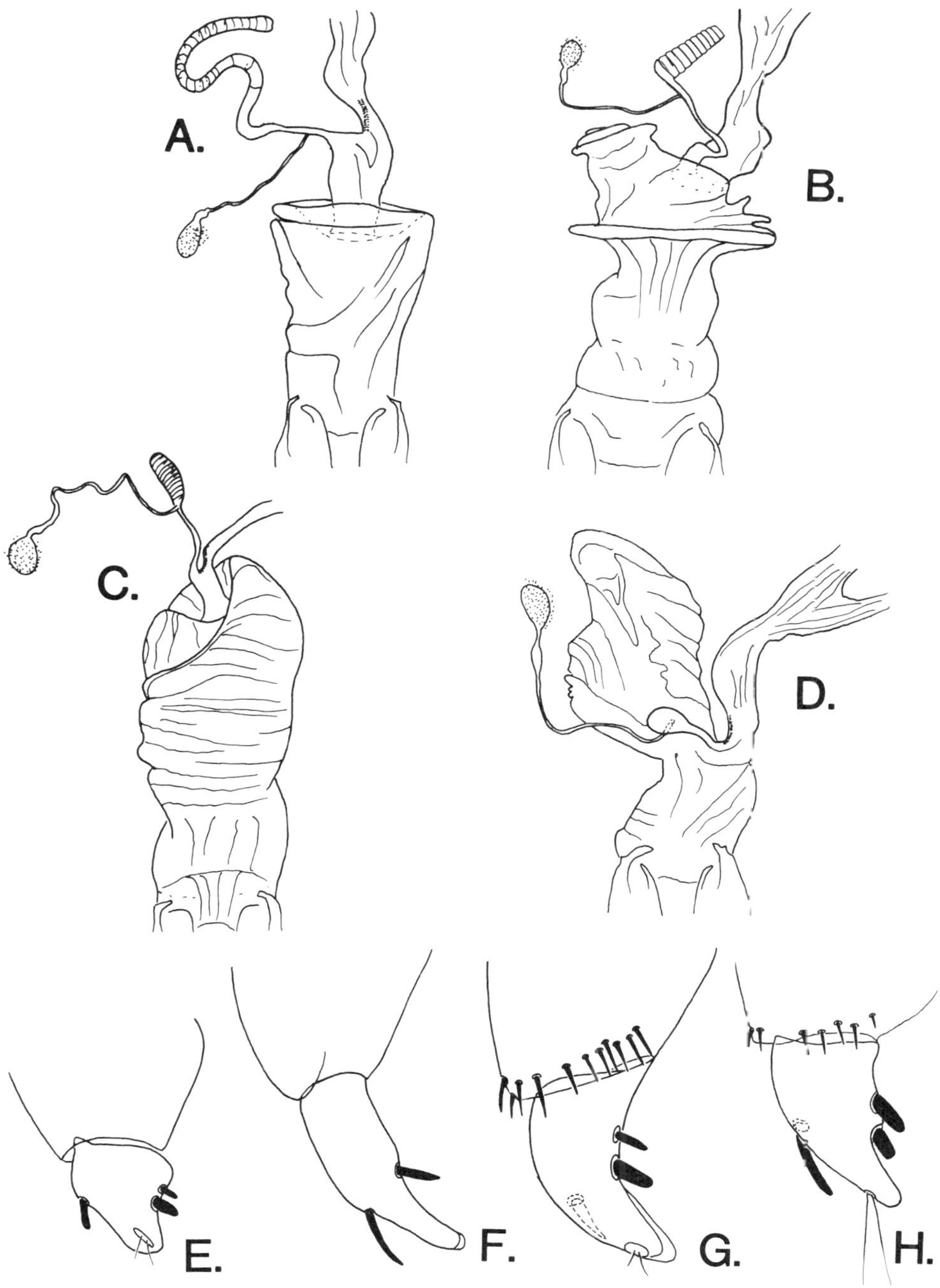

Figure 5. A–D. Bursa copulatrix, median oviduct, spermatheca, and spermathecal gland from ventral aspect, penultimate segments of gonostyli at bottom of view (spermathecal gland stippled): A. *Platynus decentis* Say; B. *Platynus ovipennis* Mann.; C.*Onypterygia tricolor* Dej.; D. *Platynus hypolithos* Say. E.–H. Left gonostylus from ventral aspect (penultimate segment bears more heavily sclerotized apical segment): E. *Calathus ruficollis* Dej.; F. *Synuchus impunctatus* Say; G. *Elliptoleus* sp. (dorsal seta with broken lines); H. *Sericoda obsoleta* Say.

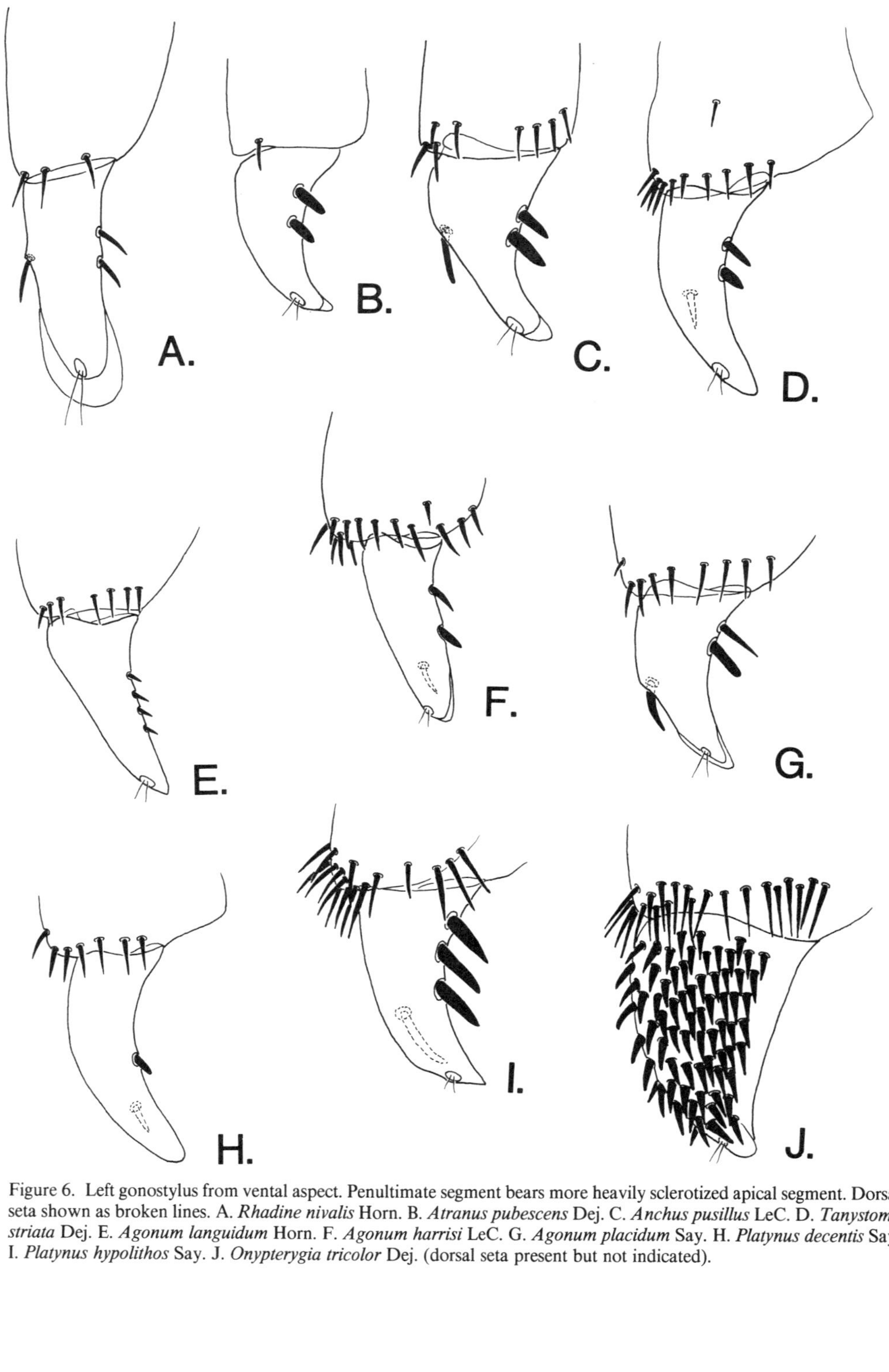

Figure 6. Left gonostylus from ventral aspect. Penultimate segment bears more heavily sclerotized apical segment. Dorsal seta shown as broken lines. A. *Rhadine nivalis* Horn. B. *Atranus pubescens* Dej. C. *Anchus pusillus* LeC. D. *Tanystoma striata* Dej. E. *Agonum languidum* Horn. F. *Agonum harrisi* LeC. G. *Agonum placidum* Say. H. *Platynus decentis* Say. I. *Platynus hypolithos* Say. J. *Onypterygia tricolor* Dej. (dorsal seta present but not indicated).

38. The mentum of Carabidae is generally lightly impressed on each side of the middle (state 0); this depression may be deeper, with the depressions taking the form of foveae (state 1).
39. The neck of Carabidae is generally gently tapering behind the temples, with little evidence of an impression on the dorsum of the head (state 0); the neck may be more strongly constricted, with evidence of a depression on the dorsum of the head. This state is encountered in a limited number of lineages and is assumed to be derived.
40, 41. The generalized carabid possesses two setae on the lateral margin of the prothorax: one at the basal angle, one at or near the midpoint of the lateral margin (state 0,0 for characters 40 and 41; figs. 31c-h; 32; 33). Either or both of these setae may be absent, although only one or the other is absent in this study set. If the anterior seta is lost, the state is coded 1,0. If the posterior seta is lost, the state is coded 0,1.
42, 43. Three setae are most commonly associated with the third elytral interval. Either two or three setae are found on or adjacent to this elytral interval over many carabid lineages. For this study set, if three setae are present, the state is coded 0,0 for characters 42 and 43. If fewer setae are present, the state is coded 1,0. If four or more setae are present, the state is coded 0,1.
44. In most carabids, setigerous elytral punctures are restricted to the third elytral interval or adjacent striae (state 0; figs. 35a,c); in a limited number of taxa, setae can be found on or adjacent to the fifth elytral interval (state 1; fig. 35b).
45. Most carabid taxa possess glabrous body surfaces, with the exception of articulated sensory macrosetae (state 0); a secondary pubescence composed of smaller, finer unarticulated microsetae may be present (state 1). This setation occurs over large portions of the body surface, taking the form of a pelage.
46. In Bembidiini, Patrobini, Harpalini and Lebiini, the foretibia has an anterior and posterior carina with shallow grooves medial to the ridges. This state is assumed to be primitive within the study set. The foretibia can be more robust, with the ridges and associated sulci effaced. This state is considered derived.
47, 48. The mesocoxa bears setae upon a ventrolateral ridge (figs. 35d-f; 36a-d). Across a variety of tribes allied to the Platynini, there is only one seta borne on the mesocoxal ridge (coded 0,0 for characters 47 and 48). The number of setae may be variable between specimens, with one or two setae found on the ridge (coded 1,0). The number of setae may consistently be two (coded 1,1).
49, 50, 51. The tarsi are generally cylindrical, without well-developed carinae or associated sulci on their dorsal surface. There may be carinae on the inner and outer lateral surfaces of the tarsi, with associated sulci. There may also be a median sulcus in the middle of the medial elevated portion of the tarsi. Based on the generalized plan of cylindrical tarsi without grooves, the presence of an outer sulcus (state 1, character 49), an inner sulcus (state 1, character 50), and a median sulcus (state 1, character 51) are considered derived. The presence or absence of sulci is judged by inspection of the basal metatarsus. Even if sulci are reduced on other tarsal segments or legs, the taxon is coded as derived based on the basal metatarsus. The presence of lateral carinae with associated sulci is considered primitive for the foretibia, and derived for metatarsi. This coding would appear incorrect, but is based on the fact that foretibial sulci are found in many more out-groups of Carabidae, including at least some member taxa of Carabini, Scaritini, Bembidiini, Opisthiini, Elaphrini, Trechini, Patrobini, *Pterostichus*, *Harpalus*, and *Lebia*. The taxa with dorsal or lateral tarsal sulci are less numerous. The occurrence of tarsal sulci may be partially correlated with smaller body size. Tarsi with a narrow diameter could retain structural strength if the dorsal surface were fluted. The independent development of lateral sulci in

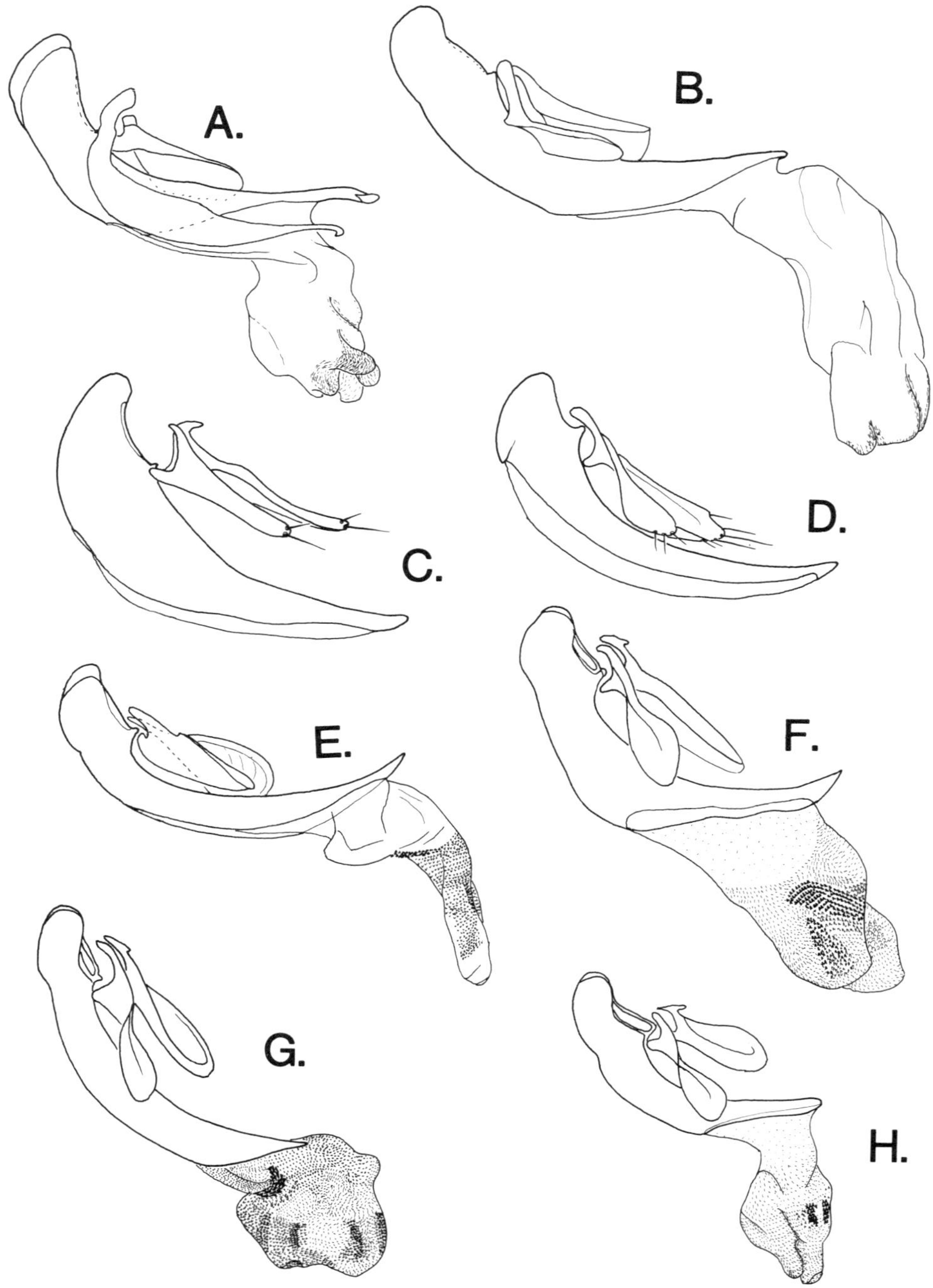

Figure 7. Aedeagal shaft and parameres viewed from ventral aspect (resting position). C and D approximately 1.3 times more enlarged than others and with sac uninflated. A. *Calathus ruficollis* Dej. B. *Synuchus impunctatus* Say. C. *Sericoda obsoleta* Say. D. *Elliptoleus* sp. E. *Rhadine nivalis* Horn. F. *Tanystoma maculicolle* Dej. G. *Tanystoma striata* Dej. H. *Agonum collare* Say.

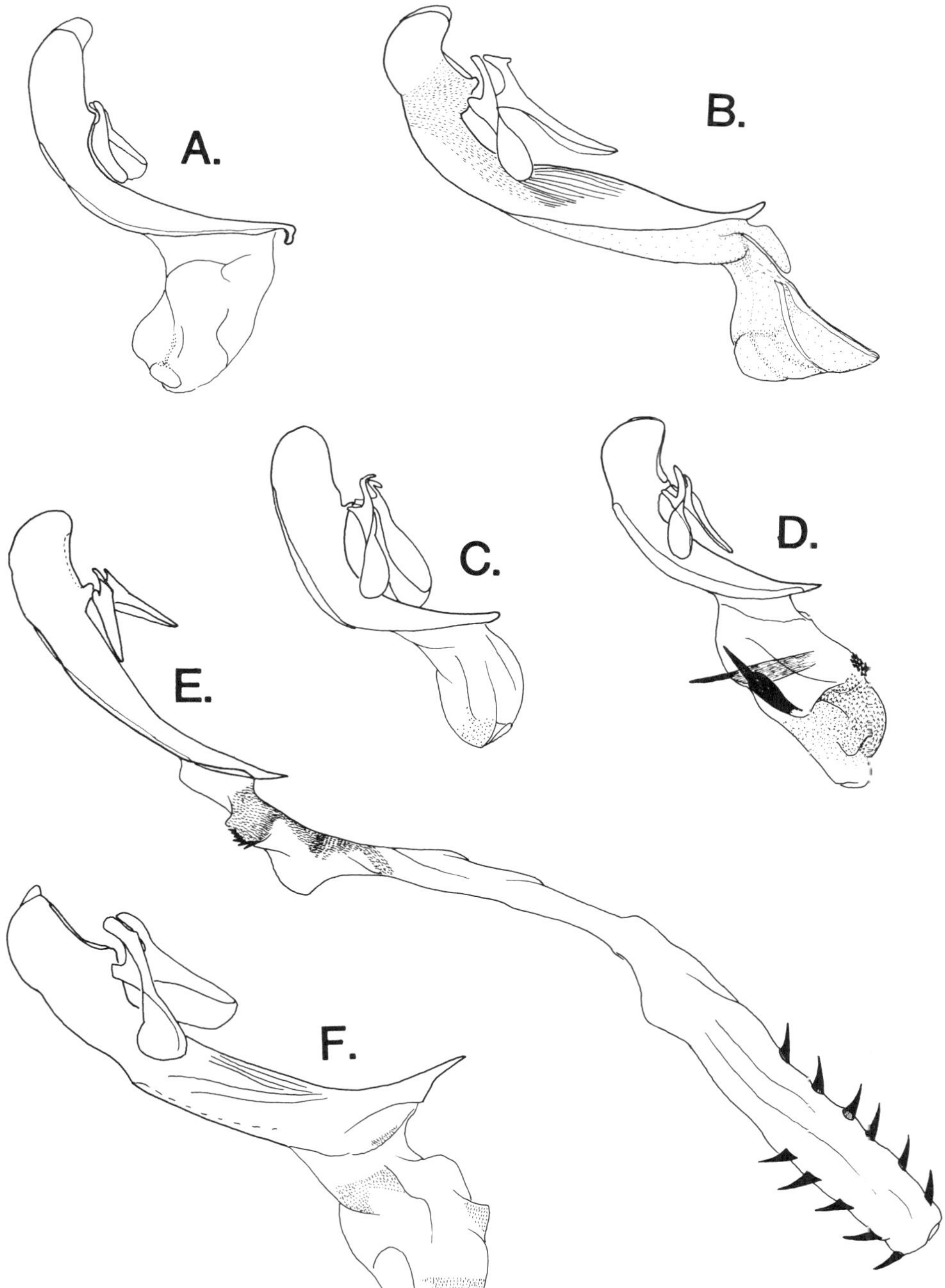

Figure 8. Aedeagal shaft and parameres viewed from ventral aspect (resting position) with aedeagal internal sac inflated. E somewhat reduced (approx. .75X others). A. *Atranus pubescens* Dej. B. *Agonum pacificum* Casey. C. *Anchomenus dorsalis* Pont. D. *Agonum languidum* Horn. E. *Onypterygia* sp. F. *Platynus ovipennis* Mann.

Elaphrus, *Bembidion*, and Platynini is a correlate of smaller body size. As such, we should expect these characters to be relatively poor indicators of phylogenetic relationships.

52. The primitive carabid tarsus has a truncate apex (state 0); in arboricolous taxa the apex of the fourth tarsal segment is emarginate to deeply incised, with two well-developed lobes (state 1). This bilobed condition is also seen in some taxa that are found at least part of the time on the ground.

53, 54. The fifth tarsal segment, or ungues, may bear two rows of ventral setae much like the four basal segments (coded 0,0); the ventral setae may be reduced in size and number, with up to four or six small setae near the apex of the segment (coded 1,0); the ventral side of the ungues may be glabrous (coded 1,1).

55. The generalized carabid tarsal claw is smooth, without modifications on the inner surface (state 0); the inner surface may bear toothlike pectinations (state 1).

56. The generalized carabid tarsal claws close against one another, not against the ventral side of the ungues (state 0); in *Onypterygia*, the claws approach a chelate condition, in which they partially oppose one another, but also close back toward the ungues (state 1).

Methods of Cladistic Analysis. Preliminary analyses were conducted using the character compatibility algorithm of Estabrook, Strauch, Jr., and Fiala (1977) and the Wagner 78 algorithm of Farris (1970). This early data set contained taxa now considered Platynini, plus four taxa from the Lachnophorini that were previously placed in the Platynini. The discovery of the larva of *Calybe (Ega) sallei* Chevrolat permits removal of *Lachnophorus, Calybe, Euphorticus*, and *Anchonoderus* from Platynini (Liebherr, 1983b). The results of these earlier cladistic analyses showed that the taxa included in the Sphodri and Synuchi of Lindroth (1956) are the sister group to the Agoni (= Platyni). Within the *Agonum-Platynus* lineage, the configuration of the spermathecal assembly proves most useful for estimation of affinities.

Based on this knowledge, an expanded data set containing 43 taxa coded for 56 unit characters is analyzed here. These characters are used to construct a synapomorphy scheme cladogram (Hennig 1966). The relatively large number of characters obtained from the female tract (23) is not an artifact of ignoring other parts of the body or the male genitalia. The female reproductive structures are extraordinarily variable across the Platynini.

The 30 internal and genitalic characters are first used to generate a cladogram using the Wagner groundplan-divergence method (Wagner 1980). This cladogram (fig. 9) groups genera into more inclusive lineages, however, resolution of 43 taxa based on 30 characters is not possible. In addition to construction of a cladogram based on 30 characters, all 56 characters can be used to generate a cladogram (fig. 10). Characters that group recognized generic level taxa are used to dictate the basic structure of this second cladogram. The external characters modify and further resolve lower level groupings.

Minimization of character state changes is the overall criterion for placement of taxa on the cladogram. If two characters conflict, meaning they dictate different nested sets of taxa, an evaluation of the suitability of each character for phylogenetic estimation can be made. In this way overall parsimony is tempered by confidence in the homology of derived character states. The decisions made between characters are discussed in the results section.

RESULTS AND DISCUSSION OF CLADISTIC ANALYSIS

Internal Reproductive and External Genitalic Character Cladogram. Based on the internal reproductive and external genitalic characters of males and females, five lineages exist within the Platynini (fig. 9). The genus *Olisthopus* is characterized by a spermatheca with a sclerotized cup-shaped duct entering the reproductive tract on the median oviduct (fig. 2a, characters 1 and 14). *Olisthopus* shares these derived states with taxa in the Lachnophorini (Liebherr 1983b). Based on internal characters, *Olisthopus* is an outlier of the Platynini.

The second lineage based on reproductive characters comprises the subtribe Sphodri. It is characterized by the absence of the fringe of setae on the penultimate segment of the ovipositor (figs. 5e, f; characters 15 and 16). The spermatheca may be uniform in diameter as in *Calathus* or *Synuchus* (figs. 2b, c), or apically expanded as in *Pristonychus*. In the taxa studied, the bursa is symmetrical (character 6). The male parameres exhibit a transition within this group, from the moderately derived *Synuchus* type (fig. 7b) to the styloid type of *Calathus* (fig. 7a) and *Pristonychus*.

The third lineage in the Platynini is characterized by a spermatheca that is basally widened, with an apical filament (figs. 2d-f, h; 3a; character 3). Both *Elliptoleus* and *Sericoda* contain species that have parameres bearing two to four apical setae (figs. 7c, d). This condition is the primitive state for Carabidae, and three interpretations are possible: (1) these two genera may be early derivatives of the platynine stock, with all later lineages possessing glabrous parameres; (2) the parameral setae may have been lost a number of times within the Platynini; (3) the condition in *Sericoda* and *Elliptoleus* may be a reversal from glabrous parameres to the primitive setose condition. The last alternative is the most parsimonious, with the reversal to the primitive state uniting these two taxa. The sharing of the basally expanded spermatheca (character 3) strongly supports this interpretation.

The other exemplars that possess a basally expanded spermatheca include *Anchomenus dorsalis, A. picticornis, Pseudanchus funebris,* and *P. quadratus*. External characters resolve the relationships among these taxa.

The fourth lineage possesses spermathecae superficially similar to the primitive condition of *Calathus* and *Synuchus*, but there are consistent differences. The spermatheca in the *Rhadine-Tanystoma* line is moderately long and medially inflated, with the narrower apical portion at an angle to the medial portion (figs. 3b-h; characters 4 and 8). *Atranus pubescens* has a bizarre spermatheca (fig. 2g), which is coded as angulate (state 1, character 8). *A. pubescens* is loosely allied to the *Rhadine-Tanystoma* lineage.

Several derived character states unite subsets of taxa within this line. *Tanystoma* spp., *Rhadine* spp., and *Agonum collare* all possess aedeagal internal sacs with some sort of spination or spicular fields (figs. 7e-h; character 27). The configuration of these fields varies greatly even between closely related species, creating difficulties in homologizing spines on specific parts of the sac. It is assumed that occurrence of aedeagal sac spination is labile enough to accept parallel derivation if other characters dictate.

Rhadine caudata and *R. nivalis* both possess a reduced setal fringe on the penultimate segment of the gonostyli. The *Rhadine jejuna* group species studied also possesses a reduced setal fringe. The combination of medially expanded spermatheca and reduced ovipositor setal fringe distinguishes *Rhadine* from the superficially similar species of *Platynus*.

The fifth, and by far the most speciose lineage suggested by reproductive and genitalic characters is the *Agonum-Platynus* lineage. This line is characterized by an apically expanded spermathecal assembly (figs. 8a-h; character 5). Within this lineage a number of groups can be circumscribed based on characters of the females. Most *Agonum* possess a spermatheca attached to a long duct (figs. 4e-h; character 7). In some species, this duct is 1.5-2.0 times the body length. Little structure is present within the clade delimited by the derived state of character 7, and

Figure 9. Cladogram including 43 exemplar taxa within the Platynini based on 30 characters of the female internal reproductive tract and male and female external genitalia. Circled numbers indicate character state change from primitive to derived state, hexagon denotes character state reversal. Small numbers next to state changes indicate how many times a character changes state on cladogram.

discussion of the affinities within this part of *Agonum* must be deferred for the full complement of characters.

Two other species groups of *Agonum* can be delimited based on internal characters. The *Agonum nigriceps* group comprises species that have spermathecal ducts of short to moderate length (figs. 4b, d) and gonostyli bearing four lateral setae (fig. 6e; character 19). Possession of the derived state for character 19 is shared with *Platynus hypolithos* and *Onypterygia tricolor*. Shared ancestry based on character 19 is not supported when external characters are added. Even on the reduced character cladogram (fig. 9) increased number of lateral setae on the gonostylus occurs five times, indicating that this character is prone to convergent development.

The species of the *Agonum ferreum* group, *A. ferreum, A. errans,* and *A. sulcipenne*, all possess spermathecae with well-developed ringlike constrictions on the reservoir (figs. 4a, c; character 13). This condition recurs in *Tanystoma maculicolle, Agonum cyanope, Platynus decentis, P. ovipennis,* and *Onypterygia tricolor* (figs. 3c, 5a-c, 38c). Ringlike spermathecal constrictions occur in other taxa, but are not as regularly concentric as in the taxa coded for it above. This condition has undoubtedly arisen a number of times in the Platynini. On this cladogram (fig. 9), character 13 unites the *A. ferreum* group with *Platynus decentis* and *P. ovipennis*. This grouping is not supported when external characters are added to the analysis.

The few North American exemplars indicate how variable the internal structures are in *Platynus*. *Platynus brunneomarginatus* possesses no derived states aside from an apical sperm reservoir. *P. decentis* and *P. ovipennis* have an angled spermatheca with ringlike constrictions, and a spermathecal gland that enters on the duct, not adjacent to the reservoir (figs. 5a, b; characters 8, 12, 13). *Platynus hypolithos* is very specialized, with the spermathecal duct reduced and the spermatheca appressed to the bursa (fig. 5d; character 11). Study of the female reproductive tract will prove extremely useful for classification of *Platynus*.

Onypterygia spp. appear to be specialized *Platynus*, with a relatively unmodified spermatheca (fig. 5c) and highly modified gonostyli (fig. 6j). The gonostyli may be used for specialized oviposition in the arboreal habitat occupied by *Onypterygia*. The internal sac of the *Onypterygia* aedeagus is greatly elongated, and bears a dorsal and ventral row of apical spines (fig. 8e).

Cladogram Using All Characters and Discussion of Classification. The cladogram based on all 56 characters (fig. 10) divides the Platynini into four lineages. In the first lineage, *Olisthopus* is united with the other sphodrine genera based on synapomorphies in apical setal number for scape and pedicel (characters 32 and 33) and dorsal elytral setal number (character 42). These characters change five, seven, and five times respectively. Grouping taxa on such homoplasious characters is unsatisfactory. LeConte and Horn (1883) suggested that *Olisthopus* shares affinity with the Lebiini based on the configuration of the ligula. The similarity of the spermatheca to that of Lachnophorini (Liebherr 1983b) supports placement of *Olisthopus* outside the Platynini, closer to the basal lebiine lineages such as Lachnophorini or Odacanthini. In the classification presented in Appendix 2, *Olisthopus* is retained within the Platynini as an unplaced genus.

The hypothesized relationships of the genera of the subtribe Sphodri remain unchanged on the full character cladogram. The unique synapomorphies of characters 6 and 24 unite *Synuchus, Calathus*, and *Pristonychus*. Pectinate claws (state 1, character 55) also unite these taxa. The only other platynine taxon in which this state occurs is *Onypterygia*. The two occurrences of pectinate claws are probably not homologous. The claws of *Onypterygia* are chelate and strongly pectinate, useful in an arboreal habitat. The claws of the terrestrial sphodrine taxa are nonchelate, with weaker pectinations.

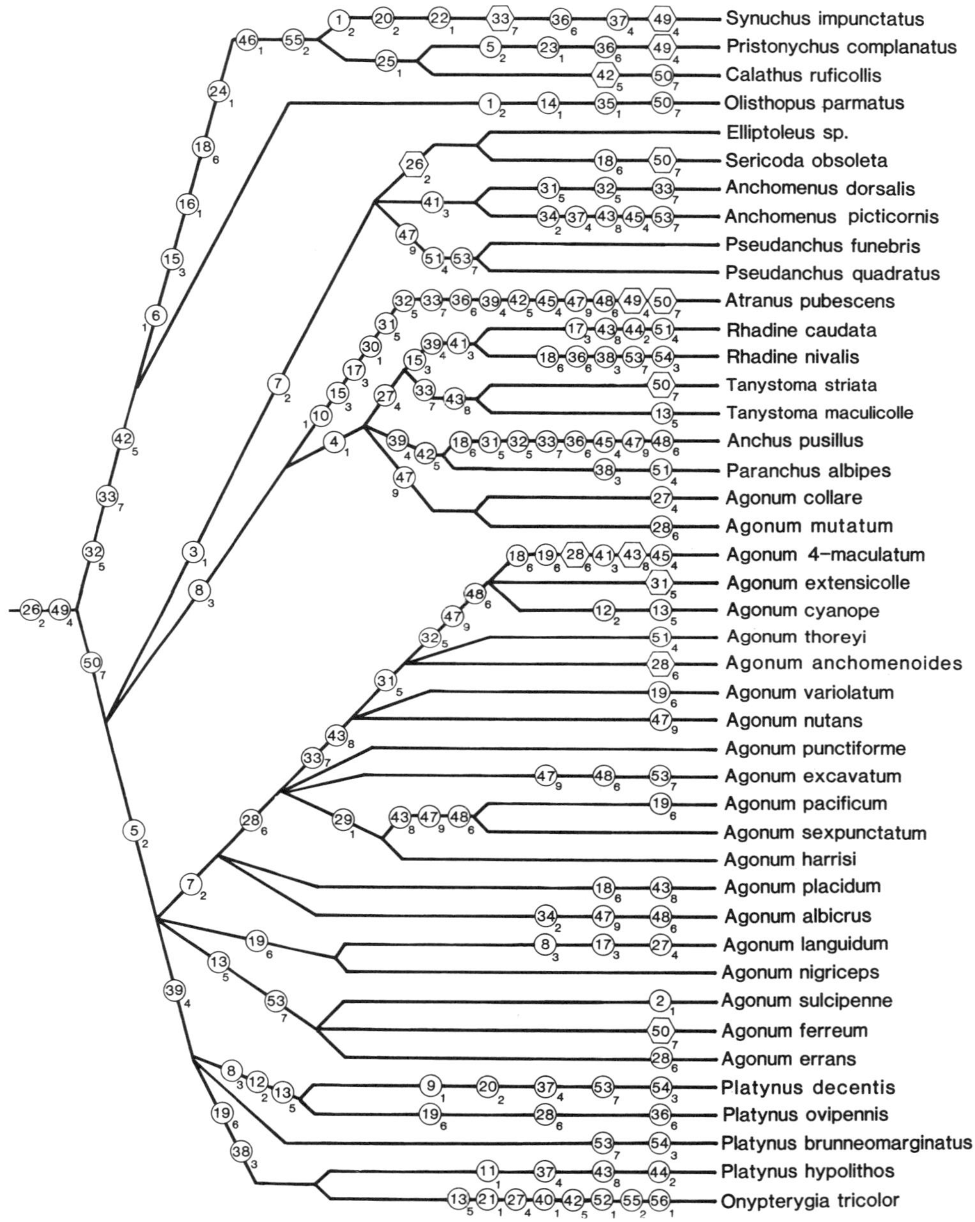

Figure 10. Cladogram including 43 exemplar taxa within the Platynini based on 56 characters: the 30 characters used in figure 9 plus 26 external characters. Conventions the same as for figure 9. Total number of character state changes on cladogram is 169.

The second major lineage in the Platynini, which is itself composed of three lineages, is the subtribe Platyni. The *Anchomenus, Rhadine-Tanystoma*, and *Agonum-Platynus* lineages all share external and internal sulci, evident at least on the metatarsi (characters 49 and 50). Several taxa within the Platyni lack internal tarsal sulci (*Sericoda obsoleta, Atranus pubescens, Tanystoma striata*, and *Agonum ferreum* of this data set), whereas several sphodrine taxa possess them (*Calathus ruficollis* of this data set). A diagnostic grouping of the Platyni based on this character is not possible. The Platyni should be viewed as a loosely knit, paraphyletic group of generalized carabids. It is composed of all Platynini excluding the more specialized and less speciose Sphodri.

The *Anchomenus* lineage contains four genera in the North American fauna. *Elliptoleus* and *Sericoda* both share the primitive condition of parameral setae present (character 26). As discussed above, it is most parsimonious based on all characters to assume that this is a shared reversal to the primitive state for this character.

The genus *Anchomenus* is diagnosed by a basally widened spermatheca with apical filament, and lack of a basal prothoracic seta (character 41). The type for this genus is *A. dorsalis* Pont., from Europe and Siberia (Csiki 1931). The only North American species allied to *A. dorsalis* is *A. picticornis* Newman, a species found in the eastern United States. *A. picticornis* shares pale apical antennal segments with *Agonum albicrus* Dejean, but the number of pale segments differs between the species, and the weight of all other characters implies this is a parallel derivation.

The genus *Pseudanchus* has previously been treated as a subgenus of *Anchomenus* (Casey 1920) and as a subgenus of *Agonum* (Gray and Hatch 1941; Lindroth 1966). Recognition of *Sericoda, Elliptoleus*, and *Anchomenus* at the generic level makes it appropriate to recognize *Pseudanchus* as a distinct genus. The three described species in this genus are streamside hydrophiles, restricted to areas west of the Rocky Mountains. *Pseudanchus* can be distinguished by a basally widened spermatheca and median tarsal sulcus.

The *Rhadine-Tanystoma* lineage is comprised of six generic level taxa in North America, one of which has been introduced. All possess a spermatheca that is angulate medially (character 8). *Atranus pubescens* has a number of uniquely derived character states. The corkscrew-shaped spermatheca (fig. 2g; character 10) and hook-shaped aedeagal apex (fig. 8a; character 30), and increased pubescence (characters 31, 32, 33, 45, 47, and 48) serve to diagnose this species. *Atranus* has been variously associated with *Anchus*, and the lachnophorine genus *Anchonoderus*. However, *Atranus* possesses a spermatheca with a duct entering near the bursa-median oviduct junction, whereas *Anchonoderus* has a bipartite spermatheca that enters on the median oviduct, as in other Lachnophorini (Liebherr 1983b). Placement of *Atranus* near *Anchus* appears suitable based on overall external appearance. Judging from the internal characters, increased pubescence has been derived independently in the two groups.

Five other groups in the *Rhadine-Tanystoma* lineage are united by a medially inflated spermatheca (figs. 3b-h; character 4). *Rhadine* and *Tanystoma* are closely related genera, characterized by aedeagal sacs with spination. *Rhadine* appear to be specialized *Tanystoma*, distinguished by elongate appendages, constriction of the neck, narrowing of the prothoracic base, and prolongation of latero-basal pronotal lobes. Both taxa are found in drier areas throughout the West. *Tanystoma* are winter active, taking advantage of winter rains to utilize the chaparral habitat. *Rhadine* can be found in caves or associated with animal burrows. Brachyptery occurs in both genera. Within both genera, *T. maculicolle* Dejean is the least specialized and most common species.

Anchus also has a medially inflated, angulate spermatheca. The secondary pubescence seen in *A. pusillus* is lacking in *A. obscurus,* implying that the shared presence in *Anchus* and *Atranus* is not a true indication of affinity.

Paranchus Lindroth (1974a) is elevated to generic rank to accomodate *P. albipes* (Fabricius). This species possesses gonostyli with a dorsal seta, whereas *Anchus* has an inner lateral seta on the gonostylus (character 18); it possesses median tarsal sulci (character 51) whereas *Anchus* does not. These genera show affinities, but the amount of divergence dictates they be recognized as distinct entities.

The final group in the *Rhadine-Tanystoma* lineage is composed of *Agonum collare* and *Agonum mutatum.* These species make up the *Agonum collare* species group. Both display the shortened, medially expanded, apically angulate spermatheca (characters 4, 8) seen in *Rhadine, Tanystoma, Anchus,* and *Paranchus. A. collare* also has spines on the aedeagal inner sac (character 27), a feature rare in *Agonum.* These are probably not homologous with the spinose fields in *Rhadine* and *Tanystoma.* Based on external characters, these species strongly resemble the *Agonum marginatum* species group. The spermathecal configuration in this case may represent parallel development of the medially expanded spermatheca from the primitive type, or reversal from the *Agonum* type. Placement of these taxa in this lineage is more parsimonious than placement in the *Agonum-Platynus* lineage.

Inclusion of the *A. collare* group in *Agonum* makes *Agonum* a polyphyletic taxon. This hypothesis can be tested against future data, possibly solving the problem of placing these species. For now, the practical problems of recognizing taxa based on accessible characters dictates retention of the *A. collare* group as the most outlying species group in *Agonum.* A worldwide survey of the *Agonum* species is required to determine the affinities of *A. collare* and *A. mutatum.*

The *Agonum-Platynus* lineage is comprised of two grades. *Agonum* is a polyphyletic assemblage composed of at least four lineages: the *A. collare* group discussed above; the *A. ferreum* group (subgenus *Stereagonum* Casey); the *A. nigriceps* group (subgenus *Platynomicrus* Casey); and the true *Agonum* lineage containing the rest of the species groups. A classification placing all species of North American *Agonum* into species groups appears in Appendix 2.

The *Agonum ferreum* group contains three species in North America, distinguished by having ringlike constrictions on the spermatheca (figs. 4a, c; character 13), reduced setation on the ventral surface of the ungues (character 53), and an unconstricted neck. *A. sulcipenne* differs from the others in having the spermatheca empty into the bursa (character 2). It may have closer affinities to exotic species, but is provisionally included in this group.

The *Agonum nigriceps* group is composed of five small, lightly sclerotized species found in moist situations. All the species examined possess spermathecal assemblies similar to those shown in figures 4b and 4d, and more than two lateral setae on the gonostylus (fig. 6e; character 19).

The remainder of the species groups in *Agonum* are characterized by an apical sperm reservoir and a long spermathecal duct (figs. 4e-h; characters 5 and 7). The species groups within this line can be differentially diagnosed using morphological characters. If one restricts analysis to those characters for which primitive and derived states can be hypothesized with some certainty, the resolution of groups is reduced. Based on this cladistic analysis, which supplements Lindroth's (1966) treatment, the following groups are recognized:

1. *A. marginatum* group. This group corresponds to the nominate subgenus. It is a diverse group in North America, containing species with latero-basal areas of the prothorax more or less

rounded and expanded, generally fewer than three setae at the apex of the pedicel of the antenna (state 0, character 33). All species except *A. placidum, A. muiri*, and *A. excavatum* have three dorsal elytral setae (state 0, character 43). This group has previously been divided into subgenera, usually on the basis of dorsal elytral seta number and metallic sheen of the integument. Neither of these characters is constant within any of the taxa previously recognized. Several isolated species with uncertain affinities are included in this group. *A. albicrus* shares plurisetose mesocoxal ridges (characters 47, 48) with *A. pacificum* and *A. sexpunctatum* but lacks the aedeagal wrinkles (character 28), and increased number of dorsal elytral setae (character 43) seen in those species. *A. placidum* has four or more dorsal elytral setae (character 43) but lacks the aedeagal wrinkles (character 28) and plurisetose mesocoxal ridge (characters 47, 48). *A. excavatum* possesses reduced ungual setation (character 53), a unique derivation within this group. Inclusion of these species in this subgenus is based on overall phenetic resemblance. Their inclusion also reflects a decision to eliminate excessive numbers of monotypic species groups. A revision and cladistic analysis of all species in *Agonum sensu stricto* would help elucidate the affinities of these species.

2. *A. punctiforme* group. This corresponds to the subgenus *Circinalia* Casey, characterized by very small, deep latero-basal pronotal depressions, narrow prothoracic lateral bead, and lack of pronotal hind angles. All species examined possess spermathecal assemblages with ducts of moderate length (fig. 4e).

3. *A. nutans* group. This corresponds to subgenus *Micragonum* Casey and contains six species. The prothorax is narrow with rounded basal angles and a linear latero-basal impression. The spermathecal duct is moderately long to very long (as in fig. 4f). *A. variolatum* was placed in a species group containing most species of the *A. nigriceps* group by Lindroth (1966). The increased setation of the pedicel (character 33), long spermathecal duct (character 7), and wrinkles on the shaft of the aedeagus (character 28) support removal of it from that vicinity, with placement in the heterogeneous group allied to *A. nutans*.

4. *A. sordens* group. The subgenus *Europhilus* has long been recognized as a distinct lineage. These species are small and piceous, with the third antennal segment apically pubescent. Other *Agonum* possess pubescence on the third antennal segment; *A. belleri*, species in the *A. extensicolle* group, and *A. quadrimaculatum*. Most of these species possess this pubescence in concert with secondary pubescence on other parts of the body. Such is not the case for the *A. sordens* group, which have glabrous bodies. *A. anchomenoides* was placed with *A. ferruginosum, A. piceolum*, and *A. bicolor* based on the possession of an indented apex on the palpi (Lindroth 1966). Lindroth gave this character more weight than the antennal pubescence, which would have placed *A. anchomenoides* in the *A. sordens* group. Examination of series of specimens from a number of species in both groups indicates that the indentation is principally an artifact of preservation. The apical membranous area may inflex upon drying, but this is inconsistently seen within and among these species. The spermathecal assembly supports rejection of palpal apex configuration as a useful character. *A. anchomenoides* is herein placed in *Europhilus* based on spermathecal configuration and antennal pubescence.

5. *A. extensicolle* group. This group has been referred to as subgenus *Stictanchus* Casey (Lindroth 1966) and is the subject of the accompanying revision. The seven species are metallic in color; uniformly possess two, or usually more, setae on the mesocoxal ridge (characters 47 and 48); four or more dorsal elytral setae (character 43); and four or more setae on the apex of the antennal scape (character 32).

6. *A. quadrimaculatum* group. *A. quadrimaculatum* lacks the basal prothoracic seta, a character

state it shares with *Anchomenus dorsalis*. This, and the coloration of *A. quadrimaculatum*, caused Lindroth to include it in the subgenus *Anchomenus*. The two species are not closely related, based on different female internal reproductive tracts. The internal tract of *Agonum quadrimaculatum* has a moderately long duct, much like that of *A. anchomenoides* (fig. 4f). The appropriate subgeneric name for *A. quadrimaculatum* is *Deratanchus* Casey (1920); type species *A. quadrimaculatum*. It can be diagnosed from the *A. extensicolle* group by the presence of only three dorsal elytral setae (character 43), lack of prothoracic basal setae (character 41), spermathecal duct shorter than *A. extensicolle* group, gonostyli with inner lateral seta (character 18), and more than two outer lateral setae (character 19).

The final lineage within the Platyni is *Platynus* and allied genera. The exemplars used in this analysis all possess constricted necks (character 39). This character serves to diagnose many *Platynus*, but is not present in all Mexican and Central American species. None of the *Platynus* examined possess the long spermathecal duct common to many *Agonum*, and this difference may serve as a diagnostic character of some utility. The presence of at least two lineages in North American *Platynus* is indicated by spermathecal configuration. *P. decentis, P. ovipennis, P. brunneomarginatus*, and *P. tenuicollis* LeConte all possess spermathecae that attach via a duct to the median oviduct at the bursa (figs. 5a, b). *P. hypolithos, P. angustatus*, and *P. cincticollis* Say have the spermatheca appressed to the bursa (fig. 5d). Such a grouping runs counter to those previously suggested (e.g. Lindroth 1966).

Onypterygia appear to be derived *Platynus*, exhibiting specialized gonostyli (fig. 6j; character 21), large conical spines on the internal sac (fig. 8e; character 27), no anterior lateral prothoracic seta (at least in *O. tricolor* and *O. fulgens*) (character 40), bilobed tarsi (character 52), and pectinate, chelate claws (characters 55 and 56). The *Onypterygia* groundplan may in fact have been derived more than once from a *Platynus*-like ancestor (D. R. Whitehead, pers. comm.).

Glyptolenus is not included in the analysis due to lack of males for dissection. The female possesses a reproductive tract in which the spermatheca enters the bursa (character 2), a short spermathecal duct with apical spermatheca (characters 5 and 6), constricted neck (character 39), foretibial anterior and posterior sulci (character 26), and internal and external metatarsal furrows (characters 49 and 50). Based on these features, *Glyptolenus* can be placed as a derivative of a *Platynus*-like ancestor.

Colpodes buchanani Hope was also examined. It possesses all the synapomorphies required to place it in the *Platynus* line (characters 5, 26, 39, 49, 50). The spermatheca is nearly identical to that of *P. brunneomarginatus*. It possesses glabrous ungues (characters 53 and 54), and a spinose aedeagal sac (character 27). Such a combination of synapomorphies precludes assignment to any group in the set of exemplars used in this analysis. It is clear that a worldwide survey of *Platynus* and *Colpodes* is required to determine generic concepts for these names.

The classification derived from this cladistic analysis is presented in Appendix 2. The Platynini is considered a distinct tribe. Its ancestor probably represents the basal stock for carabid tribes such as the lebiomorph lineages, among others. Diversification within the platynine lineage occurred without extensive modification of the platynine groundplan, confounding the use of only external characters as the basis for classification. The large amount of homoplasy existing in the cladistic data set implies that certain derived traits arose independently many times. The affinities among European and American species implies that a regional analysis will be unsatisfactory for determination of phylogenetic relationships within this group. Only a worldwide perspective will produce a stable classification.

COMPARISON WITH PRIOR CLASSIFICATIONS

The classification presented herein recognizes two major lineages in North American Platynini; the subtribes Sphodri and Platyni. Comparison with two classifications previously proposed (Lindroth 1956; Habu 1978) shows that the differences result from differing principles of classification, as well as different data.

Lindroth (1956) recognizes four groups within this tribe; Sphodri, Synuchi, Pristosiae, and Agoni (=Platyni). The Sphodri possess highly derived styloid parameres in the male genitalia, which Lindroth used to distinguish them from the Synuchi. The interpretation in this study that the styloid parameres are a derivation of the already somewhat derived parameres (character 24) of the Synuchi unites these groups. The additional shared derived character of absence of the setal fringe on the penultimate gonostylus segment (character 16) also unites the Sphodri and Synuchi. Hence, Lindroth's Sphodri and Synuchi are recognized herein as one subtribe; the Sphodri. This study's Platyni corresponds to Lindroth's Agoni. The Pristosiae is not represented in North America, and was not included in this analysis.

Habu's (1978) classification recognizes three subtribes of Platynini: his Platynina, Sphodrina, and Dolichina. Since Habu bases his classification strictly on characters of the female gonostyli, his Sphodrina includes Lindroth's Pristosiae. Habu's recognition of Sphodrina and Dolichina as distinct entities is derived from a phenetic interpretation of differences between the two types of gonostyli (as in figs. 5e, f). Habu's phenetic interpretation does not account for the loss of the preapical sensory setae (character 22) and reduction in lateral style seta number (character 20) as derived character states of *Synuchus*. The important synapomorphies of characters 16 and 24 unite these groups, whereas the autapomorphies of *Synuchus* are useful only for resolution within the subtribe Sphodri. Habu's unification of Sphodrina and Pristosiae discounts the importance of the inverted aedeagus of the latter.

Jeannel (1942) was the first to recognize the Platyni and Sphodri as sister taxa, ranking them as tribes within his Pterostichidae. Schuler (1963) analyzed the internal female reproductive tract and found that characters of the female supported Jeannel's classification of European Platynini. The utility of the male and female characters for establishment of a worldwide classification of Platynini is shown by the condordance of Jeannel's (1942) classification with that proposed above. Analysis of Habu's (1978) results using cladistic criteria likewise produces a concordant classification.

KEY TO GENERA OF PLATYNINI IN NORTH AMERICA

1. Mentum without tooth; spermathecal duct short and cuplike, sclerotized, entering on median oviduct. . . . . *Olisthopus*
   Mentum with variously developed median tooth; spermatheca ovoid to elongate, duct not heavily sclerotized. 2
2. Claws variously pectinate. 3
   Claws smooth, not pectinate. 6
3. Claws chelate; tarsi of both sexes with numerous flattened setae on underside of first four tarsal segments. . . . . *Onypterygia*
   Claws not chelate; hind and mid-tarsi with only lateral rows of spines beneath. 4
4. Tarsi without lateral sulci on upper surface; upper surfaces hairy. . . . . *Pristonychus*
   Tarsi with lateral sulci; upper surfaces glabrous. 5
5. Right paramere styloid (fig. 7a); gonostyli with two lateral setae, one median seta and preapical sensory setae (fig. 5e). . . . . *Calathus*
   Right paramere fiddle-shaped (fig. 7b); gonostyli with one lateral seta, one median seta, lacking preapical sensory setae (fig. 5f). . . . . *Synuchus*
6. Outer surface of foretibia with median canalicula, fore and hind grooves. . . . . *Glyptolenus*
   Outer surface of foretibia without groove, fore and hind grooves present. 7
7. Male genitalia with apical setae on parameres (figs. 7c, d). 8
   Male genitalia with glabrous parameres. 9
8. Dorsal surface flat, elytra with parallel sides; wings well-developed. . . . . *Sericoda*
   Dorsal surface convex, elytra broadly ovoid, brachypterous. . . . . *Elliptoleus*
9. Median sulcus well-developed on metatarsi. 10
   Median sulcus absent or only faintly visible on metatarsi (doubtful cases key both ways). 14
10. Neck not constricted, impression not visible on dorsum of head when viewed from the side. 11
    Neck constricted, at least moderate impression visible. 13
11. Third antennal segment pubescent on apical half. *Agonum thoreyi*
    Third antennal segment glabrous except for apical ring of setae. 12
12. Upper body surface bright metallic green to aeneous; tibiae testaceous. . . . *Agonum errans*
    Upper body surface dark, may be faintly metallic; legs concolorous with body. . . . . *Pseudanchus*

13. Third antennal segment subequal in length to fourth, glabrous except for apical ring of setae; hind wings long with a reflexed apex. . . . . *Paranchus albipes*
    Third antennal segment at least 1.15 times larger than fourth, often longer, often covered with sparse fine pubescence; hind wings totally reduced. . . . . *Rhadine* (in part)
14. Posterior lateral pronotal seta absent. 15
    Posterior lateral pronotal seta present. 17
15. Neck constricted; prothorax constricted at base; third antennal segment 1.15 or more times length of fourth; rufotestaceous beetles with long legs. . . . . *Rhadine* (in part)
    Neck not constricted; third and fourth antennal segments of subequal length; body at least partially piceous. 16
16. Elytra bicolored orange and black; mesocoxal ridge with more than two setae (as in fig. 36b-c). . . . . *Agonum quadrimaculatum*
    Elytra bicolored or not; mesocoxal ridge with one seta. . . . . *Anchomenus*
17. Spines or heavy spicular fields present on internal sac of aedeagus. 18
    Aedeagal sac without heavy spines or fields of stout spicules. 22
18. Neck constricted. 19
    Neck not constricted. 20
19. Gonostyli with penultimate segment bearing a fringe of setae (figs. 6h, 8). . . . . *Platynus* (in part: *Colpodes buchanani* Hope will key to this couplet; species introduced on Pacific Coast; elytra are metallic aeneous)
    Gonostyli with penultimate segment glabrous or with only a few setae (fig. 6a). . . . . *Rhadine* (in part)
20. Four or more setae at apex of second antennal segments; more than three dorsal elytral setae. . . . . *Tanystoma*
    Less than four setae at apex of second antennal segment, three dorsal slytral setae. 21
21. Elytral striae fine, impunctate; body brunneus; 6.25-7.0mm long; Arizona. . . . . *Agonum languidum*
    Elytral striae with fine but evident punctures; dorsum piceous although legs and body margins rufo-piceous; 7.2-8.7mm long; eastern United States. . . . . *Agonum collare*
22. Neck constricted; fine pubescence covering body surface. 23
    Neck constricted or not; if constricted, body glabrous. 24
23. Mentum with bifid tooth. . . . . *Atranus pubescens*
    Mentum with simple, unidentate tooth. . . . . *Anchus pusillus*
24. Neck constricted (reliable character only for *Platynus* north of Mexico). . *Platynus* (in part)
    Neck not constricted. . . . . *Agonum* (in part)

KEY TO SPECIES GROUPS OF *AGONUM* IN AMERICA NORTH OF MEXICO

The following key can be used to place *Agonum* species into the groups recognized in the classification presented in Appendix 2. All species groups, except the *A. collare* group, already possess an available subgeneric name. These are included to facilitate comparison of the North American fauna to those of Europe or Japan. The less formal species group designations are better suited for discussion limited to the North American fauna.

Lindroth's (1966) key remains the best means for identification of the species of North American *Agonum*. His key includes species of genera recognized herein: *Sericoda, Anchomenus, Pseudanchus, Paranchus, Anchus, Rhadine, Tanystoma*, and *Platynus*. His interpretation of the species of the *A. extensicolle* group was not totally correct, largely because this group is most diverse in the American Southwest. These species can be keyed using this work. The only other described species found north of Mexico not covered in Lindroth's key are *A. pacificum* Casey, *A. muiri* Liebherr, and *A. languidum* Horn. *A. pacificum* and *A. muiri* can be keyed using Liebherr (1984). *A. languidum* will key to couplet 59 of Lindroth's key, and can be recognized by modifying the couplet to include the third alternative:

59′. 1. metatarsal segment with evident furrow also internally. Pronotal lateral margin convex outside rounded obtuse hind angle. Pronotum 1.6 times head width across eyes. .... *A. languidum*

The following key to groups should be used along with Lindroth's (1966) key to identify *Agonum*. The major value of this key is that it allows an initial sorting of species into species groups, easing the burden of species identification. Groups that will prove difficult in this key include the *A. nigriceps* group and the *A. nutans* group.

1. Antennae with pubescence beginning on first, second, or third article. 2
   Antennae with pubescence beginning on fourth article, articles 1-3 glabrous except for apical setae. 5
2. Elytra with basal and subapical orange spot, otherwise piceous to black; thorax orange; head black; prothorax with latero-basal seta lacking. .. *A. quadrimaculatum* (subg. *Deratanchus*)
   Elytra and thorax not colored as above; prothorax with latero-basal seta present. 3
3. Elytra with coarse setigerous punctures on the odd numbered intervals; latero-basal depressions of pronotum deep and broad, with rugose surface; upper body surface metallic

blue green to aeneous *(A. belleri)*. . . . . *A. marginatum* group in part (subg. *Agonum*)
Elytra without course setigerous punctures on odd numbered intervals, punctures restricted to third interval; or if punctures are present on intervals 3, 5, and 7, dorsal surface is covered with fine pubescence. 4

4. Mesocoxal ridge bearing one seta. . . . . *A. sordens* group (subg. *Europhilus*)
Mesocoxal ridge bearing three or more (rarely two) setae. . . . . *A. extensicolle* group in part (subg. *Stictanchus*)

5. One large sensory seta and three or more smaller setae at apex of first antennal segment; three or more (rarely two) setae on mesocoxal ridge. . . . . *A. extensicolle* group in part (subg. *Stictanchus*)
Large sensory seta at apex of first antennal segment accompanied by fewer than three smaller setae; if three small setae present, then mesocoxal ridge with only one seta. 6

6. Ungues glabrous underneath, or bearing up to four small setae. 7
Ungues bearing six or more setae beneath. 9

7. Elytra with three foveate dorsal punctures *(A. excavatum)*. . . . . *A. marginatum* group in part (subg. *Agonum*)
Elytra with three to six nonfoveate dorsal punctures. 8

8. Spermatheca moniliform, with regular beadlike constrictions. . . . . *A. ferreum* group (subg. *Stereagonum*)
Spermatheca fusiform, without moniliform constrictions. . . . . *A. nutans* group in part (subg. *Micragonum*)

9. Latero-basal pronotal depressions nearly obsolete (least so in *A. piceolum*); body lightly sclerotized, ferrugineous; small beetles, body length 5.0-7.8mm. . . . . *A. nigriceps* group (subg. *Platynomicrus*)
Latero-basal pronotal depressions evident, at least prothorax and head heavily sclerotized and darkly pigmented; larger beetles, body length 5.4-more than 10.0mm (*A. variolatum* is smallest, most lightly sclerotized species to key to this point). 10

10. Prothorax with restricted but deep punctiform latero-basal depressions, pronotum almost circular, hind angles lost. 11
Prothorax with broader latero-basal depressions, hind angles of prothorax developed or not. 12

11. Elytra with shallow impression at middle (near second dorsal puncture); prothorax slightly wider than head, body surface faintly metallic *(A. aeruginosum)*. . . . . *A. nutans* group in part (subg. *Micragonum*)
Elytra without depression, prothorax much broader than head, upper body surface shiny but not metallic. . . . . *A. punctiforme* group (subg. *Circinalia*)

12. Prothorax narrow, hind angles obsolete, width to length ratio less than 1.30. . . . . *A. nutans* group in part (subg. *Micragonum*)
Prothorax wider, hind angles weakly to well-developed, width to length ratio generally greater than 1.30. 13

13. Spermathecal duct at least three times length of spermatheca (figs. 4g, h). . . . . *A. marginatum* group in part (subg. *Agonum*)
Spermathecal duct as wide as spermatheca, subequal in length to spermatheca (figs. 3g, h). . . . . *A. collare* group

CIRCUMSCRIPTION OF SPECIES

Taxonomic study involves grouping organisms into taxa that can be defined on the basis of their intrinsic attributes. Beyond this discrimination of kinds, an assessment of the affinities of the various kinds can be made, creating a classification that permits an understanding of the diversity of the group under study.

The preferable classificatory system is one that predicts the most about a group's attributes, given the least information. An example of such a system is the hierarchical classificatory system used in biology. The species is the most important unit in this system because it is the rank at which different kinds of organisms are universally recognized. The concept of a species has changed drastically through time, but in practice, humans can perceive the distinctiveness of organisms with equal facility whether they are illiterate, or highly trained in orthodox scientific fields (Raven, Berlin, and Breedlove 1971). With the establishment of evolutionary theory, the concept of the species as an evolutionary unit has become almost universally accepted. The definition of that evolutionary unit, however, has never been agreed upon.

The biological species concept of Mayr (1940, 1963) is one definition that views the species as more than "the typological interpretation of a 'class of objects.'" Considering species to be "groups of interbreeding natural populations that are reproductively isolated from other such groups," Mayr (1969) assumes that the species is a reproductive unit exchanging genes throughout its range. Such a definition assumes far more about the nature of a species than is discernible by the taxonomist (Ehrlich and Raven 1969).

The species definition of Simpson (1961) views the species as an evolutionary unit much as does that of Mayr, without the added stricture of constant gene flow. The evolutionary species of Simpson is "a lineage (an ancestral-descendant sequence of populations) evolving separately from others and with its own unitary evolutionary role and tendencies." As an idealized concept, this definition is consistent with what we know of the evolution of the world's biota. The ancestral-descendant sequence of populations insures genetic continuity within the species. The temporary separation of one set of populations from other sets does not diminish the evolutionary role or tendencies of the species, unless the separation is sufficient to insure divergence to full species status. This definition permits integration of taxonomic study with the fields of molecular biology and genetics. The evolutionary species definition also assumes more than is known to the systematist.

Taxonomists should set forth the criteria used for recognition of taxa in order to permit evaluation of their work. If the idealized species concepts require more information than that available for circumscription of species, those concepts cannot be used as criteria. They may still

Figure 11. Flow diagram illustrating steps involved in circumscription of species. Electrophoretic analysis can be used alone to circumscribe species (dashed lines), but in this study it is followed up by biometric analysis to assist in circumscription.

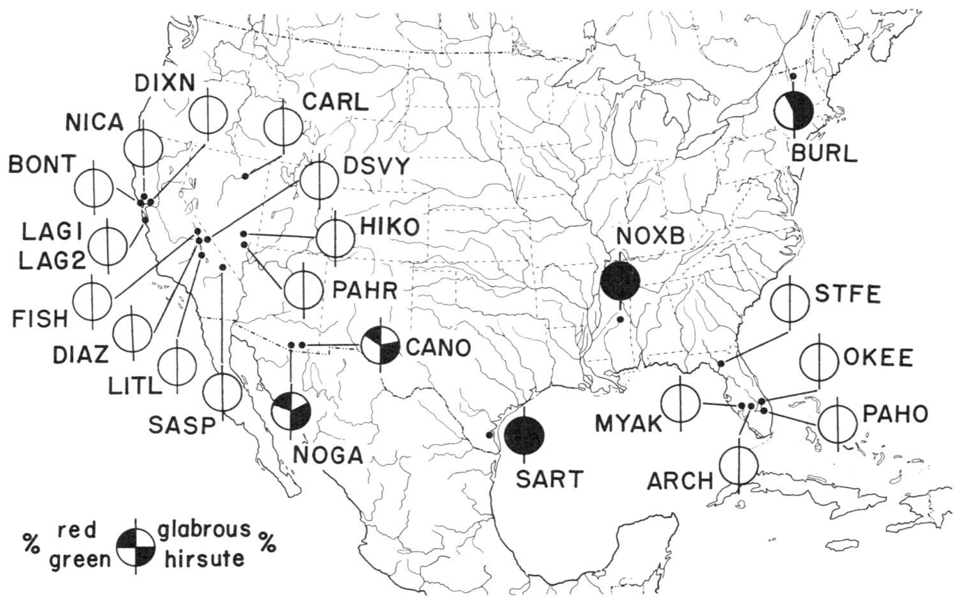

Figure 12. Sample localities for *Agonum decorum* (Say) and *Agonum elongatulum* (Dej.) used in electrophoretic analysis. Five populations in Florida represent *A. elongatulum*; all others represent *A. decorum*. Right half of circle represents percent individuals of glabrous or hirsute phenotype, left half represents percent of red or green phenotype.

be of use in formulating hypotheses concerning evolutionary mechanisms. An operational species concept must also be presented that allows examination of the methods used in species circumscription. A species criterion borrowing heavily from that of Doyen and Slobodchikoff (1974) is used in this study. Their model is a generalized scheme that recognizes species distinctness based on phenetic differences between taxa. If a study set produces ambiguous results based on an examination of overall similarity, additional criteria are then used to detect specific limits. In sympatry, reproductive isolation between two or more sets of individuals demonstrated by the existence of discrete classes of individuals permits recognition of species. If sets of individuals are allopatrically distributed, the ecological requirements of the various sets are examined. If differences in these requirements can be documented, distinct species are recognized. If differences cannot be demonstrated, the allopatric sets are considered conspecific. An operational species concept is not as generalized or elegant as the biological or evolutionary species concepts, but operational criteria are essential for the advancement of systematics. Only through critical evaluation of scientific work can progress be ensured, and operational criteria are the best means for such evaluation.

The scheme used in this study for circumscription of species is illustrated in figure 11. After taxonomic material is segregated into phenetically similar groups based on qualitative morphological characters, live samples can be obtained from as many of the phena as possible.

Electrophoresis permits sampling the gene frequencies through analysis of soluble proteins. Such analysis offers the advantage that one can be relatively certain of homology for the various loci sampled. Analysis of the genetic variability observed in most natural populations permits

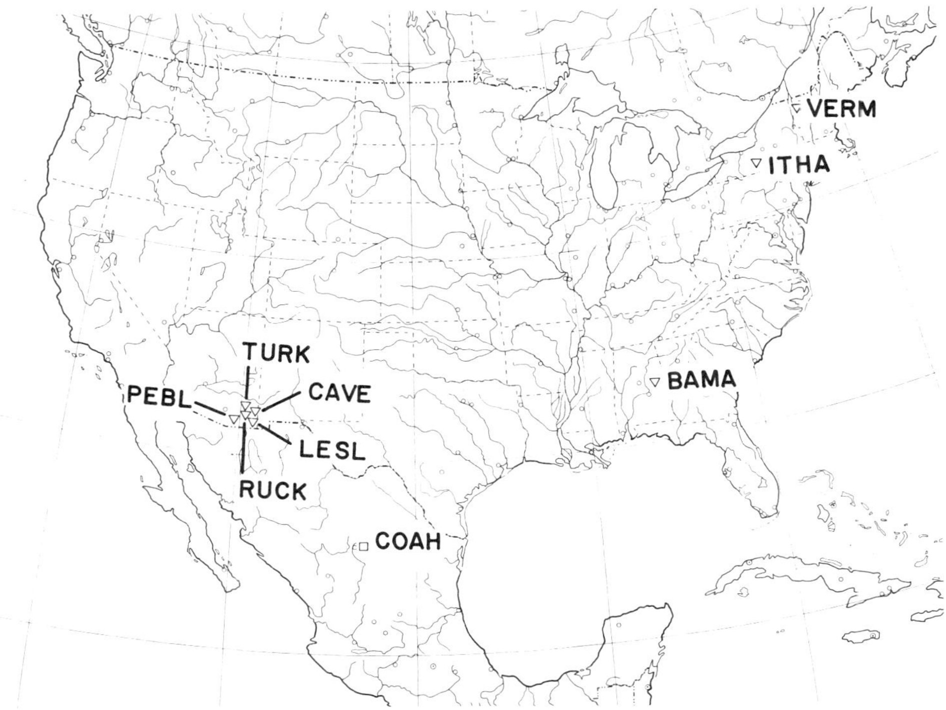

Figure 13. Sample localities of *Agonum extensicolle* (Say) (triangles) and *Agonum extimum* Liebherr, n. sp. (square) used in electrophoretic analysis.

estimation of evolutionary affinities using overall genetic similarity. Well-known difficulties with the technique include: (1) hidden variability due to insufficient resolution of alleles with very similar mobilities under generalized assaying conditions (Ayala 1982); (2) the reliance of allelomorph movement on electric charge; and (3) the possibility of amino acid replacement without change in charge (Lewontin and Hubby 1966). Electrophoretic data offer an independent data set that can be compared with hypotheses based on morphological characters. Corroboration of hypotheses is accomplished by the attainment of congruent results by the two methods.

In this study, electrophoretic analysis resulted in both corroboration and rejection of the original phenetic groupings for different species. All groups subjected to electrophoretic analysis were subsequently analyzed using biometric techniques. This analysis, using 13 morphometric characters on population samples representing all groups in the taxonomic study material, provided a third means of determining distinctiveness of population sets.

In *Agonum decorum*, both the electrophoretic and biometric analyses showed the species to be polymorphic for setation and coloration. These results dictated that this species be recognized as polymorphic. Subsequent to this recognition, crossing data from the various morphs permitted elucidation of the genetic nature of the polymorphisms (Liebherr 1983a).

Circumscription of *Agonum extensicolle* proved to be as difficult as circumscription of *A. decorum*. In this case, a substitution of alleles occurs among populations in Arizona and three

Table 1. Populations sampled and species assignment, percent polymorphic loci using 95% frequency of most common allele, and estimated ($\hat{H}$) and observed (H_o) heterozygosity with associated standard error.

Population number and code	Locality and date	Species	Percent loci polymorphic	Heterozygosity $\hat{H}$	s.e.	H_o	s.e.
1 NICA	CA:Marin Co., Nicasio Rsvr., 7 Apr 1981	*decorum*	65	.239	.059	.236	.060
2 DIXN	CA:Solano Co., 12 mi S Dixon, 4 Apr 1981	*decorum*	53	.271	.077	.244	.067
3 BONT	CA:Marin Co., Bon Tempe Lake, 7 Apr 1981	*decorum*	59	.220	.062	.199	.052
4 LAG1	CA:Sta. Clara Co., Lake Lagunitas, 17 May 1980	*decorum*	47	.234	.068	.211	.061
5 LAG2	CA:Sta. Clara Co., Lake Lagunitas, 5 Apr 1981	*decorum*	53	.278	.074	.247	.067
6 FISH	CA:Inyo Co., Fish Springs, 12 Jun 1980	*decorum*	65	.246	.065	.247	.068
7 DIAZ	CA:Inyo Co., Diaz Lake, 13 Jun 1980	*decorum*	65	.227	.060	.201	.051
‘8 LITL	CA:Inyo Co., Little Lake, 13 Jun 1980	*decorum*	65	.276	.066	.249	.059
9 DSVY	CA:Inyo Co., Deep Springs Vy., 12 Jun 1980	*decorum*	47	.230	.066	.203	.061
10 CARL	NV:Elko Co., Carlin, 18 Jun 1980	*decorum*	53	.225	.060	.173	.049
11 HIKO	NV:Lincoln Co., Hiko Springs, 16 Jun 1980	*decorum*	47	.255	.076	.217	.073
12 PAHR	NV:Lincoln Co., Pahranagat N.W.R., 16 Jun 1980	*decorum*	53	.241	.066	.245	.069
13 SASP	CA:Death Vy. N.M., Saratoga Spgs., 24 Mar 1980	*decorum*	59	.250	.069	.268	.077
14 NOGA	AZ:Sta. Cruz Co., 6 mi N Nogales, 4–5 Aug 1979	*decorum*	53	.250	.064	.157	.047
15 CANO	AZ:Sta. Cruz Co., Canelo, 6 Aug 1979	*decorum*	53	.218	.059	.174	.048
16 SART	TX:Kenedy Co., Sarita, 22 May 1981	*decorum*	53	.234	.065	.225	.061

Table 1 (cont.).

Population number and code	Locality and date	Species	Percent loci polymorphic	Heterozygosity $\hat{H}$	s.e.	H_o	s.e.
17 NOXB	MS:Oktibbeha Co., Noxubee N.W.R., 14 May 1981	*decorum*	59	.270	.057	.260	.056
18 BURL	VT:Chittenden Co., 5 mi S Burlington, 24 May 1981	*decorum*	65	.245	.059	.203	.055
19 STFE	FL:Suwanee Co., Sta. Fe R. N Bell, 23 Apr 1981	*elongatulum*	53	.236	.060	.215	.055
20 OKEE	FL:Okeechobee Co., Okee-Tantie Rec. Area, 30 Apr 1981	*elongatulum*	71	.302	.061	.257	.049
21 ARCH	FL:Highlands Co., Archbold Biol. Sta., 2 May 1981	*elongatulum*	65	.286	.061	.238	.056
22 MYAK	FL:Sarasota Co., Myakka R. St. Pk., 1 May 1981	*elongatulum*	53	.248	.063	.233	.063
23 PAHO	FL:Palm Beach Co., Pahokee St. Pk., 29 Apr 1981	*elongatulum*	59	.241	.056	.221	.052
24 ITHA	NY:Tompkins Co., Ithaca, 10 Jun 1981	*extensicolle*	53	.206	.063	.157	.054
25 VERM	VT:Chittenden Co., 5 mi S Burlington, 24 May 1981	*extensicolle*	59	.218	.059	.220	.056
26 BAMA	AL:Talladega Co., Chandler Spgs., 12 May 1981	*extensicolle*	41	.192	.078	.342	.083
27 LESL	AZ:Cochise Co., Leslie Cyn., 4500′ el, 25 Aug 1981	*extensicolle*	53	.163	.049	.235	.060
28 CAVE	AZ:Cochise Co., Cave Ck. Cyn., 5200′ el, 22 Aug 1981	*extensicolle*	47	.174	.057	.216	.052
29 PEBL	AZ:Sta. Cruz Co., Peña Blanca Lk., 29 Aug 1981	*extensicolle*	35	.142	.056	.198	.048
30 RUCK	AZ:Cochise Co., Rucker Cyn., 6000′ el, 23 Aug 1981	*extensicolle*	53	.197	.058	.291	.071
31 TURK	AZ:Cochise Co., Turkey Ck., 5200′ el, 20 Aug 1981	*extensicolle*	29	.114	.047	.078	.032
32 COAH	Mexico:Coahuila, 19.4 mi E San Pedro, 31 May 1981	*eximum*	47	.145	.053	.146	.053

populations from eastern North America. In Ward (1980), such a substitution is grounds for species recognition. Morphological criteria were subsequently found that substantiated full species status (Ward, pers. comm. April 1983). In this study supporting data were not found, and *A. extensicolle* is treated as a genetically variable species. Evidence from the biometric study, analysis of color, and wing development support this view. Conflicting data require evaluation of each type of evidence and, especially with *A. extensicolle*, this evaluation requires an eclectic approach to the analysis.

For several species, limited or nonexistent electrophoretic data reduced the circumscription of species to comparison of morphological and biometric data. In all cases, agreement between these analyses permitted species circumscription.

For each species, ecological, and biogeographic or chorological data were used in the final assessment of species limits. These data were used as corroboration for other data, and also provided information about mechanisms creating the divergence that resulted in the species observed.

ELECTROPHORETIC ANALYSIS

Materials and Methods. Two electrophoretic analyses are presented. The first contains 32 samples from 31 populations (figs. 12, 13). These include 18 samples representing 17 populations of *Agonum decorum*, five of *Agonum elongatulum*, eight of *Agonum extensicolle*, and one of *Agonum extimum* (table 1). Specimens were collected from the margins of marshes, lakes, or rivers. Populations of *Agonum* species are usually dense, and samples could be obtained from a small area of habitat. Beetles were held without food for three days, then placed in nunc vials and stored in liquid nitrogen until brought back to the laboratory. Upon return until preparation for electrophoresis, the whole beetles were stored at –70°C. Voucher specimens from all series of material analyzed in this study have been placed in the Essig Museum of Entomology, University of California, Berkeley.

Electrophoresis was conducted using methods modified from Selander et al. (1971), Ayala et al. (1972), and Harris and Hopkinson (1976). Protocols are outlined in Liebherr (1983a). Several loci that were monomorphic in that study vary in this data set. These loci include: glyceraldehyde-phosphate dehydrogenase (Gapd), one locus, monomeric structure; mitochondrial aconitase (Acon-2), monomeric; isocitrate dehydrogenase (Idh), one locus, dimeric; and mitochondrial malate dehydrogenase (Mdh), dimeric. All loci possess alleles that migrate anodally, except those for Acon-2 and Mdh-2, which migrate cathodally. A total of 17 loci were assayed for all populations (table 2). Allelic frequency data for all samples are included in Appendix 3.

Electrophoresis was conducted in 1980 and 1981, with the dates of collection listed in table 1 showing which populations were run together. The population of *Agonum decorum* from Lake Lagunitas, Santa Clara County, California was sampled both years and used as a standard. At least one individual from this population was run on each gel to aid in determination of the most common allele. After all populations were analyzed in each year, samples from individual beetles representing each allele from each gel were rerun together to permit assignment of putatively homologous alleles. Thus all alleles were standardized within years. Enough individuals were sampled from the Lake Lagunitas population in both years to permit standardization between years.

A sample of ten individuals was the minimum desired for all populations. In this way, secondary alleles could be sampled, providing the ability to cluster populations into species units. Larger samples are desirable for accurate assessment of infrapopulational genetic structure.

Table 2. Loci assayed, with mobility of most common allele under protocol of Liebherr (1983a).

Esterase-d	Est-d	32
Phenylalanine-proline peptidase	Phpr	30
Acid phosphatase	Acph	6
Glucose phosphate isomerase	Pgi	16
Phosphoglucomutase-1	Pgm-1	42
Phosphoglucomutase-2	Pgm-2	22
Mannose phosphate isomerase	Mpi	38
Aldehyde dehydrogenase	Aldh	14
Glyceraldehyde-phosphate dehydrogenase	Gapd	27
Adenosine deaminase	Ada	32
Enolase	Eno	11
6-phosphogluconate dehydrogenase	6-Pgd	10
Superoxide dismutase	Sod	20
soluble Aconitase	Acon-1	28
mitochondrial Aconitase*	Acon-2	16
Isocitrate dehydrogenase	Idh	13
mitochondrial Malate dehydrogenase*	Mdh	6

* Migrates cathodally.

For circumscription at the species level and above, and for estimation of genetic heterozygosity, sample sizes of 10-20 are considered adequate (Gorman and Renzi 1979). Accurate scoring of 10 individuals was not possible for some loci, but smaller sample sizes are included to permit use of as many loci as possible for determination of genetic distances. Although sample sizes are small for several populations, the samples are included in order to obtain as much information as possible concerning these populations. The results obtained must be evaluated in light of the sample size, and cautionary notes are included where appropriate.

The electrophoretic data were analyzed using parameters that are averaged over all loci, and on a locus by locus basis. Individual heterozygosity estimated from gene frequencies and interpopulational Nei's and Rogers' genetic distances were obtained from the Gendist program of Roychoudhury modified by Y. Tateno to account for sample size (Levene 1949; Nei 1978) and further modified by G. Barrowclough to give Rogers' distance (Rogers 1972). The interpopulational Rogers' genetic distances are graphically portrayed using a minimally connected graph, or Prim network (Sneath and Sokal 1973). The F-statistic, F_{ST}, of Wright (1965; 1969; 1978) is presented for all variable loci using the program of G. Barrowclough. This program generates F_{ST} values based on all alleles, corrected for sample size using the method of Wright (1978).

Wright's F_{ST} value is a measure of the genetic variance of subdivisions of a study set, divided by the maximum genetic variance that is possible based on overall gene frequencies. Wright (1969, 290-95) derives the equations for this index, which is calculated as:

$$F_{ST} = \frac{\frac{1}{n} \sum_{i=1}^{n} \sum_{j=1}^{a} \left(q_{ST} - q_T\right)^2}{\sum_{j=1}^{a} q_T \left(1 - q_T\right)}$$

where n = number of population subdivisions, a = number of alleles at the locus analyzed, q_{ST} is the frequency of an allele in a population subdivision, and q_T is the overall frequency of the allele in the study set. This calculation assumes random uniting of gametes within each population subdivision. The significance of F_{ST} values can be determined using the heterogeneity χ^2 test described by Workman and Niswander (1970).

The second electrophoretic analysis includes six samples representing five species (table 30; Appendix 7). Each species is represented by six specimens, with individuals representing each sample run on the same gel. This data set is drawn from material used for the first analysis, with the exception of the sample of *Agonum texanum*. The techniques used in both analyses are identical. The Rogers' genetic distances for this exemplar data set are portrayed using the Farris 78 Wagner algorithm (Farris 1970).

Results and Discussion. *Individual Heterozygosity.* The average level of heterozygosity estimated from gene frequencies for all populations sampled is shown in table 1. For populations of *Agonum decorum*, the levels of heterozygosity range from .218 to .278, averaging .245. Similar levels are seen in the five populations of *A. elongatulum*, range .236 to .302, average .263. The eight population samples of *A. extensicolle* exhibit heterozygosity levels ranging from .114 to .218, averaging .176. There is no significant difference in heterozygosity levels between *A. decorum* and *A. elongatulum*, but *A. extensicolle* has a significantly lower level of heterozygosity compared to the combined population samples of the other two species (Wilcoxon ranked sum test, $n_1 = 22$ (LAG2 sample excluded), $n_2 = 8$, $p < .001$). The single population sample of *A. extimum* possesses a heterozygosity level of .145.

These heterozygosity levels are among the highest recorded for insects (Powell 1975; Nevo 1978; Ward 1980). The high levels are not caused by selection of loci usually considered exceedingly polymorphic: only three loci of the variable substrate class (Johnson 1974) are assayed in this study (table 3). Observed heterozygosity levels of all three classes of enzymes are elevated compared to the averages for *Drosophila* species cited in Johnson (1974) and Powell (1975). In general, the results using *Agonum* parallel heterozygosities for the types of enzymes in those studies.

Only a limited number of electrophoretic studies of beetles based on a substantial number of loci have been reported. Suomalainen and Saura (1973) reported on several species of

Table 3. Observed heterozygosity for three species of *Agonum*, using locus classification of Johnson (1974). Number of loci in each class for this study in parentheses.

Class of protein	*A. decorum*	*A. elongatulum*	*A. extensicolle*	*Drosophila* spp.*
Variable substrate (3 loci)	.350	.438	.260	.240
Regulatory (7 loci)	.277	.242	.191	.190
Non-regulatory (7 loci)	.111	.138	.114	.060
Number of samples used to obtain H_o	17	5	8	13

* Values of heterozygosity from Johnson (1974).

Curculionidae. *Otiorrhynchus scaber* L. has an observed heterozygosity level of 30.9%, with 17 of 23, or 74%, variable substrate or regulatory enzymes (Johnson 1974). Another weevil, *Strophosomus capitatus* Degeer, averages 17% heterozygosity (Suomalainen and Saura 1973). Cave beetles have been relatively well studied, with different beetle families providing differing results. The leptodirid beetle, *Ptomophagus hirtus* Tellkampf, exhibits low levels of heterozygosity, averaging 4.8% heterozygosity per individual based on 13 loci (Laing, Carmody, and Peck 1976). The carabid beetle *Neophaenops tellkampfii* Erichson, exhibits moderate levels of heterozygosity, averaging 12.1% (Giuseffi, Kane, and Duggleby 1978) and 15.4% (Turanchik and Kane 1979) in the two reported studies. Avise and Selander (1972) reported heterozygosity of .218 for three loci averaged over two populations in the platynine carabid *Rhadine subterranea* Van Dyke. Based on these scanty data, it appears the three carabid genera studied possess genetic variability equal to or slightly greater than that seen in *Drosophila* species.

Inspection of heterozygosity levels for populations across the geographic distribution of *Agonum* species reveals no striking patterns. For *A. decorum*, populations in transmontane California and Nevada and the Great Basin could be considered marginal and/or isolated, since they are restricted to habitat surrounding isolated lakes, springs, and rivers. Nevertheless, heterozygosity levels for these populations (numbers 6–13) average .244 (table 1), whereas the other ten samples from more mesic life zones possess heterozygosity levels averaging .246. *Agonum extensicolle* populations from Arizona are generally less heterozygous than those in the eastern United States, but none of the differences between western and eastern populations approaches significance.

Deviations from Hardy-Weinberg Equilibrium. Most observed heterozygosity levels for *Agonum decorum* and *A. elongatulum* are lower than those expected by estimation from gene frequencies (table 1), though no differences are statistically significant. In *A. extensicolle*, the opposite tendency is represented: six of eight samples exhibit higher observed heterozygosity levels. The disparity between the observed and expected heterozygosity levels is attributed to sample size, and I assume that all populations of *Agonum* included exhibit random mating structure. Since samples were collected from dense beetle populations occupying a limited area of habitat, panmixis within population samples is expected.

Genetic Distance. A matrix of Rogers' and Nei's genetic distances between samples is presented in Appendix 4. Rogers' genetic distance is used for portrayal of populational affinities (fig. 14). Three distinct populational clusters are present on the minimally connected network. The single population of *Agonum extimum* (COAH) is separated from the *A. extensicolle* samples by a genetic distance of at least .326. The distance between *A. extimum* and *A. extensicolle* includes substitution at three loci: Est-d, Acon-1 and Mdh. The minimum distance of COAH from the populations representing *A. decorum* and *A. elongatulum* is .405 (HIKO to COAH). This distance estimate includes allelic substitutions at three loci: Acph, Est-d, and Mdh (Appendix 3).

The grouping of *Agonum extensicolle* populations reflects geographic proximity, with all populations from Arizona linked together. The Arizona populations exhibit substitution for alleles at the Pgm-1 locus (Appendix 3, Pgm-1 alleles 1.18 and 1.14). The eastern populations of *A. extensicolle* possess only alleles 1.0, .94, and .89 at this locus. Such a substitution could imply a genetic discontinuity within this set of populations, consistent with the existence of two distinct species. In this case, several overriding factors within the electrophoretic data refute such a conclusion. The genetic distance analysis shows that *A. extensicolle* is genetically heterogeneous. The Alabama population is joined to other eastern populations by a minimum genetic distance

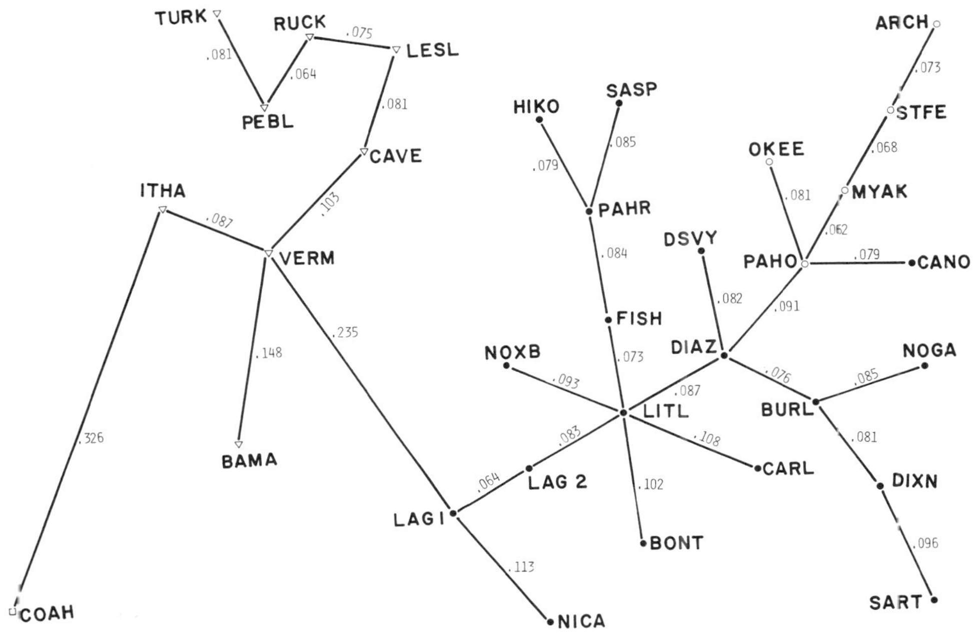

Figure 14. Minimally connected Prim network for 32 population samples in electrophoretic analysis. Sample acronyms as in table 1. Interpopulation distances are Rogers' genetic distances.

of .148. The distance between the Cave Creek Canyon, Arizona and the Burlington, Vermont samples is less than that: .103. Such large interpopulational genetic distances are consistent with the geographic distribution of samples within this species. One population sample from the Davis Mountains in Texas was initially included in the analysis, but the inability to score the Pgm-1 locus in this sample required its elimination. Clearly, sampling a transect of *A. extensicolle* populations from Arizona eastward is necessary to elucidate the nature of the allelic frequency transition for the Pgm-1 locus. The *Agonum extensicolle* group is separated from the *A. decorum-A. elongatulum* cluster of population samples by a minimum distance of .235 (VERM to LAG1), including substitution of alleles at Acph and Pgm-2.

Within the *A. decorum-A. elongatulum* cluster, the minimally connected graph links the five samples of *Agonum elongatulum* (PAHO, OKEE, MYAK, STFE, ARCH). These populations are most similar to the populations of *A. decorum* at Canelo, Arizona and Diaz Lake, California. The samples of *A. elongatulum* possess Pgm-1 allele 1.10 at frequencies of .045 to .333, and allele 1.05 at frequencies of .542 to .750. These allelic frequencies are higher than those seen in most population samples of *A. decorum.* The Ada locus supports linkage of the five Floridian *A. elongatulum* samples. The rare allele, Ada .80, is represented in four of the five samples but is seen nowhere else in the study set.

Geographically defined subsets of *A. decorum* also are connected. These include the Owens Valley, California populations at Fish Springs, Little Lake, and Diaz Lake and the nearby population at Buckhorn Springs, Deep Springs Valley. Also connected are the populations in the Amargosa and White River systems: Saratoga Springs in Death Valley and Pahranagat Lake and Hiko Springs, Nevada. The San Francisco Bay Area populations are more heterogeneous. The samples from Bon Tempe Lake in Marin County, and the Central Valley population near Dixon, California are not connected to the other three Bay Area samples. The Lake Lagunitas samples are most similar to each other, reaffirming the standardization of gel scoring using these samples as standards.

The conspicuous polymorphisms of coloration and setation occurring in *Agonum decorum* (fig. 12) are governed by two unlinked diallelic loci (Liebherr 1983a). These polymorphisms are not strongly correlated with allelic frequencies reported in this analysis, as demonstrated by the loose association of populations possessing the red and hirsute alleles: Nogales, Canelo, Sarita, Noxubee, and Burlington. The Rogers' genetic distances between these five populations average .118, higher than the distances for most other neighboring samples on the minimally connected graph. The genetic distance between the Nogales and Canelo samples is .097, a distance poorly preserved in the graphical representation. Other genetic distances within this set of five populations loosely reflect geographic distance.

F_{ST} Analysis: Genetic Heterogeneity. The F_{ST} values for the three species with sufficient samples are presented in table 4. *Agonum elongatulum* is very homogeneous genetically. Only Acph exhibits an F_{ST} value differing significantly from zero. The other 14 loci that exhibit variability within this species are genetically homogeneous throughout Florida.

Table 4. F_{ST} values for *Agonum elongatulum, A. decorum,* and *A. extensicolle.* Values tested for deviation from zero using heterogeneity chi-square test.

Locus	*A. elongatulum*	*A. decorum*	*A. extensicolle*
Pgm-1	.0033	.1370**	.6611**
Gapd	0	.3241**	0
Est-d	0	.1359**	.0671**
6-Pgd	.0100	.0982**	.0581
Pgm-2	0	.0788**	0
Mpi	0	.0522	0
Phpr	0	.0480	.0960*
Aldh	0	.0434	0
Idh	0	.0282	0
Pgi	0	.0194	.0463*
Sod	0	.0154	0
Acph	.0448*	.0098	0
Eno	0	0	0
Mdh	0	0	0
Acon-1	0	0	.0768**
Ada	0	0	0
$\overline{F_{ST}}$	.004	.062	.063
% loci with $F_{ST} \neq 0$	.067	.333	.385
number of populations	5	17	8

* P less than .05.
** P less than .005.

Agonum decorum is much more heterogeneous genetically. Five loci exhibit F_{ST} values significantly different from zero. The very high value of F_{ST} for Gapd is due to the presence of a rare allele at frequency .400 in the Noxubee National Wildlife Refuge population in Mississippi. Because the F_{ST} index is quite sensitive to the presence of a rare allele in one subdivision of a study set, such an allele frequency causes a great increase in the F_{ST} value. Overall distribution of alleles must be examined along with the F_{ST} value in evaluating genetic heterogeneity. The F_{ST} value of Pgm-1 is significantly greater than zero, owing to differentiation at this locus across the range of *A. decorum.* One of three alleles is most common in various populations. Est-d and 6-Pgd also have three different alleles most common in the various samples of *A. decorum.* Either of two alleles are most common at the Pgm-2 locus.

In the eight population samples of *A. extensicolle*, five loci possess F_{ST} values significantly different from zero. The F_{ST} value of .6611 for Pgm-1 is caused by the substitution of alleles 1.18 and 1.14 for alleles 1.0 to .89 in the five Arizona samples versus the three eastern population samples. At the Phpr peptidase locus, allele .60 is most common in the very small Chandler Springs sample, with allele .81 most common in the other seven samples. The other three loci with F_{ST} values significantly different from zero have only one allele most commonly represented in the various population samples.

The average F_{ST} value is .062 for *A. decorum* and .063 for *A. extensicolle.* The value of .062 for *A. decorum* occurs in a species known for polymorphism in color and setation. Liebherr (1983a) showed that the hirsute allele of the setational polymorphism varied erratically between central Texas, where it is fixed, to the Colorado River in California, where it is absent. The erratic fluctuation in allelic frequency implies that gene flow is not sufficient to create a smooth clinal transition for this allele. Selection or genetic drift may be causing allelic frequency change within populations, and the lack of homogenizing gene flow is apparent. The presence of complementary sets of alleles at Pgm-1 seen in eastern versus Arizonan *A. extensicolle* is suggestive of barriers to gene flow within this species. Separation of sets of *A. extensicolle* populations may be due to the same climatic factors that have fragmented the range of *A. decorum.* Other data in the biometric analyses, wing analysis, and body color analysis, illustrate the variability within *A. extensicolle.* The F_{ST} values corroborate the existence of variability at the genetic level.

BIOMETRIC ANALYSIS

Materials and Methods. Adult specimens for all recognized phena within the *Agonum extensicolle* group are included in the biometric analysis, with up to ten individuals per locality measured for 13 morphometric characters. Males and females are analyzed separately. The population localities, and numbers of male and female specimens measured are presented in table 5. All measurements were made using a calibrated ocular grid, set in a dissecting microscope ocular. Measurements were taken on body parts held horizontal in the field of view at 20 power magnification.

The measurements made are depicted in figure 15, with conventions as follows:

1. head width (hw): width across the head measured from the outer supraorbital ridges immediately above the middle of the antennal socket.

2. prothoracic length (pl): median length of pronotum, including any marginal beads.

3. apical prothoracic width (apw): width joining the two most anterior points at pronotal front angles.

4. maximum prothoracic width (mpw).

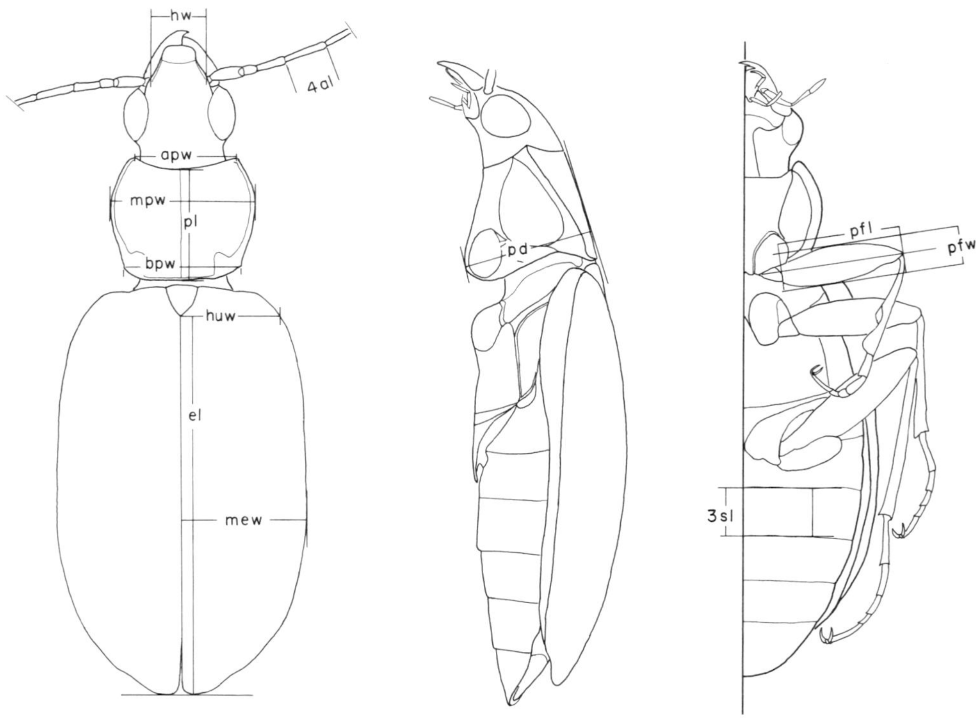

Figure 15. Method of measurement for 13 morphometric characters used in biometric analysis. Character abbreviations as in text.

5. basal prothoracic width (bpw): width of pronotum along line drawn through setae at hind angles.

6. humeral width (huw): width from tip of scutellum to elytral edge at humerus, measured perpendicular to suture.

7. maximum elytral width (mew): maximum width of one elytron, measured perpendicular to suture.

8. elytral length (el): distance from tip of scutellum to apex of elytra, measured parallel to suture.

9. length fourth antennomere (4al).

10. profemur length (pfl): distance from apex of trochanter to apex of femur, measured along median of ventral (anterior) femoral surface.

11. profemur width (pfw): maximum width of profemur measured perpendicular to axis of length measurement.

12. lateral length of third abdominal sternite (3sl): maximum length of sternite measured anywhere from median third to apical quarter of segment.

13. prothoracic depth (pd): distance from convexity of pronotum to convexity of prosternal process, measured perpendicular to line drawn along notum equalizing angle of curvature of notal surface from a line toward the front and back of specimen.

Types of Analysis. The raw measurements made on individual beetles are used to study individual variation. The overall correlation of biometric measurements, as well as the between-species body shape differences, can also be analyzed. To do these types of analyses, averages of within sample measurements can be used. Average measurements for population samples are presented in Appendix 5 and 6. (Individual measurements are too voluminous to present here; raw data are available from the author upon request.)

To analyze geographic variation within species, discriminant function (canonical variate) analysis was performed on individual scores, with population samples as groups. The SPSS subprogram DISCRIMINANT (Klecka 1975) was used. In all cases, the independent variables were utilized simultaneously for derivation of the discriminant function. The DISCRIMINANT program can include univariate analysis of characters using a univariate F-test. For each within-species analysis, such a test was conducted. In general, any character whose value did not differ significantly in at least one population sample in the analysis was excluded from the discriminant analysis. The results of the within-species geographic variation analysis are included with the species treatments in the taxonomic treatment of the *A. extensicolle* group.

Two species complexes were analyzed in an effort to determine whether biometric data could be used for species discrimination. Populations were the groups specified in the discriminant analysis, with the discriminant function portraying between-population biometric distances. In this way, population samples were unconstrained by species grouping, and sets of populations could be placed within species based on the discriminant function. A second type of analysis was also conducted on these species complexes. Individuals were grouped into species and a discriminant function derived which best separated the two species. These two types of analysis, within the *Agonum decorum-A. elongatulum* complex and the *A. extimum-A. parextimum* complex, are essential for species circumscription.

To determine the correlation between the various characters, an R-mode factor analysis of biometric characters was conducted using the FACTOR program (Kim 1975). An R-mode factor analysis considers characters as units, determining the correlation of characters taken over all samples. Within the FACTOR program, the average within-population character values are submitted to principal factoring without iteration, with the diagonal of the correlation matrix not altered, a technique known as principal components analysis. The purpose of this technique is to portray character correlations in factor space, and no rotations are utilized before final graphic portrayal of the results.

The average values for measurements taken over populations are used in a discriminant function analysis employing populations as individual samples, and species as groups. As in the geographic variation and species discrimination analyses, all variables were entered simultaneously in derivation of the discriminant function.

Evaluation of Discriminant Analysis. Discriminant function analysis is a means of summarizing multivariate data, permitting classification of individuals by means of a linear function (Selvin 1980). Klecka (1980) summarizes much practical knowledge regarding this technique. The overall task of a discriminant function is the ordination of groups such that a line is defined that maximizes the between-groups component of variance. The technique is canonical, (hierarchical), in that lower-level units can be included in a larger group. The centroids of the groups, defined by the average values of the characters for individuals within the groups, can be ordinated, maximizing between-groups variance. The technique is used here because the species in the *A. extensicolle* group are extremely variable in biometric characters. Species assignments can generally be made with confidence based on qualitative characters. A

Table 5. List of localities for morphometric data set.

Population number	Species	Locality	Number ♂	♀
1*	*decorum*	FL:Highlands Co., Archbold Bio. Sta.	9	9
2	*decorum*	GA:Ware Co., 9 mi NE Fargo	10	7
3*	*decorum*	MS:Middlesex Co., Wayland	10	10
4*	*decorum*	NB:Lincoln Co., nr. Lincoln	10	10
5*	*decorum*	AZ:Sta. Cruz Co., Peña Blanca Lake	10	10
6*	*decorum*	UT:Utah Co., east shore Utah Lake	10	10
7*	*decorum*	CA:San Diego Co., San Luis Rey R.	10	7
8*	*decorum*	NV:Elko Co., Humboldt R. at Carlin	10	10
9*	*decorum*	CA:Inyo Co., Diaz Lake	10	10
10*	*decorum*	CA:Sta. Clara Co., Lake Lagunitas	10	10
11*	*decorum*	WA:Snohomish Co., Chase Lake	10	—
11*	*decorum*	WA:Pierce Co., Tacoma	—	10
12*	*decorum*	MS:Oktibbeha Co., Noxubee N.W.R.	10	8
13*	*decorum*	TX:Kenedy Co., Sarita	10	—
13*	*decorum*	TX:Harris Co., Houston	—	10
14*	*decorum*	Mexico:Sinaloa, Rosario	10	6
15**	*decorum*	Mexico:Villahermosa, Cuernavaca, Jalapa	7	9
16	*decorum*	Cuba:Cienfuegos, Soledad	10	10
17*	*decorum*	Haiti:swamps N of Dessalines	10	10
18*	*decorum*	Jamaica:Ocho Rios	10	10
19*	*elongatulum*	FL:Suwannee Co., Sta. Fe R. at Hwy. 129	10	10
20	*elongatulum*	FL:Hernando Co., Brooksville	10	10
21	*elongatulum*	FL:Orange Co., Winter Park	10	10
22	*elongatulum*	FL:Highlands Co., Archbold Bio. Sta.	10	10
23*	*elongatulum*	FL:Manatee Co., Myakka R. St. Pk.	10	10
24	*elongatulum*	FL:Hendry Co., Clewiston	10	10
25	*elongatulum*	FL:Dade Co.	10	10
26*	*extensicolle*	Mexico:Durango, Rio Chico 15.7 mi. W Rte. 40	10	5
27*	*extensicolle*	Mexico:Aguascalientes, 5.3 mi. N	10	5
28*	*extensicolle*	AZ:Pima Co., Tucson	10	10
29*	*extensicolle*	AZ:Coconino Co., 15 mi SE Flagstaff	7	8

canonical analysis whereby the centroids average the infraspecific biometric variation, permits a generalized analysis with characteristics of very distinct populations becoming tempered by the overall characteristics of other populations sampled within the species.

A discriminant function is composed of terms comprising discriminant coefficients multiplied by variable values. When all variables are standardized to zero mean and unit variance, the coefficients in the linear discriminant function are said to be standardized. The absolute values of the standardized coefficients determine how much a variable will contribute to the total discriminant score. Although these absolute values are often equated with the characteristics of the function—the larger a particular value, the more likely the function is to reflect the corresponding character (Klecka 1975)—this assumption is somewhat oversimplified (Campbell and Atchley 1981). If variables are highly correlated within groups, those variables with the largest absolute coefficients need not be the best discriminators. Wilks' lambda values, or the ratio of the within-group sum of squares divided by the total sum of squares, determined

Table 5 (cont.).

Population number	Species	Locality	Number ♂	Number ♀
30*	*extensicolle*	NM:Grant Co., Gila R. at U.S. 260	10	7
31*	*extensicolle*	TX:Jeff Davis Co., Limpia Ck. Cyn., el. 5000′	10	10
32*	*extensicolle*	TX:Kerr Co., Kerrville	10	10
33*	*extensicolle*	CO:Boulder Co., Boulder	10	10
34*	*extensicolle*	AR:Washington Co., Farmington	10	10
35*	*extensicolle*	Canada:Manitoba, Lake Winnipeg, Victoria Beach	9	10
36*	*extensicolle*	SC:Oconee Co., Clemson	10	10
37*	*extensicolle*	VA:Fairfax Co., Potomac R.	10	10
38*	*extensicolle*	NY:Tompkins Co., Ithaca	10	10
39*	*extensicolle*	Canada:Nova Scotia, nr. Truro	10	8
40*	*extimum*	Mexico:Coahuila, 19.4 mi E San Pedro	10	10
41	*extimum*	NM:Luna Co., Deming	4	7
42	*extimum*	AZ:Cochise Co., nr. Douglas	10	10
43	*parextimum*	AZ:Pima Co., nr. Tucson	—	7
44	*parextimum*	AZ:Pima Co., Brawley Wash W Tucson	—	6
44	*parextimum*	AZ:Pima Co., 17 mi. W Quijotoa	4	—
45*	*parextimum*	Mexico:Sinaloa, Venedio/Los Mochis	8	8
46*	*cyanope*	Mexico:Durango, 8 mi. N Coyotes	6	4
47*	*cyanope*	AZ:Cochise Co., Chiricahua Mts.	8	10
48*	*cyanope*	AZ:Cochise Co., Huachuca Mts.	5	6
49*	*texanum*	Mexico:Chiapas, 5 km. E Comitan	9	10
50*	*texanum*	TX:Cameron Co., Brownsville	10	10
51	*texanum*	Mexico:Durango, Villa Lerdo	6	—
52	*texanum*	TX:San Patricio Co., Sinton	10	10
53	*texanum*	TX:Bexar Co., Ft. Sam Houston	10	10
54	*texanum*	TX:McLennan Co., Waco	10	10
55*	*texanum*	TX:Terrell Co., Dryden	10	10
56	*texanum*	TX:Jeff Davis Co., Ft. Davis	5	5
57*	*texanum*	AZ:Yavapai Co., Kirkland	10	10

* Sample included in wing analysis.
** Sample consists of pooled localities from Bates (1882, 1884).

univariately, can be used in conjunction with standardized coefficients to evaluate the importance of variables (Selvin 1980). If a character possesses a small Wilks' lambda, it possesses limited variability within groups relative to its overall variability. If it also possesses a large standardized coefficient, it is a good candidate for recognition as a character with good discriminating power. The final step in this assessment involves inspection of the raw data to determine whether the contribution of the particular characters to the discriminant function score is positively correlated with the sign of its coefficient and the direction of difference in the raw data values.

The presentation of results for all discriminant function analyses includes: (1) univariate Wilks' lambda values; (2) univariate F-ratios with assigned probabilities that the heterogeneity in character values in the various groups is due to chance alone; and (3) standardized canonical discriminant function coefficients. Assessment of the importance of various characters is based on the Wilks' lambda, standardized coefficient, and raw data values.

RESULTS AND DICUSSION

Species Group Analyses. *Principal components analysis of biometric characters.* The results are similar for the male and female data sets. For males, the first component accounts for 72.3% of the total variance (table 6). The second component accounts for 15% of the total variance, and permits separation of the biometric characters into three suites, or groups of correlated characters (fig. 16). One group is composed of appendage measurements; profemur length, profemur width, and fourth antennomere length. A second group contains mostly overall body size and length characters; elytral length, prothoracic length, third sternite length, prothoracic depth, and maximum elytral width. The third group consists of mid-body widths; apical, maximum and basal prothoracic width, and humeral width. Head width is not highly correlated with any of the other characters, and is highly separated on component three.

Table 6. Percentage of variance accounted for and eigenvalue for first three principal components for male and female species level biometric data sets.

Principal	Males		Females	
component	% variance	Eigenvalue	% variance	Eigenvalue
1	72.3	9.40	68.1	8.85
2	15.0	1.96	15.4	2.00
3	4.8	0.63	7.1	0.93
Cumulative % variance	92.1		90.6	

The female data set produces identical groupings of correlated characters: three appendage measurements, five body size and length measurements, four mid-body width measurements, and head width (fig. 17). In this case, component 1 accounts for 68.1% of the variance, component 2 accounts for 15.4%, and component 3 for 7.1% (table 6).

Graphic portrayal of the character correlations (table 7) in reduced factor space, as is possible using principal component analysis, permits comprehension of the overall correlation trends in the characters. This data summary is directly applicable only to the higher level, or species group analyses. The results of the character correlations can be used in the interpretation of the between-species discriminant function analysis detailed below.

Species group discriminant analysis using males. The discriminant function analysis for males is based on 55 population samples (table 5, Appendix 5). This analysis utilizes species as groups, with population samples assigned a priori to species according to the qualitative characters initially used to sort phena. The classificatory efficiency of the discriminant functions provides a posteriori confirmation of the initial species assignments. The first four discriminant functions account for 97.0% of the total variance in the data set, apportioned as 61.2% by discriminant function (d.f.) 1, 20.0% by d.f. 2, 9.2% by d.f. 3, and 6.6% by d.f. 4. The graphic portrayal of the first three (fig. 18) shows separation of species, with all population samples falling nearest the centroid of the a priori assigned species. Table 8 shows the classificatory efficiency of assignment of populations to groups, with the 100% correct assignment illustrating the distinctiveness of all seven species based on body shape data.

Table 7. Correlation coefficient matrix for 13 morphometric characters, calculated using entire morphometric data set. Values for females above diagonal, values for males below diagonal.

Character	hw	pl	apw	mpw	bpw	huw	mew	el	4al	pfl	pfw	3sl	pd
head width (hw)	1.00	.271	.397	.290	.191	.227	.267	.249	.205	.339	.337	.419	.325
prothoracic length (pl)	.458	1.00	.759	.741	.589	.695	.870	.774	.769	.838	.821	.879	.912
apical prothoracic width (apw)	.665	.808	1.00	.942	.888	.876	.824	.705	.317	.488	.544	.740	.820
maximum prothoracic width (mpw)	.610	.767	.937	1.00	.955	.857	.809	.682	.238	.404	.491	.667	.814
basal prothoracic width (bpw)	.553	.643	.896	.955	1.00	.809	.683	.544	.041	.213	.363	.519	.679
humeral width (huw)	.521	.786	.922	.907	.867	1.00	.865	.768	.313	.453	.380	.686	.790
maximum elytral width (mew)	.565	.905	.855	.845	.744	.924	1.00	.875	.599	.741	.628	.878	.952
elytral length (el)	.620	.873	.875	.826	.719	.897	.938	1.00	.568	.698	.560	.773	.835
length 4. antennomere (4al)	.170	.722	.327	.204	.023	.363	.587	.559	1.00	.896	.717	.714	.625
profemur length (pfl)	.329	.861	.577	.458	.303	.575	.768	.736	.874	1.00	.817	.853	.794
profemur width (pfw)	.346	.801	.492	.484	.309	.450	.645	.621	.766	.762	1.00	.729	.728
length 3. sternite (3sl)	.578	.794	.744	.677	.557	.713	.834	.789	.584	.750	.653	1.00	.877
prothoracic depth (pd)	.587	.951	.875	.836	.730	.862	.937	.922	.632	.824	.719	.825	1.00

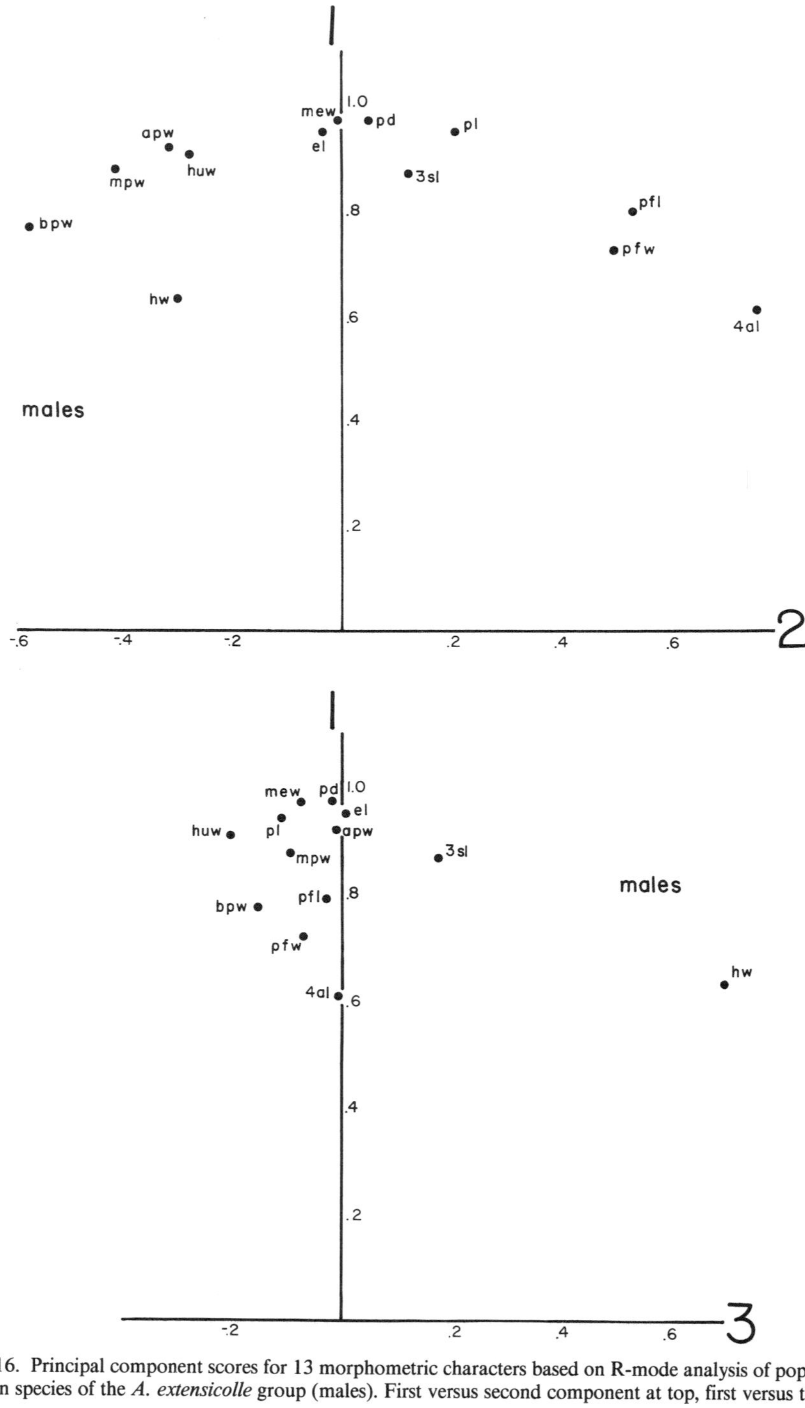

Figure 16. Principal component scores for 13 morphometric characters based on R-mode analysis of population means for seven species of the *A. extensicolle* group (males). First versus second component at top, first versus third below.

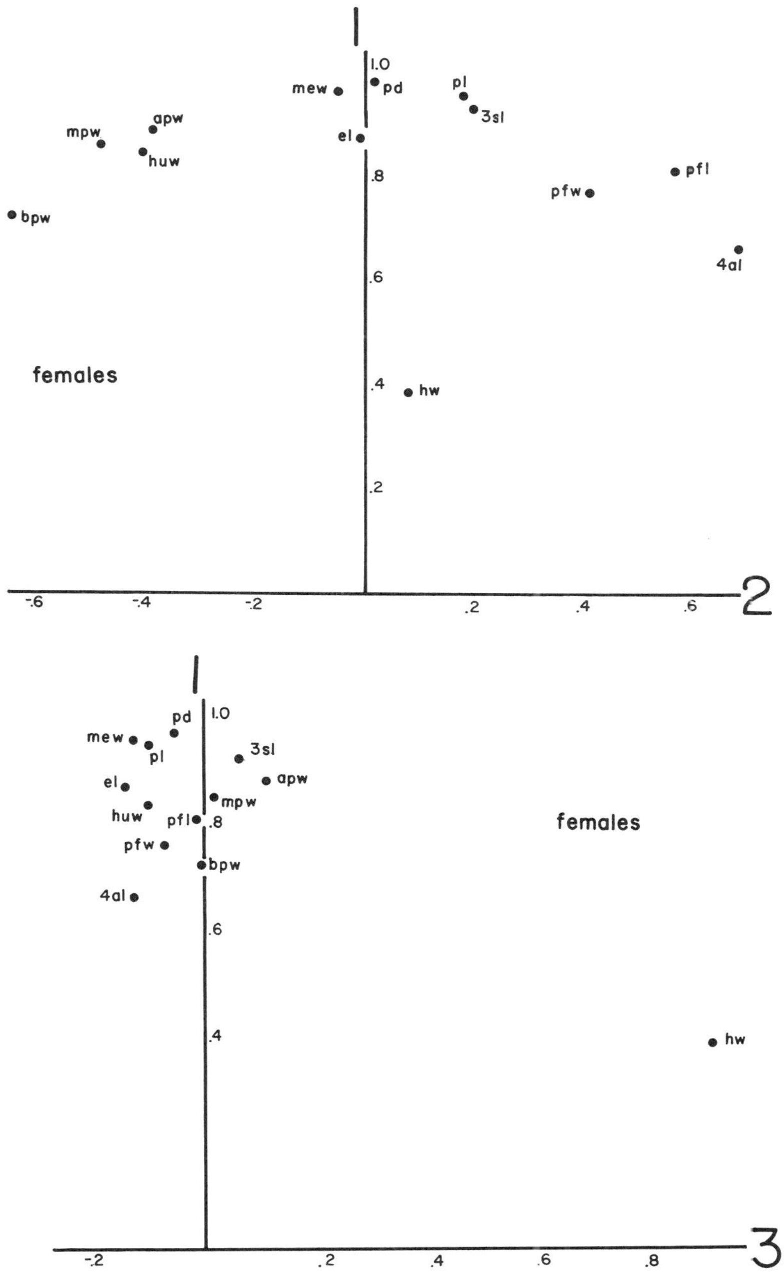

Figure 17. Principal component scores for 13 morphometric characters based on R-mode analysis of population means for seven species of the *A. extensicolle* group (females). First versus second component at top, first versus third below.

Table 8. Classificatory efficiency using first four canonical discriminant functions for males of the *Agonum extensicolle* group.

			Predicted group						
Group	Species	No. of cases	1	2	3	4	5	6	7
1	*A. decorum*	18	18	0	0	0	0	0	0
2	*A. elongatulum*	7	0	7	0	0	0	0	0
3	*A. extensicolle*	14	0	0	14	0	0	0	0
4	*A. extimum*	3	0	0	0	3	0	0	0
5	*A. parextimum*	2	0	0	0	0	2	0	0
6	*A. cyanope*	3	0	0	0	0	0	3	0
7	*A. texanum*	9	0	0	0	0	0	0	9

Agonum extensicolle population samples receive large negative scores for discriminant function 1; *Agonum texanum* samples receive large positive scores. As shown by the large standardized coefficients for d.f. 1 and small Wilks' lambda values for species as groups (table 9), maximum prothoracic width and elytral length contribute heavily to the first discriminant function. *A. texanum* possesses a wide prothorax, with maximum widths averaging 2.04-2.28mm. The average prothoracic width in *A. extensicolle* samples ranges 1.72-1.97mm. *A. texanum* is a large beetle with average elytral length a good indicator of overall size, ranging 5.29-5.99mm. *A. extensicolle* is smaller, with average elytral lengths of 4.83-5.70mm.

In addition to the contribution of maximum prothoracic width and elytral length taken alone, other characters contribute to separation on discriminant function 1. The concerted action of mensural characters can be illustrated by looking at suites of correlated characters and the shapes of structures they represent. The shape of the prothorax is the best example of how d.f. 1 represents shape. *A. extensicolle*, *A. extimum*, and *A. parextimum* all have cordate prothoraces

Table 9. Wilks' lambda values, univariate F-ratios, and standardized canonical discriminant function coefficients for males of the *Agonum extensicolle* group.

			Coefficients			
Variable	Wilks'	F*	d.f. 1	d.f. 2	d.f. 3	d.f. 4
hw	.523	7.44	–0.15	–0.33	–0.59	1.13
pl	.347	15.34	0.22	0.66	–0.19	0.08
apw	.277	21.37	–0.79	0.89	–0.52	–0.94
mpw	.241	25.65	2.11	–0.53	–0.34	1.05
bpw	.220	28.94	0.61	–0.41	0.74	0.18
huw	.263	22.91	–0.67	–0.13	0.12	0.29
mew	.221	28.78	–0.57	1.43	0.65	0.14
el	.295	19.52	1.79	–0.32	–1.99	–0.95
4al	.312	18.04	–0.35	–0.33	0.23	–0.04
pfl	.321	17.31	–0.65	–0.14	–0.90	0.83
pfw	.488	8.56	–1.01	–0.19	0.42	0.37
3sl	.359	14.57	–0.22	0.13	0.19	0.72
pd	.354	14.89	–0.75	–0.10	2.03	–1.89

* P less than .001 for all univariate tests.

(figs. 32a-h). *A. texanum* and *A. cyanope* possess prothoraces with more rounded lateral margins and little or no constriction before the hind angles (figs. 31c-h). The scores of the species on the first discriminant function reflect this difference. Both maximum and basal prothoracic width have positive standardized coefficients (table 9). Their contribution to the d.f. scores correlates positively with the broader prothorax relative to body size seen in *A. cyanope* and *A. texanum.* These two species score positively on d.f. 1. The positive coefficients for these characters, when combined with the narrow, constricted prothorax seen in *A. extensicolle*, *A. extimum*, and *A. parextimum*, permits other terms in the function to predominate, producing negative d.f. scores.

One appendicular measurement, profemur length, also contributes in a positively correlated manner on d.f. 1. *A. extensicolle* sample profemoral lengths average 1.79-2.01mm, with only 3 of 14 samples averaging 1.85mm or less. *A. texanum* samples average 1.69-1.88mm, with only 2 of 8 samples averaging more than 1.85mm. This difference, coupled with a negative coefficient (-.65), directly increases *A. texanum* scores and decreases *A. extensicolle* scores.

Discriminant function 2 separates *A. extimum* and *A. parextimum* from *A. extensicolle*, and *A. cyanope* from *A. texanum.* The principal character acting on d.f. 2 is maximum elytral width, which possesses a small Wilks' lambda (.221) and a large coefficient of 1.43 (table 9). *A. extimum* and *A. parextimum* have average maximum elytral widths ranging 1.78-1.82mm in samples analyzed, whereas *A. extensicolle* is smaller, with average maximum elytral widths of 1.53-1.75mm. The two wider species receive more positive scores on d.f. 2. *A. texanum* is wider than *A. cyanope*, with average maximum elytral widths ranging 1.72-1.88mm in the former versus 1.57-1.80 in the latter. As in the other example, the wider species scores higher on d.f. 2.

The third discriminant function largely separates *A. decorum* and *A. elongatulum* (fig. 18). The combination of characters causing this separation is ambiguous in this large data set, with the best separation between these two species achieved on an axis oblique to functions 1, 2, and 3. The two-species analysis, restricted to the *A. decorum* complex, better evaluates the mensural differences between these species.

Discriminant function 3 also separates *A. cyanope* from all other species, giving it high scores. This separation is largely due to the measurement of elytral length, with a standardized coefficient of -1.99 (table 9). *A. cyanope* is a small species, with average elytral lengths of 4.82-5.49mm, and is robust, as illustrated by the large prothorax, with prothoracic depth ranging from 1.64-1.87mm and basal prothoracic width averaging 1.47-1.73mm. The combination of short elytra with a negative elytral length coefficient, broad prothoracic base with coefficient of .74, and deep robust prothorax with a coefficient of 2.03 aid in producing high scores on d.f.3 for the samples of *A. cyanope. A. elongatulum* scores negatively on d.f.3. It has prothoracic depth values averaging, from 1.66 to 1.83mm, similar to those in *A. cyanope.* In this case, moderate prothoracic depth is correlated with larger body size. This is shown by elytral lengths averaging 5.15-5.71mm, and prothoracic basal widths averaging 1.49-1.58mm. *A. elongatulum* can be considered a larger more normally proportioned *Agonum*, whereas *A. cyanope* is a smaller and more robust *Agonum.* The other five species in the group fall more or less at the middle of the scores on d.f. 3. As this function accounts for only 9.2% of the variance in the data set, little separation of species can be expected.

The importance of the various characters used in the discriminant analysis can be evaluated in light of the R-mode principal components analysis. The separation of samples on discriminant function 1 can be traced to large contributions by maximum prothoracic width and elytral length, with apical prothoracic width and basal prothoracic width of secondary importance. These measurements represent two of the four correlated groups. The presence of the two most strongly discriminating characters in different groups of correlated characters is not surprising, as these account for a large percent of the variance of the data within their correlated groups. The other measurements in their groups can univariately separate samples, but their high correlation with these two strong characters precludes significant contribution to the discriminant function.

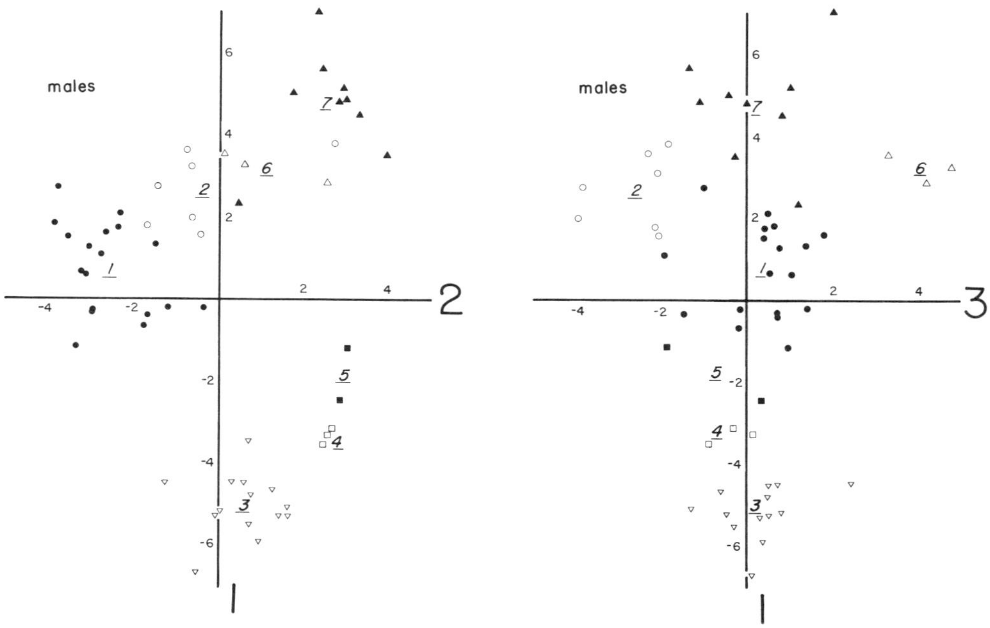

Figure 18. Discriminant function scatterplots for functions 1 and 2, and 1 and 3, for males of population samples of species in the *A. extensicolle* species group. Species 1 (●), *A. decorum*; species 2 (○), *A. elongatulum*; species 3 (▽), *A. extensicolle*; species 4 (□), *A. extimum*; species 5 (■), *A. parextimum*; species 6 (△), *A. cyanope*; species 7 (▲), *A. texanum*.

Appendicular measurements, profemur width and length, and fourth antennomere length contribute little to separation of species on any of the discriminant functions. Such features are relatively variable within species, and more or less uniform throughout the species group. This limits their between-species discriminating power. Head width is also a poor means for discrimination of species. Within-species variation in head width is greater than any between-species differences. The selective pressures requiring retention of an efficient, generalized mouthpart system in these generalized omnivores would retard extensive changes in mouthpart configuration. The variation that does exist is correlated with overall size.

Species group discriminant analysis using females. The canonical discriminant function analysis based on 56 population samples of female beetles produces a scattergram based on discriminant functions 1, 2, and 3 much like that seen for males (fig. 19). The first four discriminant functions account for 98.9% of the variance in the data set, apportioned as 77.7% accounted for by d.f. 1, 11.0% by d.f. 2, 6.6% by d.f. 3 and 3.6% by d.f. 4. Overall, 94.6% of the population samples are classified into the a priori assigned species groups (table 10). Two pairs of closely related species similar enough in body shape that some misclassification occurs are discussed below: *A. decorum* and *A. elongatulum*, and *A. extimum* and *A. parextimum*.

Table 10. Classificatory efficiency using first four canonical discriminant functions for females of the *Agonum extensicolle* group.

			Predicted group*						
Group	Species	No. of cases	1	2	3	4	5	6	7
1	*A. decorum*	18	17	1	0	0	0	0	0
2	*A. elongatulum*	7	1	6	0	0	0	0	0
3	*A. extensicolle*	14	0	0	14	0	0	0	0
4	*A. extimum*	3	0	0	0	2	1	0	0
5	*A. parextimum*	3	0	0	0	0	3	0	0
6	*A. cyanope*	3	0	0	0	0	0	3	0
7	*A. texanum*	8	0	0	0	0	0	0	8

* 94.6% of population sampes correctly assigned.

As in the male data set, discriminant function 1 is characterized by maximum separation of *A. texanum*-*A. cyanope* with positive d.f. scores, and *A. extensicolle*, with negative scores. In females, this separation is based on measurements different from those seen in the male data set. Prothoracic length, apical prothoracic width, and profemur length all possess relatively small Wilks' lambda values and large negative standardized coefficients (table 11). Prothoracic depth possesses a relatively small Wilks' lambda and a large positive standardized discriminant coefficient.

Profemur length is positively correlated with d.f. score. Profemur length averages from 1.82-2.05mm in *A. extensicolle* samples and 1.60-1.91mm in *A. texanum* samples. The standardized coefficient of -1.61 insures a larger negative contribution to d.f. score for *A. extensicolle* and a smaller negative contribution to d.f. scores of *A. texanum*.

Prothoracic depth has a standardized coefficient of 1.56. For *A. texanum*, this measurement ranges from 1.92-2.03mm, except for the small sized individuals from Comitan, Chiapas, Mexico. *A. extensicolle* is smaller, with prothoracic depth averaging 1.73-1.92mm. The negative values for *A. extensicolle* are positively correlated with the contribution of this measurement.

Prothoracic length and apical prothoracic width are less clearly correlated with d.f. scores for the first function. Prothoracic length averages 1.61-1.85mm in *A. extensicolle* whereas it averages 1.54-1.87mm in *A. texanum*. Such a large overlap indicates that the standardized coefficient of -1.07 is contributing little discriminating power to this particular between species comparison. Apical prothoracic width has a standardized coefficient of -1.72. The values of this measurement average 1.32-1.52mm in *A. extensicolle*, and 1.39-1.68mm in *A. texanum*. Since *A. texanum* receives high scores on d.f. 1, the contribution of this character actually runs counter to the total scores observed.

Head width displays a standardized coefficient of -1.30 on d.f. 1 in the females (table 11) but the Wilks' lambda is .695, indicating that much of the variance accounted for by this character is found within groups. This is the same trend seen in this measurement for the male data set.

Basal prothoracic width has a coefficient on d.f. 1 of .68, with d.f. scores positively correlated with its contribution. *A. extensicolle* samples average 1.27-1.46mm basal prothoracic width, whereas *A. texanum* samples average 1.58-1.82mm. With this large difference, and a moderately sized positive coefficient, this measurement supplies a significant level of discrimination on d.f. 1.

Table 11. Wilks' lambda values, univariate F-ratios, and standardized canonical discriminant function coefficients for females of *Agonum extensicolle* group.

			Coefficients			
Variable	Wilks'	F*	d.f. 1	d.f. 2	d.f. 3	d.f. 4
hw	.695	3.59*	–1.30	–0.65	–0.38	–0.54
pl	.432	10.73**	–1.07	0.32	0.41	0.60
apw	.442	10.33**	–1.72	1.11	–0.57	–0.26
mpw	.336	16.14**	0.07	–0.90	–0.69	–0.69
bpw	.301	18.96**	0.68	–0.85	–0.02	0.79
huw	.434	10.67**	–0.63	–0.25	0.94	0.30
mew	.325	16.95**	–0.17	1.31	–0.47	–2.03
el	.456	9.74**	0.41	–0.17	0.37	–0.66
4al	.376	13.57**	–0.13	–0.34	0.49	0.54
pfl	.400	12.26**	–1.61	0.26	–0.67	–0.64
pfw	.649	4.41*	–0.13	–0.08	0.48	–0.25
3sl	.363	14.30**	0.07	0.65	–1.42	1.20
pd	.444	10.20**	1.56	–0.99	1.98	1.53

* P less than .005.
** P less than .001.

The above analysis points out several weaknesses of relying exclusively on standardized coefficients for determination of character contributions in a particular discriminant function. All these characters possess moderate Wilks' lambda values. Their within-group sum of squares approaches half or more of the total sum of squares. Two of the measurements contribute to the d.f. scores in correlation with the magnitude of the average sample values (prothoracic depth, profemur length); two do not provide discriminating power (prothoracic length, head width); one actually contributes in opposition to the magnitude of the average sample values (apical prothoracic width). As in the male data set, the two best discriminating characters on d.f. 1 belong to different groups of correlated characters. In summary, d.f. 1 represents body size acting in a positive manner, with *A. texanum* a larger species. Appendicular length acts in a negative manner, with *A. extensicolle* possessing somewhat longer legs.

Discriminant function 2 contains one character with both a large standardized coefficient and small Wilks' lambda. Maximum elytral width has a coefficient of 1.31, and a lambda value of .325. Species with positive d.f. scores on function 2 include *A. texanum*, *A. parextimum*, *A. extimum*, and *A. extensicolle*. Over these species, maximum elytral width averages 1.57-2.00mm, with only 4 of 28 samples averaging less than 1.75mm. *A. decorum* population samples have negative scores in d.f. 2, and they have narrower elytra, averaging from 1.49-1.74mm. This character exhibits a contribution to d.f. score that is positively correlated with the raw values.

Another character whose contribution to d.f. 2 is positively correlated with overall values is lateral length of the third sternite. *A. decorum* averages from .59-.72mm for this character, whereas the four species with positive d.f. scores average from .67-.82mm. Only one sample of these species averages less than .72mm (the Chiapas sample of *A. texanum*). With a standardized coefficient of .65, larger sternite lengths produce larger d.f. scores.

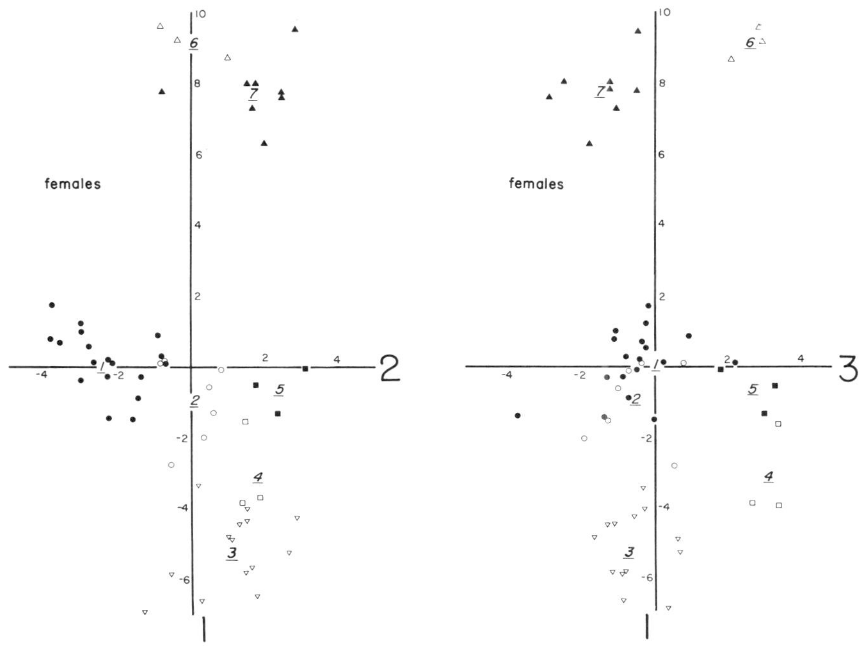

Figure 19. Discriminant function scatterplots for functions 1 and 2, and 1 and 3, for females of population samples of species in the *A. extensicolle* species group. Species 1 (●), *A. decorum*; species 2 (O), *A. elongatulum*; species 3 (▽), *A. extensicolle*; species 4 (□), *A. extimum*; species 5 (■), *A. parextimum*; species 6 (△), *A. cyanope*; species 7 (▲), *A. texanum*.

Discriminant function 3 accounts for only 6.6% of the overall variance of the data set. Nevertheless, significant discrimination of *A. cyanope* from *A. texanum*, and *A. extimum* and *A. parextimum* from *A. extensicolle* is accomplished by this function (fig. 19). Length of the third sternite possesses a coefficient of -1.42. *A. cyanope* has shorter sternite lengths than *A. texanum*: .62-.76mm versus .67-.83mm. This character helps *A. cyanope* receive positive d.f. scores and *A. texanum* negative scores.

Humeral width and prothoracic depth vary between two groups—*A. extensicolle* versus *A. extimum* and *A. parextimum*. The contribution of characters to the d.f. scores is positively correlated with the raw data values of these measurements. *A. extensicolle* has humeral width values averaging 1.14-1.43mm, whereas the other two species possess humeral widths averaging 1.38-1.47mm. As the standardized coefficient is .94, the wider humeri of *A. extimum* and *A. parextimum* cause larger d.f. scores.

The prothoracic depth measurements of *A. extensicolle* average 1.73-1.92mm, whereas *A. extimum* and *A. parextimum* samples average from 1.89-1.95mm. Based on the coefficient value of 1.98, this measurement contributes to discrimination of *A. extensicolle* from the other two species.

Summary of Species Group Analyses. The various species recognized a priori using qualitative characters can be successfully classified based strictly on biometric data. The general distribution of species on the discriminant function scatterplots is the same for males and females (figs. 18, 19). The character correlations are also similar for both sexes, based on principal components analysis (figs. 16, 17). Nevertheless, there are differences between the sexes in the types of contributions various characters make to discriminant scores. For males, function 1 has maximum prothoracic width and elytral length contributing heavily, with basal prothoracic width and profemur length of secondary importance. In females, prothoracic depth and profemur length are strong discriminators, with basal prothoracic width of secondary importance. The distribution of characters varies between the sexes; however, strong discrimination in both cases is based on characters from three of the four groups of correlated characters. The fourth group in the character correlation is head width, poorly correlated with most characters. This measurement contributes little to discrimination because of its large within group variability.

Discrimination by function 2 in both sexes is primarily based on maximum elytral width. In females, third sternite length is of secondary importance.

The third discriminant accounts for little variance, and the characters useful for discrimination vary between the sexes. Even within each sex, discrimination of different sets of closely related species is based on different characters. Discriminant analysis permits interpretation of such parallel trends based on examination of the raw data in light of the magnitude and sign of the standardized discriminant coefficients.

Species Complex Analyses. *A. decorum complex. Agonum decorum* and *A. elongatulum* are sympatric in northern Florida, with the electrophoretic analysis indicating a genetic discontinuity between *A. elongatulum* and eastern red-hirsute populations of *A. decorum*. Green-glabrous Florida populations assigned to *A. elongatulum* are very similar morphologically to green-glabrous populations in the Antilles and in western North America.

A. decorum is polymorphic for color and setation across North America, with red-hirsute populations intergrading to green-glabrous populations from east to west in the American Southwest (Liebherr 1983a). The purpose of biometric analysis in this species complex is twofold: (1) to ascertain whether *A. elongatulum* is biometrically distinct from *A. decorum*, and (2) to assign island populations from the Caribbean area to one of the two taxa.

For this analysis, 20 population samples are used to derive discriminant functions. Included are seven green-glabrous populations from peninsular Florida (populations 19-25, table 5), green-glabrous populations on Cuba, Jamaica and Hispaniola (nos. 16, 17, 18), and 10 populations of *A. decorum* from mainland North America (nos. 1-8, 11, 14). A southern Mexican sample (no. 15) is pooled from *Biologia* material housed at the British Museum (Natural History). This sample is treated as an unknown in the analysis. Discriminant scores of its individuals are determined by functions derived from the other 20 samples. In this analysis, the populations are treated as groups, with species assignment determined by population function scores.

Population level analysis of A. decorum complex males. The results of the first three discriminant functions show that *A. elongatulum*—the glabrous Florida populations—cluster together. These samples are relatively distinct from all others (fig. 20). The cumulative variance of the first three discriminant functions totals 72.1%, apportioned as 32.1% accounted for by d.f. 1, 24.9% by d.f. 2 and 15.1% by d.f. 3. The best separation of *A. elongatulum* is on d.f. 1, with elytral length possessing a small Wilks' lambda and a large coefficient (table 12). This measurement is the largest one made on the body, and is highly correlated with overall size.

The three population samples from the Antilles fall clearly within the other populations of *A. decorum*. The southern Mexico pooled sample (no. 15) has scores on d.f. 1 that are characteristic of other *A. decorum* population samples. The red-hirsute populations from the area of sympatry between *A. decorum* and *A. elongatulum* (nos. 1 and 2) are maximally separated from the samples of *A. elongatulum*. The biometric analysis shows a discontinuity in body size between these two sympatric forms, corroborating the genetic discontinuity observed in the electrophoretic analyses.

Table 12. Wilks' lambda values, univariate F-ratios, and standardized canonical discriminant function coefficients for males of *A. decorum* and *A. elongatulum*.

Variable	Wilks'	F*	Coefficients d.f. 1	d.f. 2	d.f. 3
hw	.550	7.70	0.00	0.78	–0.80
pl	.509	9.09	–0.61	–0.32	0.26
apw	.595	6.42	0.24	0.24	0.28
mpw	.604	6.17	–0.52	0.31	0.77
bpw	.572	7.05	0.25	0.26	0.08
huw	.465	10.85	0.44	–0.09	0.63
mew	.479	10.23	–0.04	0.27	0.19
el	.382	15.24	1.35	–0.60	–1.01
4al	.469	10.68	–0.34	–0.59	–0.42
pfl	.462	10.99	–0.11	–0.60	0.49
pfw	.541	8.01	–0.68	–0.08	–0.14
3sl	.573	7.01	0.27	–0.09	–0.13
pd	.616	5.87	–0.18	0.09	–0.36

* All F-ratios significant at P less than .001.

On all three functions, the seven green-glabrous Florida population samples possess similar function scores, while there is much more heterogeneity between the population samples of *A. decorum*. Function 2 causes great separation, with northern populations (nos. 6, 8, 11) receiving large positive d.f. scores. Within this data set, these populations comprise the smallest individuals, characterized by small elytra, averaging 4.51-4.68mm; short antennae, with fourth antennomere averaging .43-.48mm; and short legs, with profemur length averaging 1.49-1.55mm. The coefficients for these three characters are -.60, -.59, and -.60 (table 12), all acting to increase the scores on d.f. 2 for the smaller beetles.

The smaller body size and shorter appendages in the northern and western region of *A. decorum*'s range are independent of the color and setational polymorphisms. The Utah Lake population (no. 6) is made up of red-glabrous individuals, whereas populations from Carlin, Nevada (no. 8) and Chase Lake, Washington (no. 11) are composed of green-glabrous individuals. The trends in biometric measurements corroborate findings of Liebherr (1983a) concerning the conspecificity of the various morphs of *A. decorum*.

Population level analysis of A. decorum complex females. The first three discriminant functions (fig. 21) show some overlap between *A. elongatulum* population samples and populations 4 and 5 of *A. decorum*. The first three functions account for 73% of the variance, apportioned 32.8% to d.f. 1, 28.7% to d.f 2, and 11.5% to d.f. 3.

As in the male data set, elytral length receives a large standardized coefficient on d.f. 1 (table 13). In this case the Wilks' lambda value is .50, indicating that half of the sum of squares is within groups. The raw data show that *A. elongatulum* has a larger average elytral length (5.39-5.71mm) than *A. decorum* (4.82-5.47mm). At the level of species, this character is a good discriminator.

Table 13. Wilks' lambda values, univariate F-ratios, and standardized canonical discriminant function coefficients for females of *A. decorum* and *A. elongatulum*.

Variable	Wilks'	F*	Coefficients d.f. 1	d.f. 2	d.f. 3
hw	.671	4.39	0.11	0.39	–0.03
pl	.623	5.41	–0.85	0.29	–0.82
apw	.602	5.92	0.55	0.35	0.08
mpw	.614	5.62	–0.46	0.09	–0.09
bpw	.624	5.40	–0.23	0.60	0.88
huw	.627	5.33	0.47	0.66	0.63
mew	.600	5.96	0.19	–0.68	–0.33
el	.501	8.91	1.15	–1.08	–0.48
4al	.437	11.51	–0.61	–0.54	0.39
pfl	.486	9.47	–0.32	–0.23	1.25
pfw	.611	5.70	–0.42	0.12	–0.44
3sl	.637	5.09	0.11	–0.17	0.06
pd	.640	5.03	0.19	–0.04	–0.37

* All F-ratios significant at P less than .001.

The populations most similar biometrically to the Florida *A. elongatulum* are those from Lincoln County, Nebraska (no. 4) and Peña Blanca Lake, Arizona (no. 5). These are the largest beetles in this analysis, with elytral lengths averaging 5.46-5.47mm. These are two of the three population samples whose elytral length averages fall within the range of averages of *A. elongatulum*.

The three Antillean population samples (nos. 16-18) and the southern Mexican pooled sample (no. 15), are distinct from the samples of *A. elongatulum*. As seen in the males, the sympatric red-hirsute *A. decorum* and green-glabrous *A. elongatulum* samples are maximally distinct biometrically.

The trends of scores on d.f. 2 parallel those seen in the males. The northern and western *A. decorum* samples (nos. 6-8, 11) have high positive scores on d.f. 2. Three measurements, all with small Wilks' lambda values and negative coefficients, contribute to this separation (table 13). Elytral length averages 4.82-5.45mm in *A. decorum* samples 6-8 and 11, versus 5.15-5.38mm for all other samples. Fourth antennomere length averages .41-.51mm in samples 6-8 and 11, versus .52-.57mm in other samples. Profemur length averages 1.50-1.80mm in samples 6-8 and 11, versus 1.67-1.86mm in other samples. These differences coupled with coefficients of -1.08, -.54, and -.23, aid in producing positive d.f. scores for the smaller beetles.

Species level analysis of A. decorum complex. With the restriction of *Agonum elongatulum* to peninsular Florida, and the recognition of green-glabrous Antillean populations as island representatives of *A. decorum*, the question arises of how best to discriminate the green-glabrous forms within this complex consistent with the overall species circumscription. Because the red and hirsute polymorphisms within *A. decorum* can be used to discriminate those morphotypes

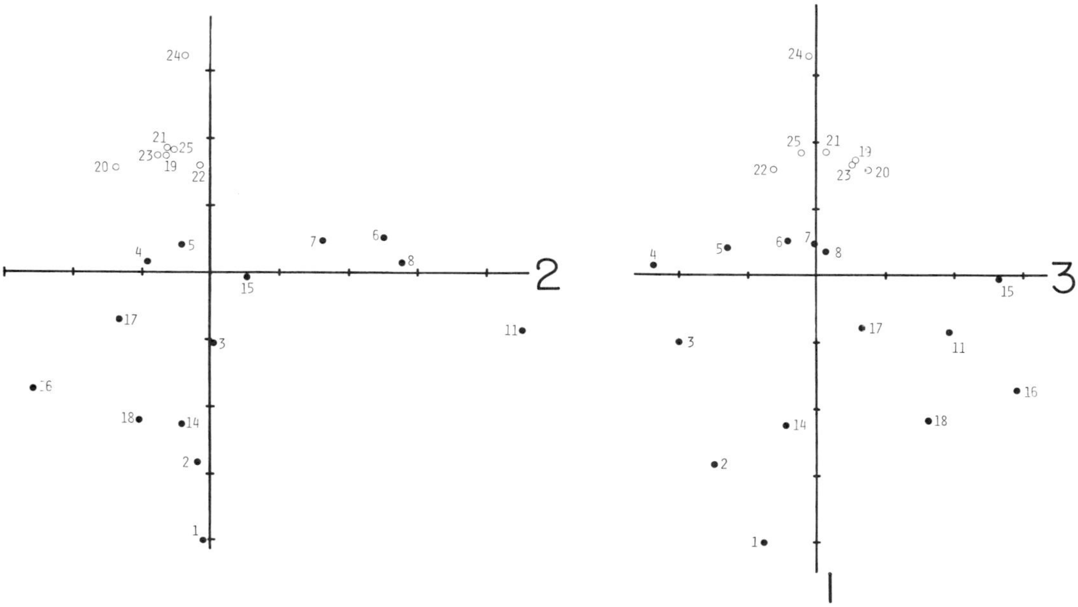

Figure 20. Scatterplots for first three discriminant functions obtained from male population samples representing *A. decorum* (●) and *A. elongatulum* (O).

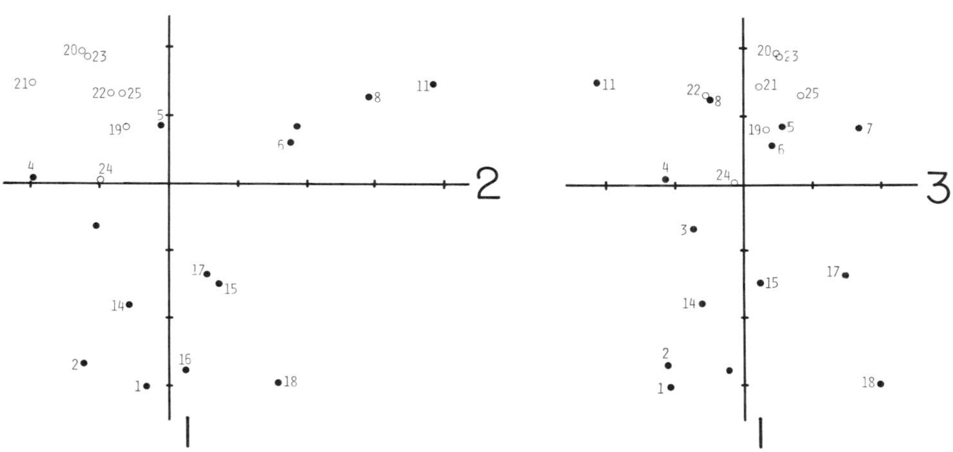

Figure 21 Scatterplots for the first three discriminant functions obtained from female population samples representing *A.decorum* (●) and *A. elongatulum* (O).

from *A. elongatulum*, this analysis deals only with green-glabrous individuals. The seven population samples of *A. elongatulum* used in the previous analysis are included. For *A. decorum*, population samples from monomorphically green-glabrous populations are included. Samples are those from populations 7-11 and 14-18 (table 5). The purpose of this analysis is the derivation of a discriminant function useful for species identification.

The two group discriminant analyses produce the standardized coefficients for male and female data sets in table 14. Discrimination of these two species based on morphometric features is the primary concern, with the results used in a dichotomous key; therefore, the maximum separation using the minimum number of characters is the ideal. In addition, the biometric characters diagnosing these species had to fulfill three other criteria: (1) the same measurements should be used to separate both sexes, (2) measurements should be unambiguous, (3) to insure use of the discriminant function the measurements should be easily made.

Table 14. Wilks' lambda values and univariate F-ratios for characters based on populations of *A. elongatulum* and green-glabrous populations of *A. decorum*.

	Males		Females	
Character	Wilks'	F	Wilks'	F
hw	.897	19.26***	.962	6.45*
pl	.885	21.88***	.968	5.36*
apw	.842	31.56***	.916	14.82***
mpw	.900	18.69***	.968	5.35*
bpw	.958	7.40**	1.000	0.02(n.s.)
huw	.810	39.41***	.888	20.44***
mew	.810	39.50***	.771	48.09***
el	.598	113.00***	.604	106.30***
4al	.910	16.62***	.914	15.32***
pfl	.850	29.71***	.881	21.83***
pfw	.990	1.64(n.s.)	.999	0.17(n.s.)
3sl	.763	52.02***	.792	42.52***
pd	.880	22.95***	.882	21.63***

* P less than .05.
**P less than .01.
***P less than .001.

Three of the characters evaluated for discrimination (table 14) proved suitable. Elytral length is a good discriminator in both sexes, based on small Wilks' lambda values. This measurement can be made rather precisely if the specimen is aligned horizontally in the field of view. Another character, visible in dorsal view, that possesses a relatively small Wilks' lambda value is maximum elytral width. This character is measured with the specimen set in the same position required for determination of elytral length. Prothoracic length is the third measurement included in the discriminant function for identification of these two species. Its inclusion is justified because it is easily measured from dorsal view. Trials with other characters used in conjunction with maximum elytral width and elytral length did not give better classification than trials with prothoracic length, elytral length, and maximum elytral width. The ease and precision of measurement proved to be the overriding criterion for character choice. Expansion of the discriminant function to six or seven characters resulted in only marginal increases in classificatory efficiency (3-4%). For males, the discriminant function to be used for separation of green glabrous *A. decorum* from *A. elongatulum* is:

d.f. score = -10.8 - 12.9(prothl) - 4.1(maxelwid) + 7.3(elyleng).

The specimen should be considered *A. decorum* if the score is less than .213, *A. elongatulum* if greater than .213. This discriminant function correctly classifies 87.7% of the individuals in this study set (table 15).

For females the discriminant function to be used for separation of these two species is:

d.f. score = -11.6 - 10.7 (prothl) + 1.7 (maxelwid) + 4.9(elyleng).

A specimen should be considered *A. decorum* if the score is less than .199, *A. elongatulum* if greater than .199. This function correctly classifies 85.4% of the individuals in this study set (table 15).

Table 15. Classification results for discriminant function used to discriminate between *A. elongatulum* and green-glabrous *A. decorum*. Males above diagonals, females below.

		Predicted species	
A priori species	*n*	*A. decorum*	*A. elongatulum*
A. decorum	100 / 96	86(86%) / 80(83%)	14(14%) / 16(17%)
A. elongatulum	70 / 68	7(10%) / 8(12%)	63(90%) / 60(88%)

The discriminant scores for individuals of the two species are portrayed in figure 22. Several unknown specimens were classified using these functions. The Cuban specimen is a paratype male of *A. extensicolle cubanum* Darlington = *A. decorum* Say, with label data: Central Soledad, Cuba, 1-VII-1932, B.B. Leavitt; M.C.Z. paratype *cubanum* Darl. (MCZ). This specimen is classified as *A. decorum* corroborating the prior analyses. Three males from the Bimini Islands are classified as *A. elongatulum* by this function. The label data for these specimens are: British West Indies, South Bimini Isl. (2), North Bimini Isl. (1); VI-1951, M. Cazier, C. & P. Vaurie (AMNH). Colonization of the Biminis by *A. elongatulum* may have been effected by man, or this short distance may have been traversed naturally.

Throughout northern peninsular Florida, *A. decorum* and *A. elongatulum* exist sympatrically without any evidence of hybridization. *A. decorum* is monomorphically red-hirsute in this area, whereas most *A. elongatulum* are green-glabrous. Rare individuals of *A. elongatulum* possess red prothoraces, but this red morph is not homologous with the red morph of *A. decorum*. In *A. decorum* the red morph has the mesosternum and mesepisternum rufous, whereas these sclerites are piceous in individuals of *A. elongatulum* with red prothoraces. No green-hirsute individuals have been observed from Florida.

Far outside the range of *A. elongatulum*, a series containing green-hirsute individuals has been taken. The label data are: NC: Ocracoke Isl., 12-VIII-1972, A. Smetana (5 specimens), (CNC). In this series there are two green-glabrous beetles (male and female), one dark red-glabrous female, one dark red-hirsute male and one green-hirsute female. The dark red character state is

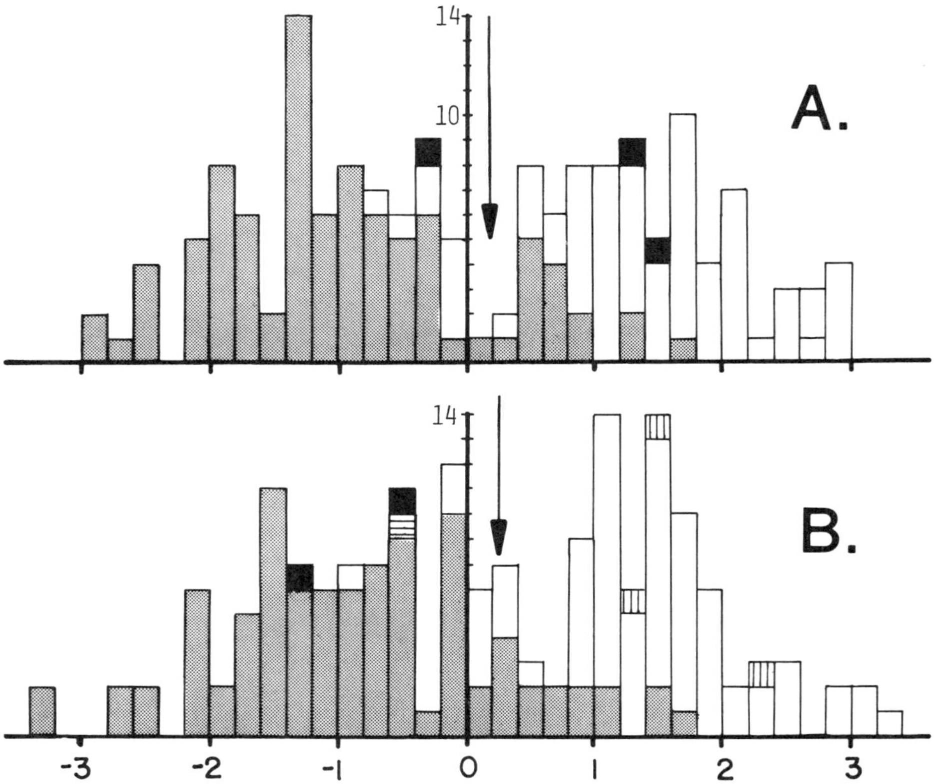

Figure 22. Distribution of discriminant function scores for two-group analysis based on green-glabrous *A. decorum* (shaded bars) and *A. elongatulum* (hollow bars): A. Females; B. Males. Vertically striped portions of bars represent individuals from South Bimini Island inserted as unknowns in discriminant function; horizontally striped portions represent paratype of *A. extensicolle cubanum* Darlington inserted as an unknown; solid portions of bars represent putative hybrids from Ocracoke Island, NC, also inserted as unknowns. Arrows mark scores that best divide two groups.

intermediate between green, in which the prothorax is metallic green dorsally and ventrally and the legs piceous, and red, in which the prothorax is rufous dorsally and ventrally and the legs rufotestaceous. The dark red state has the background prothoracic color dark red, with granules of metallic microsculpture denser on the centers of the pronotum and proepisterna. The legs are rufotestaceous.

Green-hirsute individuals of this complex have not been recorded in any areas outside the region near Lake Champlain and the southwestern United States. The absence of any red-hirsute individuals from this polymorphic sample is unique to this series. It can be hypothesized that these individuals represent hybrids between *A. decorum* and *A. elongatulum.* To test this hypothesis using the biometric data, measurements of these individuals were submitted to the three-character discriminant function. The measurements and scores are presented in table 16, with resultant positions indicated in figure 22. Two individuals are classified as *A. elongatulum* and three are classified as *A. decorum.* Based on this evidence, this series of individuals is assigned hybrid status.

Table 16. Values for three biometric measurements and discriminant scores for putative *A. decorum* × *A. elongatulum* hybrids.

Specimen	Sex	Morph*	pl	el	mew	d.f. score
1	female	GG	1.68	5.10	1.82	1.49
2	female	RG-G	1.60	5.00	1.75	1.25
3	female	GH	1.53	5.05	1.67	–.39
4	male	GG	1.47	4.73	1.47	–1.26
5	male	RG-H	1.67	5.30	1.67	–.50

* First letter G = green, RG = reddish green; second letter G = glabrous, H = hirsute.

Restriction of hybridization of *A. decorum* and *A. elongatulum* to outside of the normal range of sympatry implies one of three things. There may be secondary reinforcement of reproductive isolation in the area of sympatry. This would have acted to reduce gene flow after establishment of sympatry between these two species. This is the standard explanation of reproductive character displacement, in which hybrids are absent in areas of sympatry, but are produced when allopatric populations are brought together (Ehrman 1965). A second possibility involves the clinal transition of a factor inhibiting mating between these two species. If *A. decorum* populations in Florida possess this inhibiting characteristic and it is absent further north, the Okracoke Island hybrids are due to an intrinsic property of *A. decorum*. This alternative would parallel the cases cited in Grant (1972). Because of its very homogeneous genetic structure (table 4), clinal change in reproductive behavior is unlikely to exist in *A. elongatulum*. The third and most likely explanation assumes that the hybrid series is the progeny of a single founder. *A. elongatulum* is often taken in flight by black light traps. A dispersing individual is likely to be unmated, assuming that *A. elongatulum* expresses the oogenesis-flight syndrome of Johnson (1969). If transport of such individuals on storm fronts is rare, a successfully transported *A. elongatulum* could only hybridize with *A. decorum* individuals.

Agonum extimum complex. A. extimum and *A. parextimum* are exceedingly similar beetles. *A. extimum* can be distinguished by a pelage of fine setae on the dorsum and venter, and by 8 to 13 articulated dorsal setae on or adjacent to the third elytral interval. *A. parextimum* has a glabrous body surface, except for normally positioned articulated macrosetae, with four to six dorsal setae on or adjacent to the third elytral interval. Although such differences normally would be accepted as diagnostic for species, in this group acceptance should not be given hastily for two reasons: (1) *A. decorum* has been shown to be polymorphic for hirsuteness, with the presence or absence of a dorsal pelage governed by one allele (Liebherr 1983a), and (2) *A. extimum* and *A. parextimum* are distributed allopatrically (fig. 41). *A. extimum* is a lowland species whose distribution lies for the most part east of the Sierra Madre Occidentale. *A. parextimum* is distributed from near Tucson, to the west and southward on the west coast of Mexico. No series of specimens examined contained both types of beetles, so no locality could be sampled to obtain females for rearing as was done for *A. decorum* (Liebherr 1983a). In the absence of data on reproductive isolation in sympatry, the two qualitative setational characters must be used to recognize these two forms.

A biometric analysis was conducted on these two sets of populations to determine whether biometric differences would support species recognition. The first analyses are population level analyses of males and females; discriminant analysis is conducted with populations as groups. A

Table 17. Wilks' lambda values, univariate F-ratios, and standardized canonical discriminant function coefficients for males of *A. extimum* and *A. parextimum* analyzed with populations as groups.

Variable	Wilks'	F*	Coefficients d.f. 1	d.f. 2	d.f. 3
hw	.897	0.89	–0.99	–0.95	–0.02
pl	.903	0.84	0.77	–0.13	–0.62
apw	.879	1.07	2.23	–0.26	–0.66
mpw	.849	1.38	–0.23	1.43	1.61
bpw	.892	0.94	–0.06	0.34	–0.44
huw	.938	0.51	–1.37	1.01	0.60
mew	.977	0.19	1.78	–1.77	–0.42
el	.942	0.48	–0.40	–0.44	1.06
4al	.737	2.77**	1.09	0.19	–1.33
pfl	.881	1.05	–0.67	–0.13	0.71
pfw	.915	0.72	–0.15	–0.48	0.72
3sl	.748	2.62	–0.88	1.26	–0.22
pd	.885	1.01	0.26	0.25	–1.22

* P n.s. except as noted.
** P less than .05.

second pair of analyses treats the glabrous and hirsute forms as groups, with all individuals in one of the two groups.

Population level analysis of A.extimum complex. Discriminant analysis was conducted using five population samples, three of *A. extimum* (nos. 40-42) and two of *A. parextimum* (nos. 44, 45) (table 5). The first three discriminant functions accounted for 94.4% of the total variance, apportioned as 39.8% accounted for by d.f. 1, 31.9% by d.f. 2, and 22.7% by d.f. 3. Most characters exhibited high Wilks' lambda values, with only fourth antennomere length differing significantly in at least one population (table 17). All measurements were retained in the discriminant function, even though most were not significantly different in any of the population samples.

The two species separate on d.f. 1, although scores vary greatly within *A. parextimum* (fig. 23). The two characters that produce the strongest discrimination on d.f. 1 are fourth antennomere length and profemur length. Fourth antennomere length averages .63-.69mm in *A. extimum* versus .62-.64mm in *A. parextimum.* Profemur length averages 1.94-1.98mm in *A. extimum* versus 1.89-1.90 mm in *A. parextimum.* These differences coupled with standardized coefficients of -1.09 and -.67 (table 17) contribute to positive scores on function 1 for *A. parextimum.*

The trends in the population level analysis of females parallels those of the males. The first three functions account for 93.5% of the total variance, apportioned as 48.2% by d.f. 1, 26.9% by d.f. 2, and 18.4% by d.f. 3. Again, the only character which exhibits heterogeneity among population samples is fourth antennomere length (table 18). Populations of *A. extimum* possess antennae with average fourth segment lengths broadly overlapping those of *A. parextimum*: .63-.69mm versus .59-.67mm. The legs are also slightly longer in *A. extimum,* averaging 1.98-2.02mm versus 1.92-1.98mm in *A. parextimum.* The standardized coefficients of these two characters are -1.16 and -.55 (table 18) respectively, giving *A. parextimum* generally higher d.f.

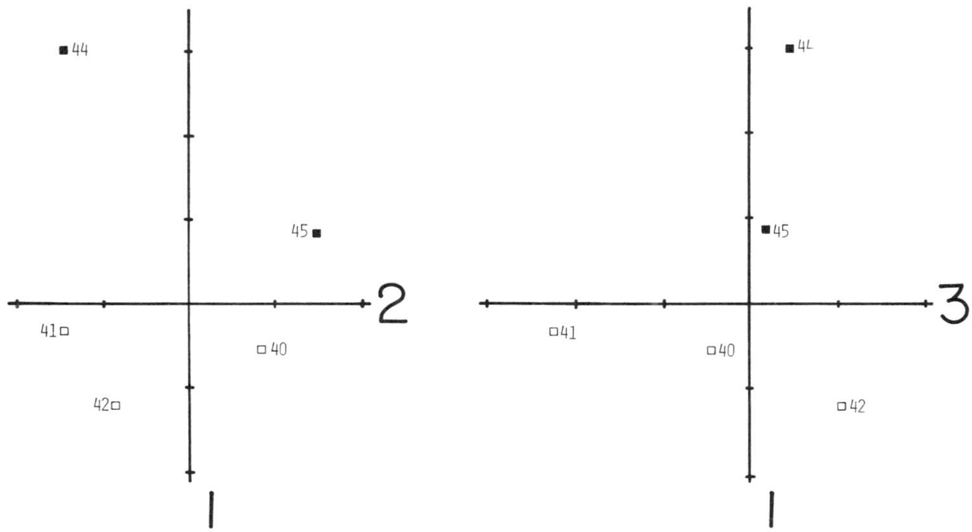

Figure 23. Scatterplots of first three discriminant functions for male sample of *A. extimum* (■) and *A. parextimum* (□).

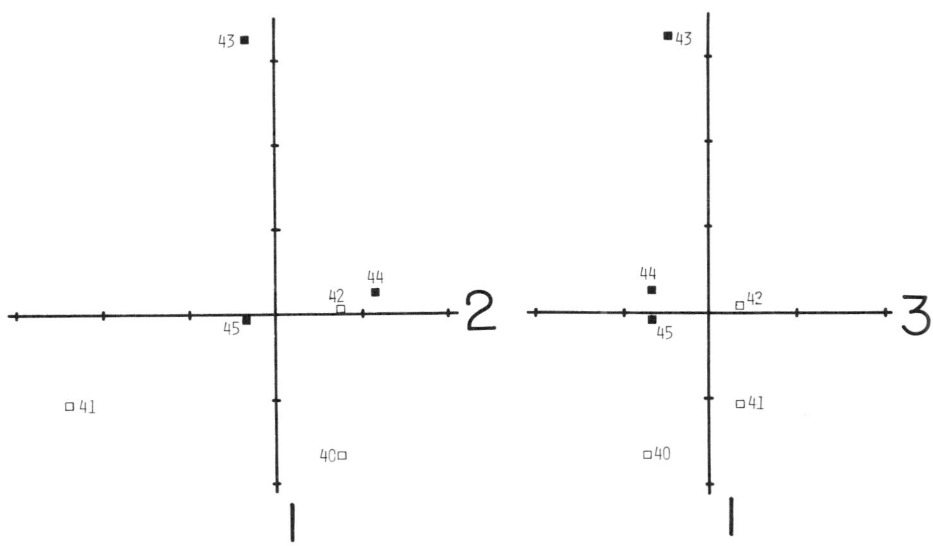

Figure 24. Scatterplots of first three discriminant functions for female samples of *A. extimum* (■) and *A. parextimum* (□).

scores on function 1 (fig. 24). There is significantly more overlap of populations between these two species in the females compared to the males. The overlap involves the Douglas, Arizona population of *A. extimum* and the Brawley Wash, Arizona and Sinaloa, Mexico populations of *A. parextimum.* The similarity of these populations does not support a clinal transition from east to west through the lowlands of southern Arizona. The very high scores of the Tucson population (no. 43) and its separation from other southern Arizona populations also argues against explanation of the variation as an east-west clinal transition in appendage length or body size.

Table 18. Wilks' lambda values, univariate F-ratios, and standardized canonical discriminant function coefficients for females of *A. extimum* and *A. parextimum* analyzed with populations as groups.

			Coefficients		
Variable	Wilks'	F*	d.f. 1	d.f. 2	d.f. 3
hw	.793	2.19	−1.36	1.04	0.88
pl	.949	0.45	−0.46	1.18	−0.19
apw	.949	0.45	0.04	1.46	−0.39
mpw	.932	0.61	0.43	−1.20	−0.43
bpw	.978	0.19	1.00	−0.04	−0.49
huw	.831	1.71	0.60	1.14	0.18
mew	.964	0.31	0.28	−1.78	−1.01
el	.977	0.20	0.26	0.57	0.18
4al	.634	4.85**	−1.16	0.30	−0.96
pfl	.884	1.10	−0.55	−1.13	1.00
pfw	.962	0.33	0.47	−0.39	0.11
3sl	.833	1.68	0.19	0.62	0.71
pd	.944	0.50	0.02	−1.23	0.45

* P n.s. except as noted.
** P less than .01.

A. extimum and *A. parextimum* differ biometrically; the differences cannot be explained by clinal transition of characters. In all cases the concerted action of all characters in a discriminant function is required to achieve separation of the two species. Univariate character analysis fails to distinguish these species due to large intrasample variability.

Species level analysis of A.extimum complex. In an effort to distinguish *A. extimum* and *A. parextimum* biometrically, individuals from the various samples were pooled, and subjected to a two group species level discriminant analysis. The resultant function scores for individuals are shown in figure 25.

For males, the standardized discriminant coefficients (table 19) are very similar to those in the population level analysis (table 17). Fourth antennomere length and profemur length possess negative standardized coefficients, and the smallest (but by no means small) Wilks' lambda values (table 19). These two characters should be considered the best discriminators of these species. For males, the discriminant function classifies 90.9% of the specimens into the a priori assigned species (table 20). Two of the four individuals of *A. extimum* that are incorrectly classified are from San Pedro, Coahuila , Mexico, as are two from near Douglas, Arizona. Like the population level analysis, these data do not support east-west clinal transition in biometric measurements as an explanation of variation in this sister species pair.

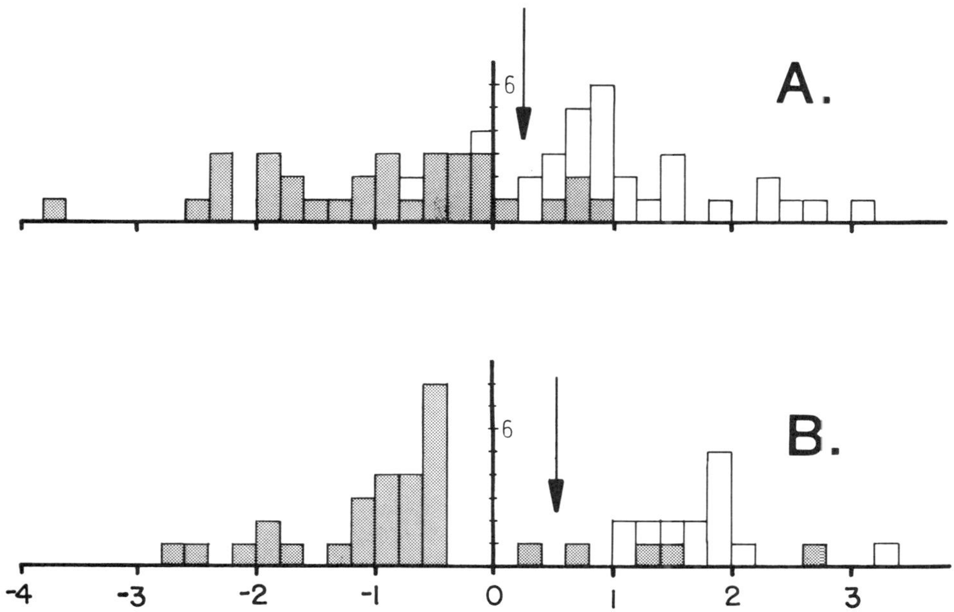

Figure 25. Distribution of discriminant function scores for two-group analysis based on *A. extimum* (shaded bars) and *A.parextimum*: A. Females; B. Males. Arrows mark scores that best divide two groups.

Table 19. Wilks' lambda values, univariate F-ratios, and standardized canonical discriminant function coefficients for *A. extimum* and *A. parextimum* analyzed with species as groups.

	Males			Females		
Variable	Wilks'	F*	Coefficient	Wilks'	F*	Coefficient
hw	.959	1.80	–0.84	.957	2.50	–0.99
pl	.994	0.26	0.29	.986	0.80	0.10
apw	.967	1.45	1.38	.942	3.47	0.19
mpw	.982	0.77	1.14	.922	4.71	1.39
bpw	.984	0.68	–0.03	.958	2.47	0.08
huw	.991	0.39	–0.47	.993	0.42	0.20
mew	.990	0.42	0.46	.996	0.21	0.32
el	.984	0.68	–0.25	.991	0.49	0.25
4al	.915	3.91	–1.23	.991	0.52	–0.17
pfl	.924	3.47	–0.66	.961	2.30	–1.29
pfw	.998	0.08	–0.12	.976	1.36	0.30
3sl	.987	0.53	–0.35	.965	2.02	0.13
pd	.991	0.39	0.16	.997	0.17	–0.49

* All F-ratios n.s.

For females, the overall correct classification into a priori assigned groups is 89.7%. The distribution of scores in this case (fig. 25) shows several individuals of each species misclassified (table 20). Of the four *A. extimum* classified as *A. parextimum*, one each is from Coahuila, Mexico, and Deming, New Mexico, and two are from Douglas, Arizona. The two misclassified individuals of *A. parextimum* are from the Sinaloa sample. Again, populations geographically intermediate in the combined species ranges are not more similar than more distant populations.

As in the males, an appendicular character is one of the strongest discriminators between the two groups. Profemur length has the third smallest Wilks' lambda and second largest coefficient (table 19), and is of moderate discriminating ability. In these samples, female *A. parextimum* exhibit slightly wider prothoraces, which are reflected in this discriminant function: Wilks' lambda of .922 and standardized coefficient of 1.39 for maximum prothoracic width (table 19). Intrasample variability within these species samples limits the usefulness of any univariate analysis. Only the discriminant function utilizing all characters provides the ability to separate these two species.

Table 20. Classification results for discriminant function used to distinguish *A. extimum* and *A. parextimum*. Males above diagonals, females below.

		Predicted species	
A priori species	*n*	*A. extimum*	*A. parextimum*
A. extimum	31 / 32	27(87%) / 28(88%)	4(13%) / 4(12%)
A. parextimum	13 / 26	0(0%) / 2(8%)	13(100%) / 24(92%)

Summary of Species Complex Discriminant Analyses. Species in this group are biometrically distinct. The ambiguities in the female data set suggested the more detailed analyses of the *A. decorum*-*A. elongatulum* and *A. extimum*-*A. parextimum* species pairs. Within the *A. decorum* complex, the greater biometric similarity of the Greater Antillean populations to mainland Mexico and western United States populations of *A. decorum*, supports a circumscription of *A. decorum* that includes these populations. *Agonum elongatulum* is comprised of populations restricted to peninsular Florida and the Biminis (fig. 58).

Darlington recognized the distinction between *A. elongatulum* and *A. decorum* in this geographic area and cited *A. floridanum* (LeConte) = *A. elongatulum* (Dejean) as occurring on the Biminis (Darlington 1953). He assigned Antillean populations to *A. extensicolle cubanum* Darlington (1934) on the basis of their similarity to Mexican material determined by Henry Bates as *A. extensicolle*, but Bates' concept of *A. extensicolle* was wrong. He wrote of *A. extensicolle*: "All the Mexican examples I have seen have dark legs and antennae, and dull bronzed elytra, thus agreeing with Dejean's description of his *A. elongatulus*"(Bates 1882, 93). What Bates was referring to is green-glabrous *A. decorum*, not *A. extensicolle*, which has bright metallic blue elytra and testaceous legs throughout Mexico. Thus, confusion of the forms adjacent to the Caribbean rests on Bates' misidentification.

Biometric analysis of *A. extimum* and *A. parextimum* shows that discrimination between the two kinds can be accomplished using mensural characters. As the distributions of the setose *A. extimum* and glabrous *A. parextimum* are allopatric, reproductive isolation in sympatry cannot be used as a species criterion. Species recognition in this case rests on the subjective opinion of the reviser. I propose recognition of two distinct species based on the biometric differences and diagnostic differences in pelage. The test of this interpretation of specific limits can be made if the two kinds are found sympatrically. The area in which to look for this is somewhere south of Tucson, east to Douglas, Arizona. Rearings of wild mated females from such a population would determine the extent of reproductive isolation.

ANALYSIS OF DISPERSAL CAPABILITIES

Agonum species tend to be good dispersers. Their colonization of habitats such as disturbed streamsides, or early successional marshes, necessitates a well-developed dispersal ability. Not all carabid species rely on a colonization strategy as strongly as do *Agonum* beetles. Varying degrees of brachyptery of the flight wings has evolved independently many times throughout the family. In extreme cases, degeneration of the sclerites used for flight can occur (Darlington 1936; Smith 1964). Wing length can be monomorphically long or short in a species, or the species may be polymorphic for wing length, with discrete macropterous and brachypterous forms. Goldschmidt (1940) summarized earlier literature on the developmental processes leading to brachyptery.

Lindroth (1946) was the first to attempt elucidation of the genetic basis for wing length polymorphism, using *Pterostichus anthracinus* Illiger. His results were consistent with the long-winged or macropterous state being recessive to the brachypterous state. Boer et al. (1980) discussed genetic data for wing polymorphism in *Calathus mollis* Marsh which agree with the mechanism in *P. anthracinus*. Carter (1976) studied *Agonum retractum* (LeConte) with results suggesting the possibility of a more complicated mechanism. In this species, short- and long-winged females gave rise to mixed broods.

Darlington (1943) correlated flight wing development with the stability of the habitat type used by a carabid species. He found that lowland hydrophilic and arboreal species tend to retain fully developed flight wings consistent with a strategy of colonizing disturbed or temporary habitats. Resident species of mountainous or highland areas are much more likely to have evolved shortened flight wings, since the fauna of such areas is composed principally of geophilic species—the ones most likely to undergo wing reduction at any altitude.

Lindroth (1949) showed that in species polymorphic for flight wing length, populations most likely to have recently colonized an area have a higher percentage of macropterous individuals than do populations of earlier origin and more continuous existence. This phenomenon is best illustrated by comparing populations located outside of areas glaciated in the Pleistocene to populations established in glaciated areas since that time.

Erwin (1979b) expanded Darlington's ideas concerning the development of flightless faunal elements, coining the term *taxon-pulse* for a model of diversification in the Carabidae. In his scenario, tropical lowland faunas are the source of generalized, vagile lineages that become fragmented through evolutionary movement into higher latitudes, or higher altitudinal or arboreal biotopes. Fragmentation and resultant diversification of lineages on mountains is caused by the tendency of such species to be wingless and thus more likely to speciate. The driving force for this model is intense competition in the lowland tropical habitats that forces taxa into peripheral areas, making this model a direct expansion of the taxon cycle of Wilson (1961). Documentation of the intense competition necessary to drive this model is not at hand, but the taxon-pulse model appears to fit the facts, *ex post facto*. It is not the purpose of this study to evaluate the taxon-pulse model; however, analysis using the *A. extensicolle* group can help answer questions relating to the taxon-pulse scenario. Specifically, is type of habitat utilized by species in this group correlated with dispersal ability? Secondly, can data quantifying amount of genetic heterogeneity within a species be correlated with dispersal ability? In other words, are the more vagile species in this clade (assuming there are differences in vagility) less heterogeneous genetically? Comparison of all species in a monophyletic group is judged to be the most rigorous protocol for evaluation of these questions. A monophyletic group consists of species with varying degrees of shared ancestry. This common ancestry insures the highest likelihood of common genetic bases for observed evolutionary phenomena. No other type of grouping can insure a better experimental analysis of species differences and similarities.

Development of Flight Apparatus. Individuals in the *A. extensicolle* species group possess fully developed flight wings. To quantify wing development relative to overall body size, five males and five females, or as many specimens as were available, were evaluated from the population samples shown in table 21. Wing development was assessed using the ratio of flight wing length to elytron length (Palmén 1944). Flight wing length is considered the distance from the basal junction of the radius and cubitus, to the apex of the posterior sector (Ward 1979). Elytral lengths are those used in the biometric analysis. This relative flight wing length ratio has the advantage that it is a simple ratio. Boer et al. (1980) advocate use of the ratio (flight wing length x flight wing width)/(elytral length x elytral width). Because it relates flight wing area to body size and biomass, this ratio may be more effective when distantly related taxa are compared. For study of closely related species, a simple ratio is sufficient, since body size and shape are similar among the species.

The population averages for flight wing length (table 21), and averages for elytral length (Appendix 5 and 6), produce the ratios of relative wing length given in figure 26. All species except *A. extensicolle* have wing length ratios greater than 1.20. Specimens of *A. texanum* are among the largest seen for this group, and this species has the largest flight wing ratio. *A. extensicolle* has the shortest relative wing length. Dividing the species into two groups based on a relative wing length less than or greater than 1.20, we can compare *A. extensicolle* to all other species. The difference between the two groups is significant at a p-value much less than .001 (chi-square value = 250; d.f. = 1).

A. cyanope individuals possess a wing length to elyton ratio greater than that of *A. extensicolle* individuals, but somewhat less than those observed in the other five species. The small number of populations available for *A. cyanope* precludes quantification of between-species differences.

Two individuals of *A. extensicolle* exhibit a discrete brachypterous condition in which wing length is .8-.9 times elytal length (fig. 26). One female from Clarke County, Virginia has an elytral length of 5.58mm, with wing length of 5.08mm (fig. 36h). For comparison, a second Clarke County, Virginia female has an elytral length of 5.38mm and wing length of 6.92mm (fig. 36g). Wing length reduction in the two cases observed does not involve major wing vein loss.

Other individuals of *A. extensicolle* did exhibit abnormalities in wing venation. Eight of the 166 specimens (5.8%) possess an oblongum cell with an incomplete basal vein (fig. 36f). This

Table 21. Average wing lengths in mm and standard deviations for males and females of the *A. extensicolle* group.

Population number	Species	Males $\overline{X}$ (s.d.)	Males *n*	Females $\overline{X}$ (s.d.)	Females *n*
1	*decorum*	6.46(.09)	5	7.06(.39)	5
3	*decorum*	7.21(.24)	3	7.03(.46)	5
4	*decorum*	7.09(.34)	5	7.55(.26)	5
5	*decorum*	6.90(.31)	5	7.11(.14)	5
6	*decorum*	6.49(.40)	5	6.78(.33)	5
7	*decorum*	6.41(.35)	5	6.87(.25)	5
8	*decorum*	6.03(.42)	5	6.54(.44)	5
9	*decorum*	6.18(.36)	5	6.38(.28)	5
10	*decorum*	5.96(.39)	5	6.45(.27)	5
11	*decorum*	5.92(.29)	5	6.21(.16)	4
12	*decorum*	6.42(.24)	5	6.87(.12)	5
13	*decorum*	6.71(.39)	5	7.24(.23)	5
14	*decorum*	6.39(.21)	5	6.55(.43)	4
17	*decorum*	6.84(.26)	10	7.18(.21)	10
18	*decorum*	6.76(.24)	10	7.23(.25)	10
19	*elongatulum*	6.71(.41)	10	7.21(.38)	10
23	*elongatulum*	7.01(.23)	10	7.42(.31)	10
26	*extensicolle*	5.71(.49)	5	5.77(.31)	5
27	*extensicolle*	5.78(.65)	5	5.59(.83)	5
28	*extensicolle*	5.58(.24)	5	5.85(.42)	5
29	*extensicolle*	5.34(.68)	5	5.27(.42)	3
30	*extensicolle*	5.76(.10)	5	5.86(.21)	5
31	*extensicolle*	6.49(.36)	5	6.72(.36)	5
32	*extensicolle*	6.55(.37)	5	7.18(.23)	5
33	*extensicolle*	6.10(.20)	5	6.75(.33)	5
34	*extensicolle*	6.61(.67)	5	6.92(.61)	5
35	*extensicolle*	6.30(.19)	5	6.43(.19)	5
36	*extensicolle*	6.10(.47)	5	5.74(.31)	5
37	*extensicolle*	5.51(.45)	5	6.07(.72)	5
38	*extensicolle*	5.98(.39)	5	6.39(.52)	5
39	*extensicolle*	6.07(.29)	5	6.28(.40)	4
40	*extimum*	8.08(.30)	5	8.07(.24)	5
45	*parextimum*	7.08(.19)	5	7.37(.30)	5
46	*cyanope*	5.56(.33)	8	5.75(.25)	8
47	*cyanope*	6.88 —	1	7.17 —	1
48	*cyanope*	6.57 —	1	6.06 —	1
49	*texanum*	7.24(.32)	5	7.20(.94)	5
50	*texanum*	8.00(.47)	5	6.19(.44)	5
55	*texanum*	7.86(.21)	5	6.18(.62)	5
57	*texanum*	7.80(.40)	5	6.66(.56)	5

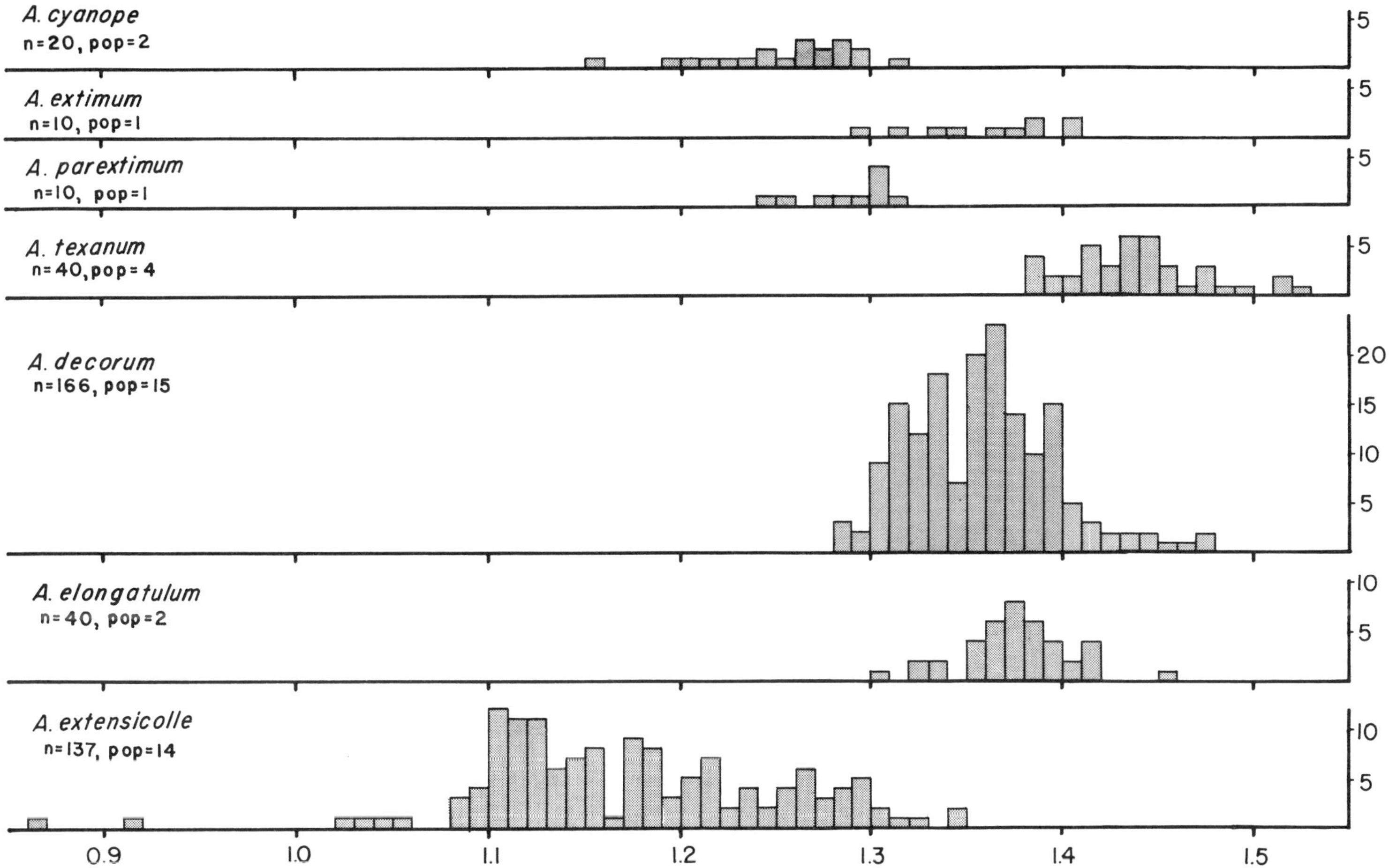

Figure 26. Frequency distribution of ratios of flight wing length to elytral length for species of the *A. extensicolle* group. Populations sampled are listed in Table 21.

abnormality is not correlated with significantly shorter than average flight wings. Wings of individuals of all other species are of normal venation.

Smith (1964) investigated the morphological ramifications of flightlessness in Coleoptera, and reported that metanotal phragmata may be reduced independently of wing reduction, or correlated with it. To investigate this in the *A. extensicolle* group, median longitudinal sections were made through individuals of six of the seven species. These individuals had been preserved in Kahle's solution.

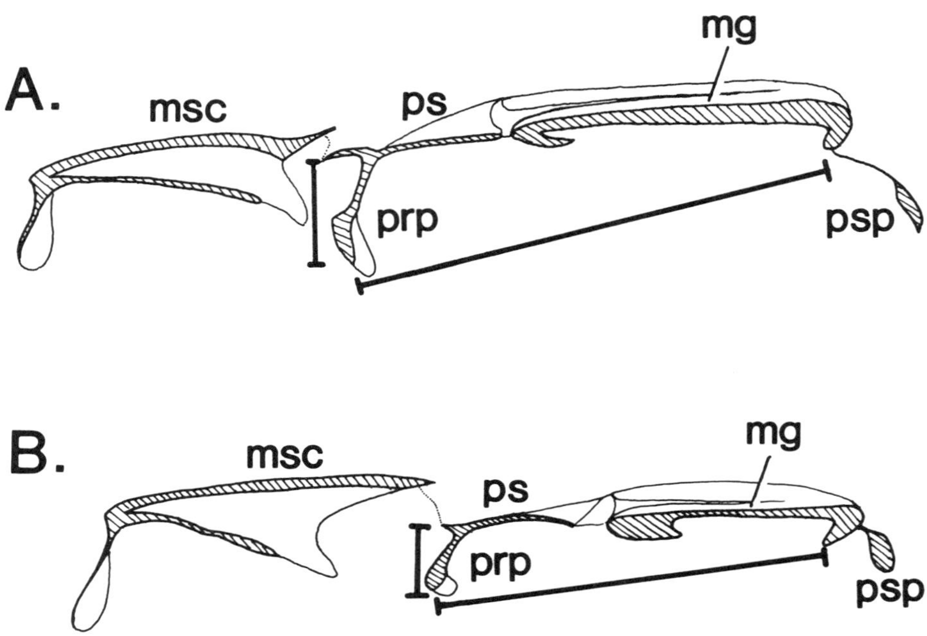

Figure 27. Diagrammatic median longitudinal section of meso- and metanotal tergites. msc = mesoscutellum; prp = metanotal prephragma; ps = metanotal prescutum; mg = metanotal median groove; psp = metanotal postphragma. Terminology follows that of Smith (1964). A. *A. elongatulum.* B. *A. extensicolle.*

Generally the metanotum has well-developed phragmata associated with it, as seen in *A. elongatulum* (fig. 27a). The area between the anterior and posterior phragmata is packed with dorsal longitudinal muscles in individuals that fly. Extent of muscle development depends on the period of the life cycle sampled. Study of European carabid species shows flight muscle autolysis commonly occurs when ovarioles are producing eggs, whereas flight muscles are well-developed during the preovipositional dispersal period (Boer et al. 1980). This phenomenon has been termed the *oogenesis-flight syndrome* by Johnson (1969).

Regardless of whether flight muscles are reduced, possession of well-developed phragmata indicates that flight capability is retained. To quantify the development of the metanotal phragmata, two measurements were made (figs. 27a,b). The vertical depth of the anterior phragma and the distance from the median apex of that phragma to the recurved base of the metascutellum were measured using a calibrated ocular grid (fig. 28). *Agonum extensicolle* and

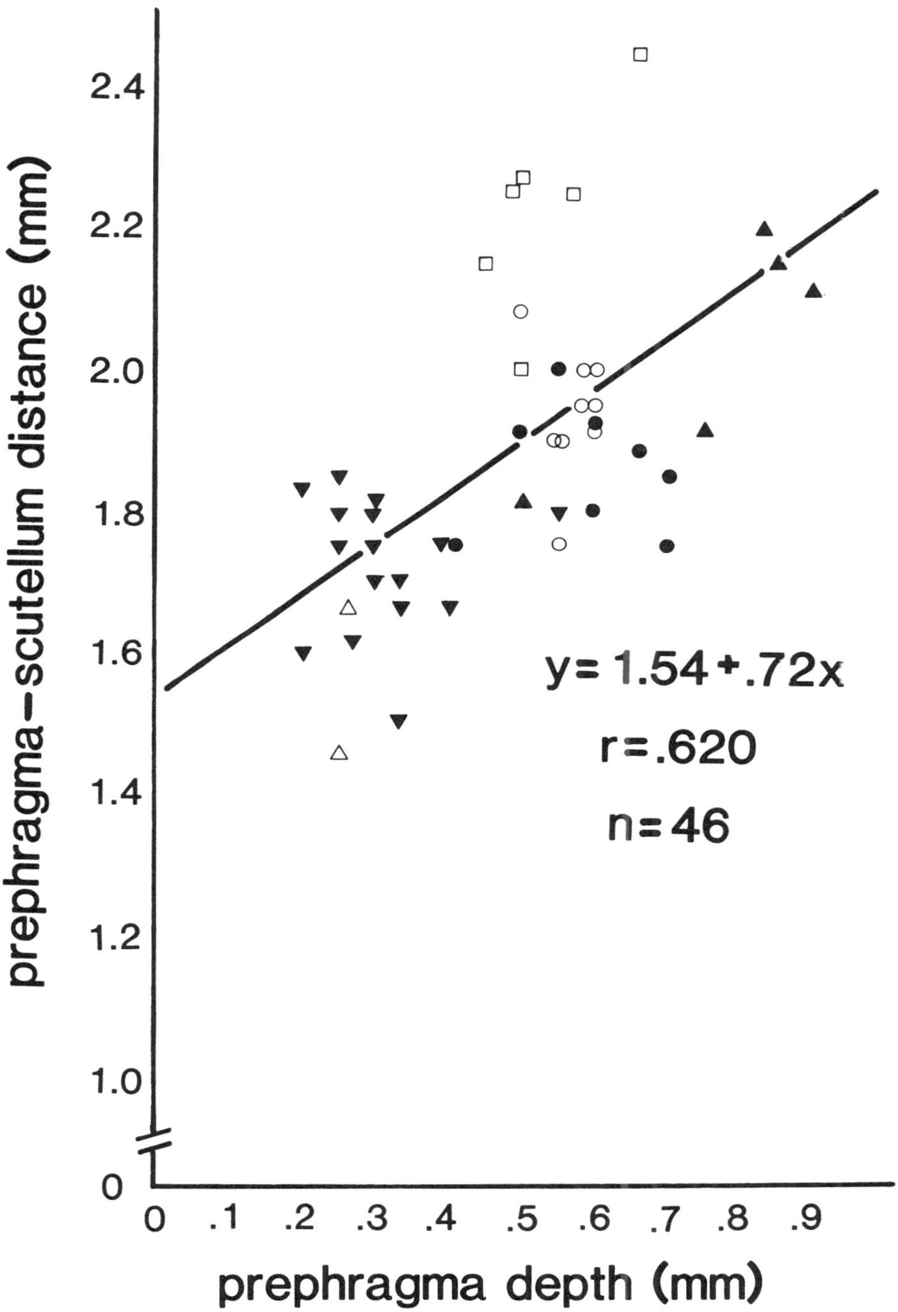

Figure 28. Distance from medial apex of metanotal prephragma to posterior of metascutellum at median groove, regressed against prephragma depth. Individuals of six species in *A. extensicolle* group: *A. cyanope* (△), *A. extensicolle* (▼), *A. texanum* (▲), *A. decorum* (●), *A. elongatulum* (O), *A. extimum* (□). Correlation coefficient significantly differing from zero a P less than .001.

A. cyanope possess a less developed anterior phragma and shorter metanotum (fig. 27b); the other four species sampled all possess a larger anterior phragma and longer metanota (fig. 27a). The data indicate reduction in metathoracic development in *A. extensicolle* and *A. cyanope*. The relationship is significant statistically, based on a t test of the correlation coefficient obtained when phragma-metascutellum distance is regressed on vertical phragma depth (fig. 28).

To summarize, interspecific differences in metathoracic development exist among the species of the *A. extensicolle* group. Reduction of the metathorax is exemplified by reduction of flight wing length, increase in wing venation abnormalities, reduction of the anterior metanotal phragma, and reduction in overall metanotal length. Correlation of all these attributes indicates that the degree of vagility within this species group depends not only on wing length, but on overall development of the metanotal flight apparatus.

Specimen Captures in Flight. Evidence of dispersal in carabids can be obtained by sampling with light traps, window traps, or other traps designed to capture aerial insects. Erwin (1979a) reports capture of one *A. decorum* in a light trap placed on an oil drilling rig 18 miles off the coast of Louisiana.

To evaluate the propensity for flight in this group, labels of museum specimens were examined to determine whether flight was indicated. The best criterion available for this determination is whether the specimens were taken at light, or in light traps. Various ambiguities enter this type of data. If specimens were taken at light sheets, they may have run to the light source. Alternatively, "at light" or "light trap" capture may not always have been noted on date-locality labels. Also, local environmental conditions determine whether flight will occur (Van Huizen 1979), making geographic distribution another determinant of flight activity. The number of specimens taken at light is thus a rough estimate of flight activity, but it is the best criterion available.

The percentage of individuals taken at light, based on date-locality labels, varies between species (table 22). *A. texanum*, found in lowland areas of the western Gulf Coast to Arizona, and possessing long flight wings (fig. 26), is commonly taken at light. Almost half the specimens examined were so taken. This high percentage is probably a function both of flight capability and suitable environmental conditions for the initiation of flight. *A. elongatulum* is also found in semitropical lowland habitats and possesses relatively long flight wings (fig. 26). Approximately one-fourth of the individuals examined for this species were taken at light.

The influence of environmental conditions on flight can be illustrated using the various morphs of *A. decorum* (table 22). The percentage of individuals noted from light is 8.6% overall, but only 3-4% for the green-glabrous or red-glabrous morphs. The majority of specimens of these two morphs are from habitats in the Great Basin desert and on the Pacific Coast from California to British Columbia (figs. 51, 53). Evenings are generally cool in these areas, limiting crepuscular and night flights of dispersing individuals. Dispersal may occur during the day, as noted by Van Huizen (1979) for European species. Sampling over 24 hours with window traps, as done in Europe, would permit a better estimate of dispersal capabilities. The green-hirsute and red-hirsute morphs are found in more southern areas of Arizona and Mexico and throughout eastern North America. The percentage of light trap captures is greater in the southern United States—a trend seen in most black-light sampling of wide-ranging carabid species. The overall percentage of *A. decorum* individuals taken at light is 11.5% for the eastern part of its range. Limiting analysis to Florida, 45.8% are captured at light. This indicates that any differences in propensity to disperse in *A. decorum* are mediated by external factors.

A significant fraction of *A. extimum* and *A. parextimum* individuals have been taken at light

Table 22. Total number of individuals and percentage of total individuals taken at light or in light traps.

Species	Total *(n)*	Taken in flight *(m)*	m/n (%)
A. cyanope	157	0	0
A. extimum	240	36	15.0
A. parextimum	62	5	8.1
A. texanum	787	345	44.1
A. extensicolle	4300	6	0.1
A. decorum	4629	400	8.6
green-glabrous	2929	121	4.0
red-glabrous	129	4	3.1
green-hirsute	216	26	12.0
red-hirsute	1303	149	11.4
r.-h. (Florida)	62	28	45.2
A. elongatulum	777	193	24.8

(table 22). The percentage for *A. parextimum* may be artificially low, since 26 of the 62 specimens were taken before 1930, when label data were less adequate than today.

The two species possessing a reduced flight apparatus, *A. cyanope* and *A. extensicolle*, are rarely taken at light (table 22). No *A. cyanope* have been noted at light, although localities listed from Arizona are commonly used for light collecting. The flight activity for this species may be greater than the data indicates if label data have been incomplete. For *A. extensicolle*, the records indicate limited capture in flight. Of the 4300 specimens for which data was taken, only six were noted as taken at light. Lindroth (1966) mentions the occurrence of macropterous *A. extensicolle* in drift material at Point Pelee, Ontario. The light trap records from this study include captures at Hudsonville, Michigan (1 ex.); College Station, Texas (3 exx.); Victoria, Texas (1 ex.); Washington County, Arkansas (1 ex.). This species can fly, but the amount of such activity appears to be less than in species with well-developed flight mechanisms.

Correlation of Flight Activity and Habitat. One hypothesis inherent in the taxon-pulse model of Erwin, derived directly from Darlington (1943), states: "Flightlessness is a feature developed among taxa occurring at mid to high elevations and mid to high latitudes" (Erwin 1979b). Darlington (1943) showed that correlation of habitats with brachyptery results from the relative rarity of streamside hydrophiles in upland areas. Hydrophiles usually retain flight capability because they must recolonize unstable habitats. In Darlington's model, upland faunas are dominated by geophilic species, which reside in limited, stable habitats, making flight unnecessary.

Darlington's model assumes the species is a homogeneous genetic unit and that selection acting on individuals causes evolutionary change. According to his view: (1) lack of habitat for hydrophiles in upland areas restricts the fauna to geophilic species; (2) geophilic species can maintain dense stable populations in which dispersal is not necessary; (3) there is an intrinsic advantage to being wingless; (4) residents of stable limited habitats become brachypterous.

The question of whether the necessity to disperse is absolutely correlated with the amount of dispersal is raised by Boer (1970). He studied a brachypterous woodland carabid, *Carabus problematicus* Herbst, which colonizes unsuitable heath habitat by running, leaving apparently suitable woodland to do so. This maladaptive dispersal strategy causes loss of individual fitness for dispersers who enter the heath. Why is this strategy tolerated by the species? If other selection pressures require dispersal before successful mating, or link dispersal with increased fecundity throughout the species, local disadaptation can be tolerated, even expected. In this case, species-level selection pressures may be counteracting individual selection.

Several of the species in the *A. extensicolle* group have extremely broad geographic ranges. For example, *A. extensicolle* is found from 5,000 to 8,000 feet in elevation in Mexico, whereas it occurs near sea level from southern Texas to Nova Scotia. The best means to assess habitat choice in this group is to restrict discussion to areas in which many of the species are sympatric, in the broad sense. Six of the seven species are found in Arizona and Mexico, and discussion of habitat utilization will concentrate on this region. The hypothesis to be tested is whether there is a correlation between reduced flight apparatus and utilization of an upland habitat.

The two species in the *A. extensicolle* group with reduced flight apparatus are *A. cyanope* and *A. extensicolle* (figure 28). *A. cyanope* is a species of upper Sonoran habitats. In the Chiricahua Mountains, specimens have been taken in Pridham Canyon and near Rucker Lake (S. McCleve, pers. comm.). Both habitats are between 5,500 and 6,000 feet elevation. In Mexico this species is recorded from elevations between 5,500 feet (Chihuahua, Rancho Primavera) and 8,800 feet (Durango, El Salto). It is associated with oak-pine forest at these higher elevations.

A. extensicolle occurs in streams from 3,800 to 5,000 feet elevation in southeastern Arizona. At Luna Lake, Apache County, Arizona, elevation 8,000 feet, it was the only *Agonum* taken. I have observed it elsewhere at elevations from 3,800 to 6,000 feet, usually in sheltered mesic canyons. Mexican records are from habitats between 5,300 and 7,600 feet.

The other species with longer metanota and larger metathoracic phragmata are principally lowland species. *A. extimum*, *A. parextimum* and *A. texanum* are found in lowland streamside habitats. These species all possess well-developed metathoraces and have been collected at light. *A. decorum* can be taken in lowland marshes. It is also taken along streams and rivers in the Southwest, but only in habitats with well-developed vegetation, or a layer of decaying vegetation. Typical elevations for *A. decorum* collections in Arizona range from 3,800 to 5,200 feet. *A. elongatulum* is restricted to peninsular Florida, residing in a variety of habitats, all lowland riparian sites.

A. extensicolle utilizes lowland habitats throughout eastern North America. As a streamside hydrophile, it must disperse at least enough to find safe overwintering spots. The streamside habitats it utilizes are continuous geographically, and this species can apparently rely on dispersal by means other than flight.

A. decorum is associated with riparian marshes that are more patchily distributed than linear creekside habitats. Very early successional shoreline habitat is unacceptable for this species. Marshes are especially patchy in western North America, where populations are sometimes restricted to desert springs or marshes on shores of lakes. Colonization of this type of habitat requires longer-distance dispersal.

Correlation of Flight Activity and Genetic Heterogeneity. The genetic variability within a species can be partitioned into hierarchical components. The within-individual genetic variability can be estimated using the percentage of heterozygous loci per individual organism. Electrophoretic analysis of soluble proteins and substrate-specific enzymes permits this estimation for a portion of the genome. The genetic variance among individuals of a population is the next hierarchical level of genetic analysis. The electrophoretic data shows the populations sampled to be composed of randomly breeding individuals, so this level of analysis is not pursued further.

The third level of genetic variability, and the one of interest in relation to dispersal capabilities, is the genetic heterogeneity among populations within a species. The degree of genetic heterogeneity among populations can be quantified using Wright's F_{ST} index or the ratio of the deviations of allelic frequencies in each population from the average frequency for the study set to the maximum variance in allelic frequencies based on overall average allelic frequencies. Wright's F_{ST} is "the correlation between random gametes within subdivisions, relative to gametes of the total population" (Wright 1969). Values of F_{ST} equal to 1.0 indicate that all populations are fixed for alternate alleles.

The relation of flight wing development to genetic heterogeneity over space tests the hypothesis that less vagile species will be more genetically heterogeneous over their range. This hypothesis assumes nothing regarding the cause for population differentiation except a difference in interpopulational gene transfer. It does assume equal time for divergence within and among populations. Divergence between populations may be due to differential selection unimpeded by gene flow (Ehrlich and Raven 1969), population bottlenecks and reassortments of the genome due to transilience events (Templeton 1980), or active changes in genotype mediated by mechanisms intrinsic to the genetic system, such as molecular drive (Dover 1982). At this level of hypothesis testing, the question is: are less vagile species more heterogeneous genetically?

Based on the electrophoretic data available in the *Agonum extensicolle* species group, three species can be compared: *A. elongatulum*, *A. decorum* and *A. extensicolle* (tables 4, 22). The most homogeneous species is *A. elongatulum*, which has only one locus, Acph, with an F_{ST} value significantly different from zero (table 4). Based on sampling 20 loci, such a result is expected at probability .05. *A. elongatulum* often flies to light, with 24.8% of the specimens examined labelled from light trap or "at light."

Agonum decorum possesses an average F_{ST} value of .062. Of these, Gapd has the highest F_{ST} value, .324, as a result of a rare allele found only in one population sample from Oktibbeha County, Mississippi. One-third of the variable loci in this species possess F_{ST} values deviating significantly from zero. Over its entire range, *A. decorum* is less apt to disperse by flight—only 8.6% of total individuals taken at light or in light traps (table 22). The percentage is higher in eastern North America (11.4%), and much higher (45.8%) in Florida. Even though *A. decorum* flies in Florida, environmental conditions over much of the range apparently limit this activity. This is illustrated by the existence of two polymorphisms, each determined by single alleles, within this species (Liebherr 1983a). The apparent introgression of the red-hirsute morph from the east and the green-glabrous morph from the west has produced an area in the Southwest where all four morphs occur (fig. 69). Gene flow has not been sufficient to introduce (or maintain) all alleles into all populations throughout the range of this species.

The electrophoretic samples of *A. extensicolle* consist of three eastern population samples and five in Arizona. The F_{ST} for these eight samples averages .063. The Pgm-1 locus exhibits alternate alleles between the Arizona and eastern U.S. populations, resulting in an F_{ST} value of .661. Of the 13 variable loci, 38.5% have F_{ST} values deviating significantly from zero. Based on average F_{ST} and percent of loci with significant F_{ST} values, *A. extensicolle* is the most genetically heterogeneous species of the three. It is also the species showing the greatest reduction in flight apparatus.

This three-species analysis corroborates the hypothesis that lowland, more vagile species are more homogeneous genetically than upland, less vagile species. Data available are limited, but predictions concerning the other species in the group can be made. *A. cyanope* possesses a flight apparatus less well-developed than the lowland species, and it occurs in cooler upland habitats. Hence I predict appreciable genetic heterogeneity occurring over the species range.

A. texanum lies at the other end of the spectrum: a lowland species that inhabits temporary mesic to xeric habitats and often is taken in flight. This species should be very homogeneous genetically. The only known candidates for divergent populations within *A. texanum* include those from western Arizona, and Chiapas, Mexico (fig. 43). If these have been separated from the central body of populations, they may have diverged genetically. Biometric data supporting this possibility is included in the infraspecific analysis of geographic variation.

A. extimum and *A. parextimum* also should be relatively homogeneous genetically, based on their well-developed metathorax and their utilization of temporary habitats. The synergistic action of environmental factors suitable for flight, and the necessity to colonize temporary habitats would result in genetic mixing within these two species.

Summary of Dispersal Capabilities

1. Species with shorter relative wing length exhibit a correlated reduction of the anterior metanotal phragma and metanotum.
2. Development of flight apparatus is positively correlated with the likelihood of in-flight capture of individuals in light traps or at light. The propensity for capture at light is partly controlled by environmental factors.
3. Species found in higher elevation habitats possess less well-developed metanota and shorter relative flight wing lengths.
4. Genetic heterogeneity, determined using Wright's F_{ST} statistic based on 16 electrophoretic loci, is inversely correlated with dispersal capability measured on a species-wide scale.

TAXONOMIC TREATMENT

This section includes a listing of subgeneric synonyms applied to the *A. extensicolle* group, a species group diagnosis, a key to species, and individual species treatments. Each treatment includes a synonymy, diagnosis, description, a section on biology and habits, and etymology of the specific epithet. In presentation of the type locality and disposition data, the symbol ! means that I have examined the type specimen.

For several of the species, additional discussion of infraspecific geographic variation is included. The foundation for these analyses is drawn from the electrophoretic and biometric circumscription of species discussed earlier.

Dates and localities are presented for all specimens of *A. cyanope, A. extimum*, and *A. parextimum*. A listing of all date-locality records is available from the author or from the Essig Museum of Entomology, University of California. Complete date-locality data is presented for all Mexican captures.

Agonum extensicolle Species Group

(figures 29-36)

subg. *Pseudanchus* Casey, 1920, *Memoirs on the Coleoptera* 9:46 (in part).
subg. *Stictanchus* Casey, 1920, *Memoirs on the Coleoptera* 9:54 (in part); Gray and Hatch, 1941, *Univ. Wash. Publ. Biol.* 10:17; Lindroth, 1966, *Opusc. Entomol. Suppl.* 29:625 (*A. extensicolle* designated as type species.)

Diagnosis. Moderate to large, often with metallic reflection, prothorax more or less constricted at base. Distinguished from other *Agonum* by ring of 6 to 12 setae at apex of second antennal segment (figs. 29h-o); mesocoxae with 3 to 4 (rarely 2) primary setae on coxal ridge (figs. 35d-f, 36a-d); elytra with 4 or more dorsal punctures in or adjacent to third elytral interval (figs. 34, 35a-c).

Description. *Head.* Moderately slender, elongate; eyes well-developed and moderately prominent; frons smooth, ocular grooves lightly impressed, extending from anterior supraorbital seta to clypeus; single seta at anterior corners of clypeus; 2 supraorbital setae; submentum with 2 setae at anterior corners.

Antennae. Length of flagellar segments at least three times width; scape 2.5 times as long as wide, widest at middle; scape with large seta dorsally at apex, smaller setae irregularly surrounding apex; second segment shortest, 1/3 to 1/2 length of scape, ring of 6 to 12 setae at

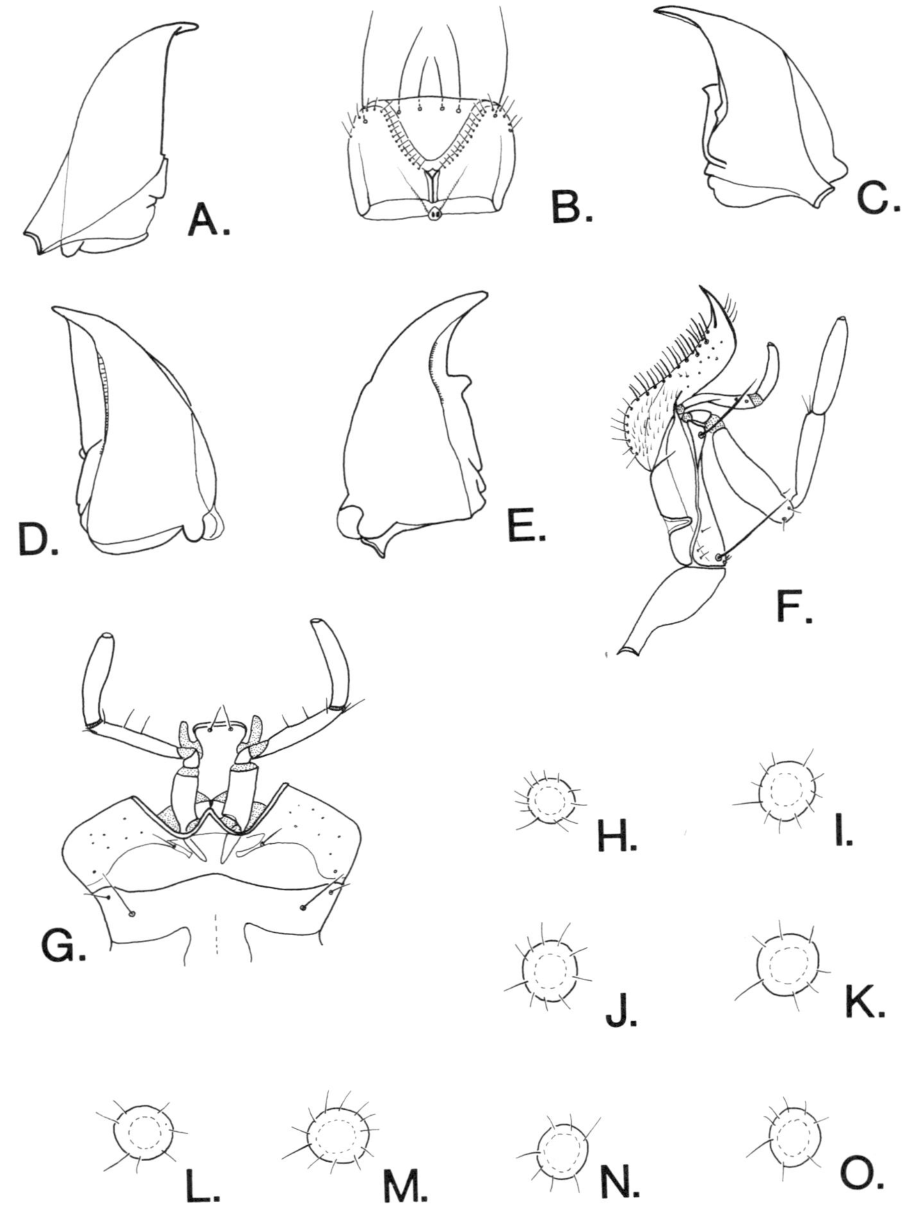

Figure 29. A–G. Mouthparts of *A. extimum* Liebherr, n. sp.:A. Left mandible, dorsal view; B. Labrum, ventral view; C. Right mandible, dorsal view; D. Left mandible, ventral view; E. Right mandible, ventral view; F. Left maxilla, ventral view; G. Labium,mentum, and submentum, ventral view. H–O. End view of apex of second antennal segment showing configuration of apical setae (broken line shows insertion of third antennal segment; larger outer sensory seta always shown at lower left): H. *A extimum*; I. *A. parextimum* Liebherr, n. sp.; J. *A. cyanope* (Bates); K. *A. texanum* (LeConte); L. *A. extensicolle* (Say); M. *A. decorum* (Say),red-hirsute form, VT:Shelburne; N. *A. decorum*, green-glabrous form, CA:Palo Alto; O. *A. elongatulum* (Dejean).

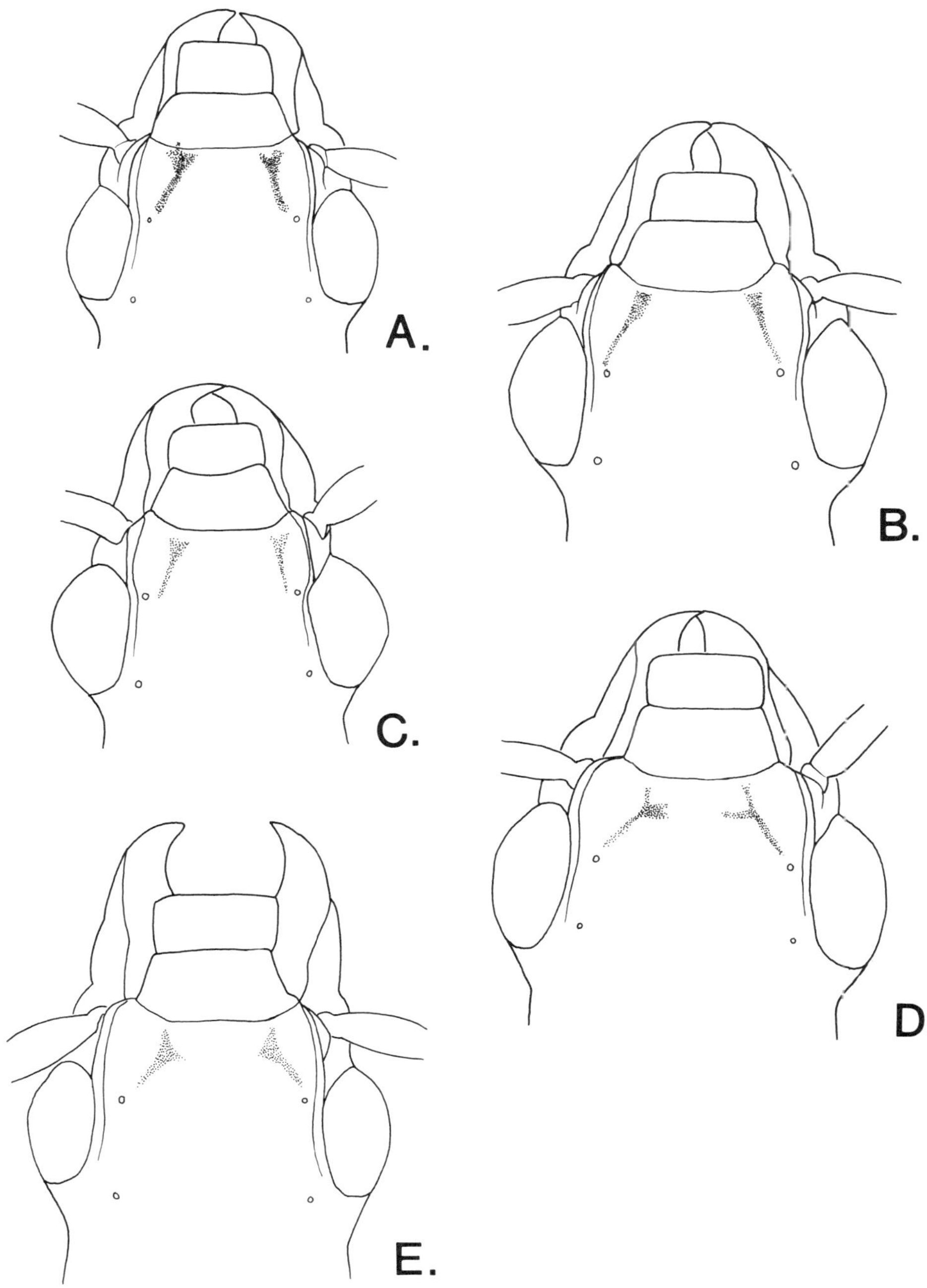

Figure 30. Dorsal view of head showing proportions of head region. All heads drawn to same scale. A. *A. cyanope*, Arizona. B. *A. extimum*, MX:Coahuila, nr. San Pedro. C. *A. parextimum*, MX:Sinaloa, Los Mochis. D. *A. texanum*, TX:Brownsville. E. *A. extensicolle*, AL:Talladega Co.

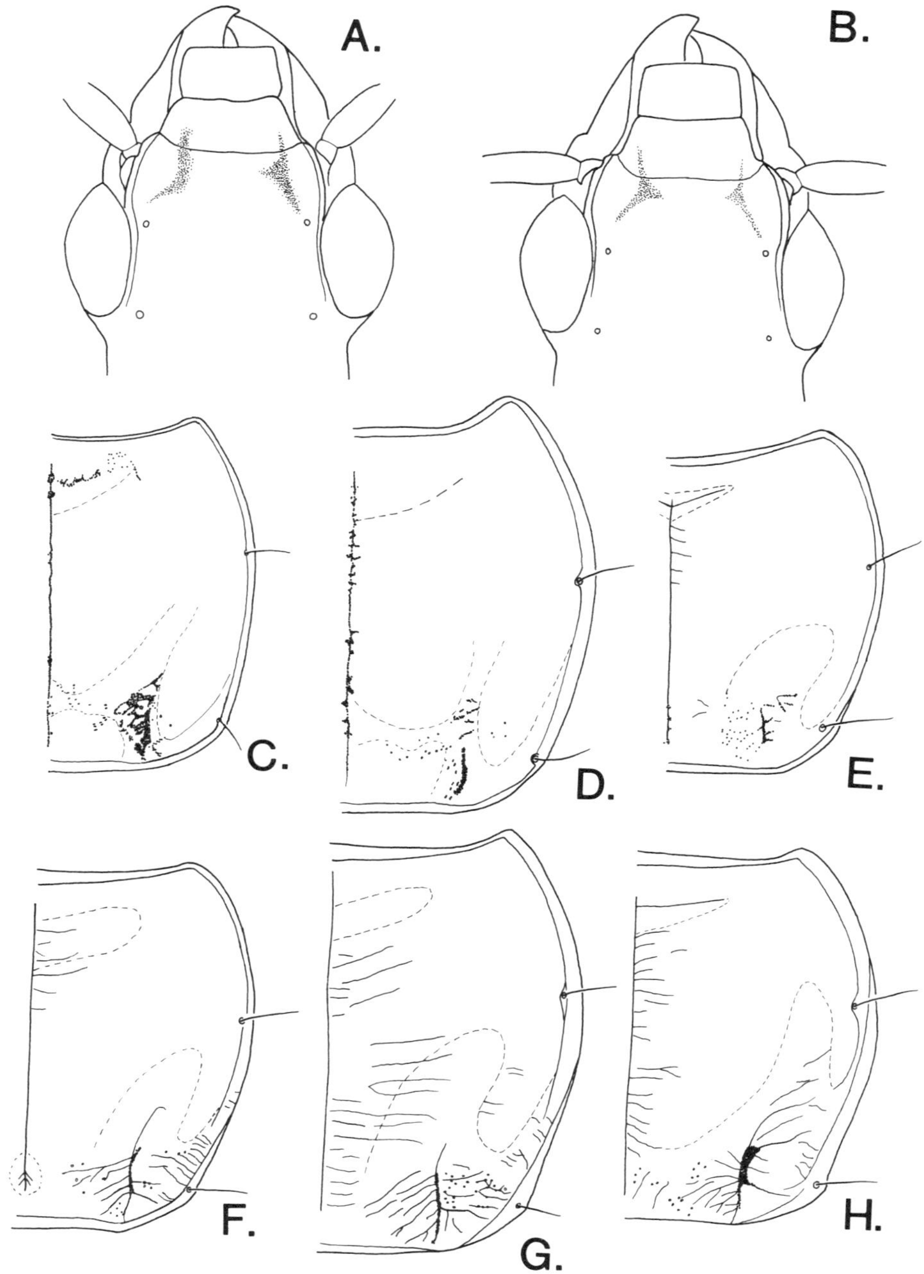

Figure 31. A–B. Dorsal view of head showing proportions of head region, drawn to same scale as figure 30: A. *A. decorum,* AZ:Sta. Cruz Co., Peña Blanca Lk.; B. *A. elongatulum,* FL:Palm Beach Co., Pahokee St. Pk. C–H. Dorsal view of right half of pronotum: C. *A. cyanope* male, AZ:Cochise Co., Pridham Cyn.; D. *A. cyanope* female, same locality; E. *A. cyanope* male, MX:Durango, 8 mi. N. Coyotes; F. *A. texanum* male, 5 km E. Comitan; G. *A. texanum* female, TX:Cameron Co., Los Fresnos; H. *A. texanum* male, AZ:Yavapai Co., Kirkland.

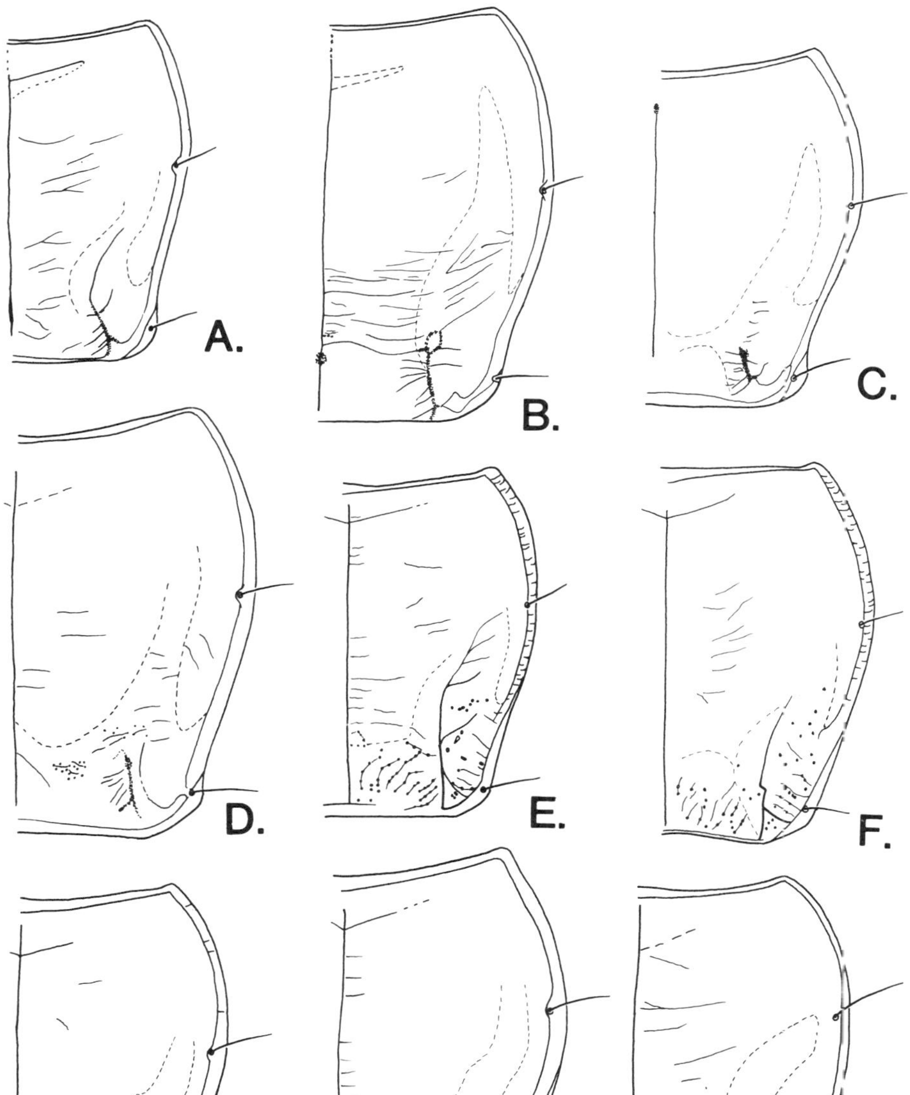

Figure 32. Dorsal view of right half of pronotum, drawn to same scale as in figure 31. A. *A. extimum* male, TX:San Diego. B. *A. extimum* female, MX:Coahuila, nr. San Pedro. C. *A. parextimum* male, MX:Sinaloa, Los Mochis. D. *A. parextimum* female, MX:Sinaloa, Los Mochis. E. *A. extensicolle* male, MX:Durango, 16 mi. W. Durango. F. *A. extensicolle* female, MI:Newaygo Co. G. *A. extensicolle* female, NC:Smokemont. H. *A. extensicolle* male, Nova Scotia:Greenfield. I. *A. decorum* female, MI:Berrien Co.

apex (figs. 29h-o), ventral sensory seta larger than other setae on apex; apical half of third segment with or without fine secondary setae; fourth antennal segment longest, segments 4-11 bearing fine secondary setae.

Mouthparts. Labrum quadrate, 6 setae along apical margin (fig. 29b); 17 to 20 setae on each side situated along lateral margin and inner ridge. Mandibles with acuminate apical tooth (figs. 29a, c-e); left mandible with terebral tooth smaller than on right, projecting dorsally, retinacular tooth ventrad of terebral tooth; right mandible with terebral tooth projecting as rounded knob on dorsal mandibular surface, retinacular tooth apical of terebral tooth well-developed; premolar area equally developed on both mandibles. Maxilla with 4 segmented palpus, penultimate and apical segments subequal in length (fig. 29f), apical segment fusiform with acuminate tip, glabrous; stipes externally unisetose, palpifer unisetose; galea with apical segment subequal to basal segment; lacinia with dorsal surface bearing numerous fine setae, most dense in basal half, stouter setae along inner surface, apex with single acuminate tooth. Labium with quadrisetose submentum (fig. 29g); mentum bisetose, apical margin biconcave with simple median tooth, ventral surface with shallow depressions laterad of setae, epilobes evenly rounded; ligula broad with truncate anterior edge, bisetose; paraglossae membranous lobes extending slightly beyond ligula; labial palpus 3 segmented, segments 2 and 3 subequal; second segment bisetose with smaller setae located at apex.

Thorax. Pronotal anterior margin weakly concave, apical corners weakly projecting; two pairs of lateral setae, one pair located at or slightly before middle of lateral margin, one pair at or near basal angles; lateral bead thin, well defined; depression alongside margin narrow; latero-basal depression well-developed, linear, parallel to side margin; median line impressed; anterior transverse depression weakly developed; basal bead variously developed medially, usually at least indicated immediately inside basal lateral setae. Prosternal projection more or less parallel-sided with rounded tip, unmargined. Metathorax moderately to well-developed, metepisterna elongate; flight wings full, only rarely shortened, never completely reduced.

Legs. Prothoracic trochanter with single seta; prothoracic femur with numerous setae along dorsal margin; protibia with carinae on anterior and posterior faces, a broad, weakly developed depression inside of posterior carina; anterior tarsi of male dilated, first and second segment with weakly developed external and internal lateral furrows on dorsum, third and fourth segments with convex dorsal surface; female with external and internal furrows present on segments 1 to 3 of foretarsi; mesocoxa with single seta on ventral surface and 3 to 4 (rarely 2) setae along anterior ridge (figs. 35d-f, 36a-d); mesotrochanter unisetose; mesofemur with numerous dorsal (=anterior) spines; mesotibia with well-developed lateral furrows on outer surface, carina between furrows with poorly developed median groove; metacoxa with 2 setae, on anterior edge and ventral to trochanteral articulation; metatrochanter ovoid-elongate, 3 times as long as wide, unisetose; metafemur long and slender with numerous dorsal spines; tibia with well-developed lateral furrows and strong median carina on outer surface, carina with median furrow; meso- and metatarsi with well-developed external and internal furrows in segments 1 to 4; fifth tarsus of all legs with 3 to 4 pairs of well-developed ventral setae, bearing a pair of slender nondenticulate claws.

Elytra. Elytron elongate, parallel-sided (figs. 34, 35a-c); striae distinct, lightly punctulate; elytral interneurs flat to moderately convex; humeri rounded but well-developed, with rounded to slightly angulate basal gutter, ninth stria close to lateral margin, lateral gutter narrow; apex slightly sinuate at junction of eighth and ninth striae; four or more dorsal setae in or adjacent to third interval; scutellar striole present, seta located at base of first stria; 15 to 18 umbilicate

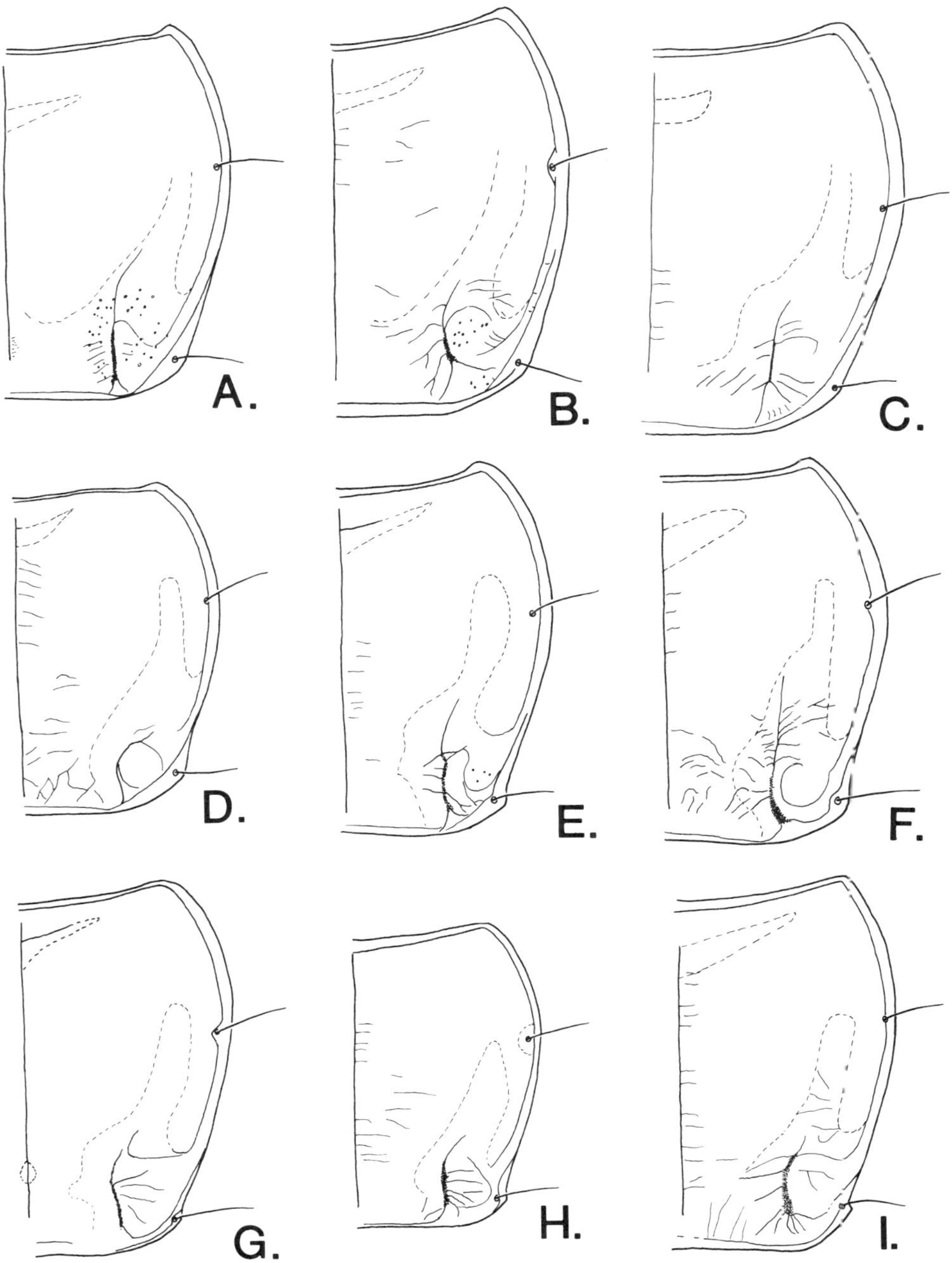

Figure 33. Dorsal view of pronotum, drawn to same scale as in figures 31 and 32. A. *A. elongatulum* male, B.W.I.:N. Bimini Isl. B. *A. elongatulum,* female, FL:Myakka R. St. Pk. C. *A. elongatulum* female, FL:Suwanee Co. D. *A. elongatulum* male, FL:Suwanee Co. E. *A. decorum* male, Cuba:Cienfuegos, Soledad. F. *A. decorum* female, Cuba:Cienfuegos, Soledad. G. *A. decorum* male, FL:Highlands Co. H. *A. decorum* male, CA:Palo Alto. I. *A. decorum* female, UT:Millard Co., Scipio Rsvr.

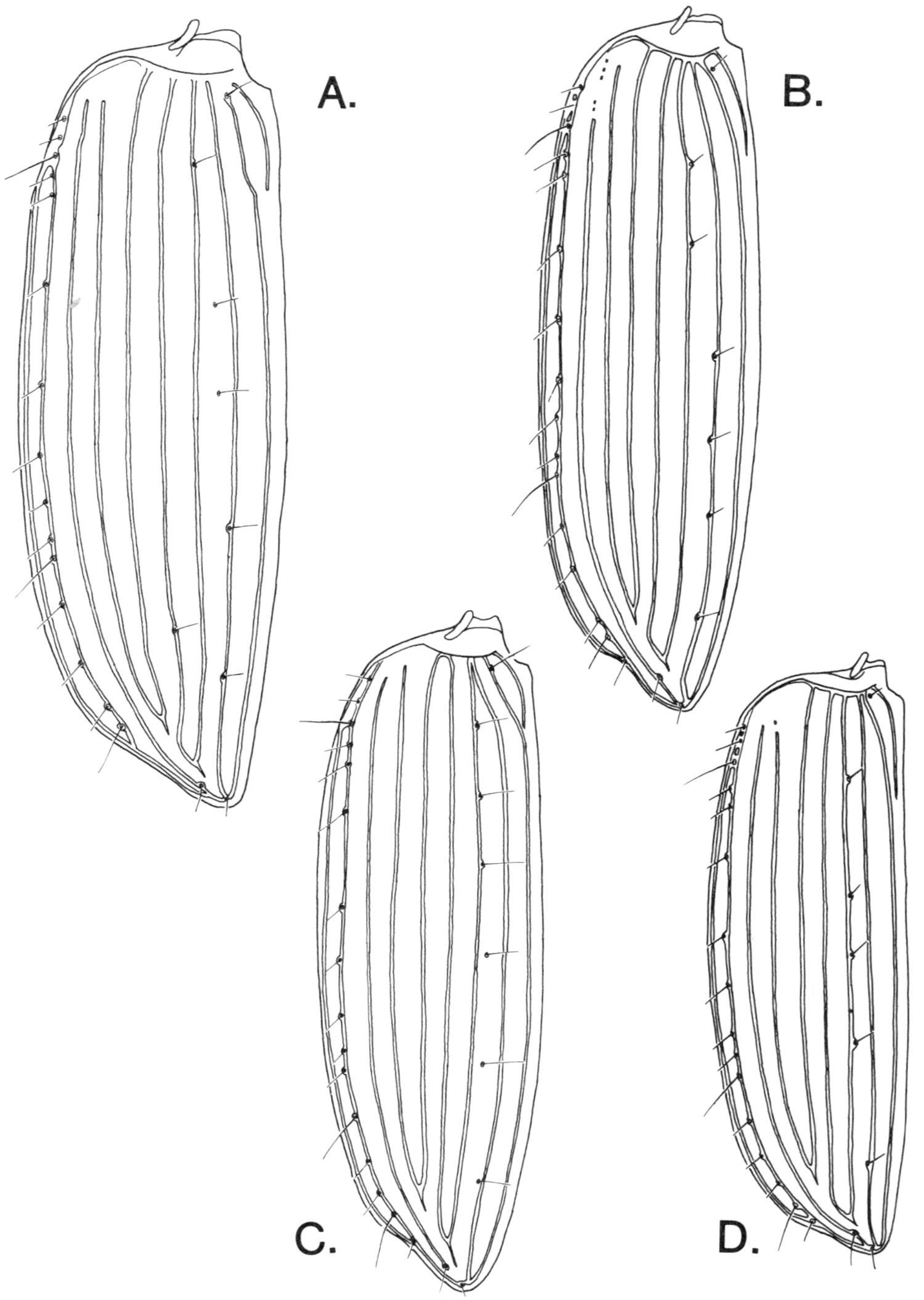

Figure 34. Left elytron, dorsal view. A. *A. texanum,* TX:Brownsville. B. *A. elongatulum*, FL:Myakka R. St. Pk. C. *A. extensicolle*, MI:Gratiot Co. D. *A. decorum*, CA:Palo Alto.

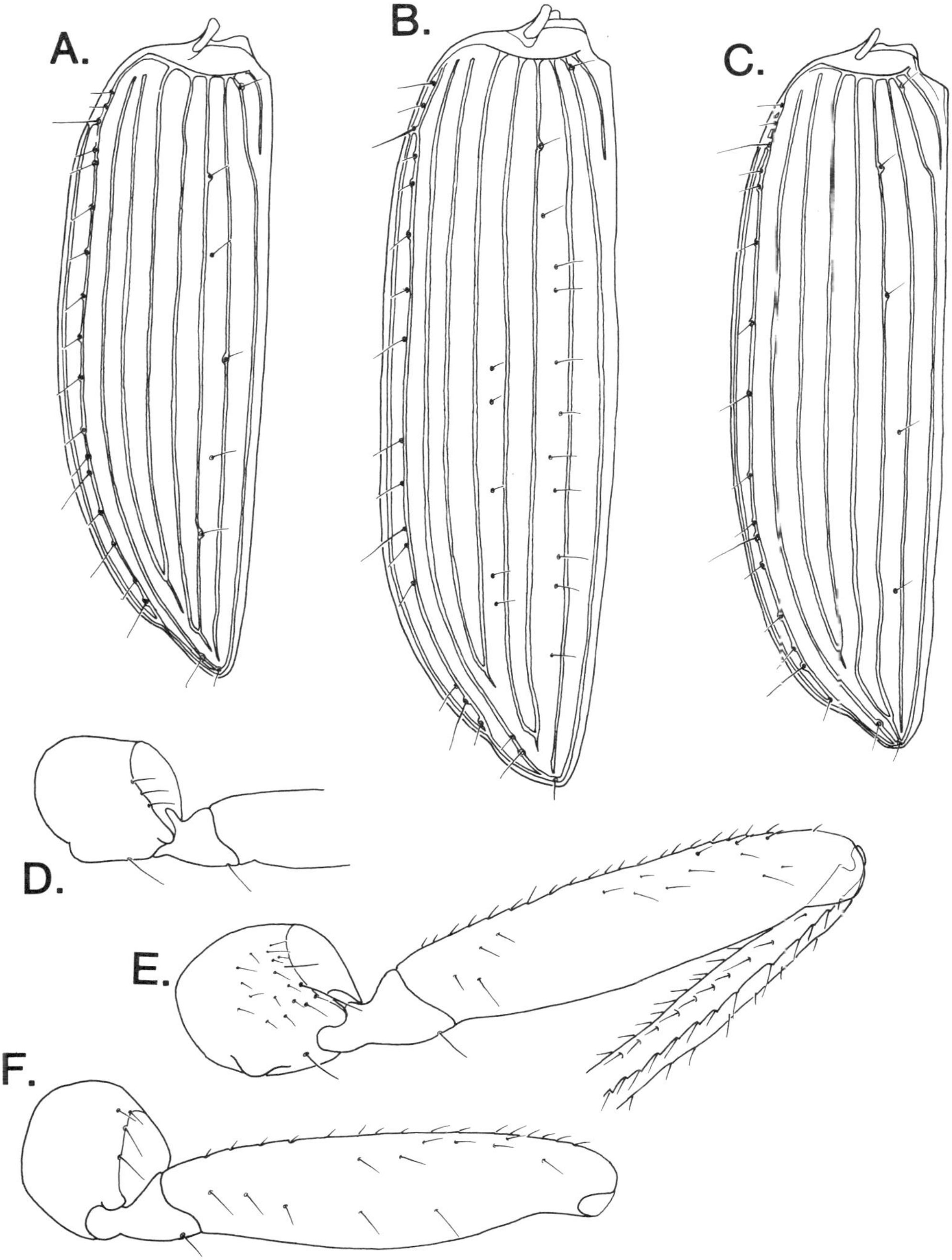

Figure 35. A–C. Left elytron, dorsal view: A. *A. cyanope*, Arizona; B. *A. extimum*, MX:Coahuila, nr. San Pedro (pelage not shown); C. *A. parextimum*, MX:Sinaloa, Los Mochis. D–F. Coxa, trochanter, and femur of meso-leg, ventral view: D. *A. cyanope*, Arizona; E. *A. extimum*, MX:Coahuila, nr. San Pedro; F. *A. parextimum*, MX:Sinaloa, Los Mochis.

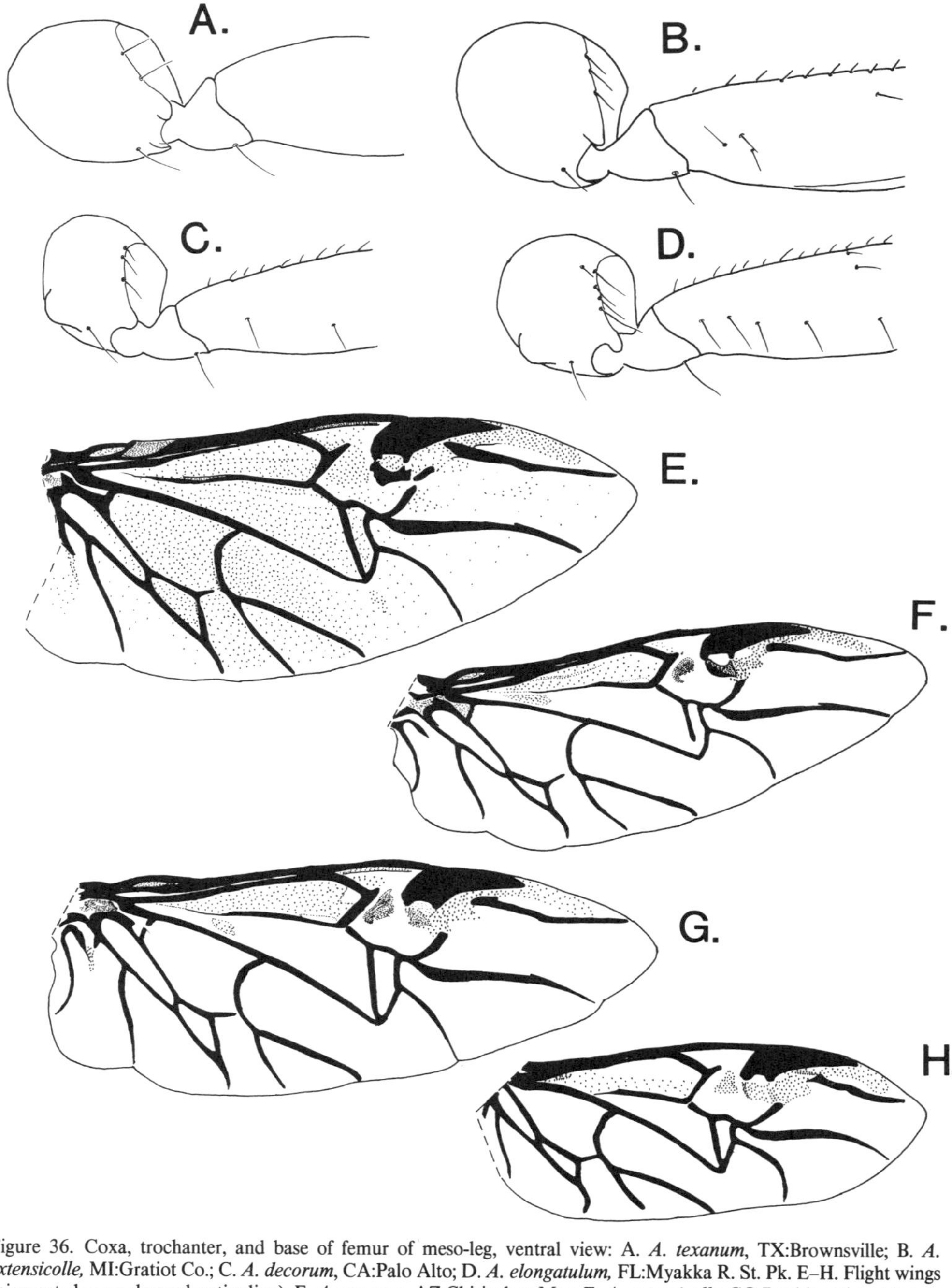

Figure 36. Coxa, trochanter, and base of femur of meso-leg, ventral view: A. *A. texanum*, TX:Brownsville; B. *A. extensicolle,* MI:Gratiot Co.; C. *A. decorum*, CA:Palo Alto; D. *A. elongatulum,* FL:Myakka R. St. Pk. E–H. Flight wings (pigmented areas shown by stippling): E. *A. cyanope,* AZ:Chiricahua Mts.; F. *A. extensicolle,* CO:Boulder (elytral length 5.05mm, wing length 6.07mm); G. *A. extensicolle,* VA:Clarke Co. (elytral length 5.38mm, wing length 6.92mm); H. *A. extensicolle,* VA:Clarke Co. (elytral length 5.58mm, wing length 5.08mm).

punctures bearing setae located in eighth elytral stria; 2 setae located at elytral tip associated with seventh stria; setae sometimes present on or adjacent to fifth and seventh elytral interval.

Abdomen. Third to fifth visible abdominal sternites with 2 setae on apical margin; apical segment of male with 2 apical setae; female with 4 apical setae.

Male Genitalia. Aedeagus with closed basal bulb and variously developed sagittal crest (fig. 49); most heavily sclerotized at junction with parameres, otherwise lightly pigmented; internal sac simple, without well-developed lobes; parameres subequal, right (=ventral) slightly smaller and more twisted than left (=dorsal), basal portions heavily pigmented, piceous.

Female Reproductive Tract. Terminal segment of gonostylus with convex ventral surface, outer upper edge concave, spadelike, 2 lateral and 1 dorsal setae, internal face of apex with pair of fine setae set in ovoid depression (figs. 38-40, 42, 45, 50, 57); basal segment of gonostylus with apical fringe of setae; bursa copulatrix membranous, saclike; spermathecal duct entering bursa at same point as does median oviduct; spermathecal duct very long relative to length of spermatheca; spermatheca long and cylindrical, curved basally where narrowed to meet spermathecal duct; spermathecal gland with globose reservoir, atrium located proximad of reservoir.

Microsculpture. Upper surface of head, prothorax, and elytra with isodiametric microsculpture, granulate in some species; ventral surface with less well-developed isodiametric microsculpture.

KEY TO SPECIES

The extensive infraspecific morphological variation in this group has impeded taxonomic work. In the following key I have attempted to account for the variability in color and setation observed in *Agonum decorum* and *A. elongatulum*. These two species are very similar morphologically, necessitating separation of the green-glabrous morph of *A. decorum* from *A. elongatulum* using a three-character discriminant function. The derivation of this function is presented in the discriminant analysis of the *A. decorum* complex. The values of the three measurements, in millimeters, can be inserted in the discriminant functions to obtain a discriminant function score. The value of this score determines the species identity. Measurement of several individuals of a series will insure a proper identification.

The cuticular microsculpture is a very useful diagnostic character. The terms used follow Lindroth (1974b). To adequately assess microsculpture, good light and greater than 50 power magnification is required.

The male genitalia are generally of limited use in this group. The unique lateral sclerites on the internal sac of *A. texanum* (fig. 42e) are one exception. This species can be reliably determined using the male genitalia. Figures 49d-j give a representation of the variability seen in aedeagi of some of the other species.

Body length is the distance from mandibular apex to elytral apex. Accurate body length measurement is often not possible due to the pinned condition of the specimen. Ranges of lengths are given only for *A. decorum*, *A. elongatulum* and *A. parextimum*.

1. Dorsum covered with fine microsetae in a pelage; normal macrosetae also present. 2
   Dorsum glabrous except for articulated macrosetae. 4
2. Eight to 16 erect dorsal macrosetae in or adjacent to third elytral interval (fig. 35b); ventral surface of body with pelage of fine setae. . . . . *A. extimum* n. sp.
   Fewer than eight dorsal elytral setae; ventral surface of body with only a few secondary setae. 3

3. Prothorax ferruginous; legs testaceous. .... *A. decorum* red-hirsute morph
   Prothorax piceous to metallic green, sometimes with aeneous tinge; legs piceous. ....
   .... *A. decorum* green-hirsute morph
4. Pronotum rugosely punctate medially at base (figs. 32e-h); upper body surface metallic blue, green or aeneous bronze; gutter along lateral margin of prothorax often with fine crenulations (high power) (figs. 32e-g). .... *A. extensicolle*
   Median area of pronotal base smoother, not rugosely punctate, although small punctures may be present in latero-basal depression; color of dorsum variable; marginal gutter of prothorax not crenulate. 5
5. Body and leg color cyanotic blue to black; flight wings dark piceous; microsculpture on pronotum granulate, isodiametric. .... *A. cyanope*
   Body color metallic green to piceous, legs piceous to testaceous; flight wings unpigmented; if pronotum with granulate isodiametric microsculpture, legs are testaceous...............6
6. Body broad, maximum width of pronotum at least 1.25 X median length; elytral striae faint at base where they join elytral groove; aedeagal shaft straight, internal sac with two well-developed lateral sclerites (fig. 42e). .... *A. texanum*
   Body more slender, width of pronotum less than 1.25 X median length; elytral striae strongly joining base where they join basal groove; aedeagal shaft evenly arcuate, internal sac with at most partially sclerotized lateral areas (figs. 40d, 50d, 57d). 7
7. Prothorax ferruginous red. 8
   Prothorax metallic green to piceous. 9
8. Mesosternum and mesepisternum paler than metasternum, usually ferruginous. ....
   .... *A. decorum* red-glabrous morph
   Mesosternum and mesepisternum piceous (a form taken in low frequency with typical dark form). .... *A. elongatulum*
9. Pronotum with granulate isodiametric microsculpture, surface not shiny; elytra piceous, with poorly developed metallic sheen; legs testaceous; overall length 8.0-10.1 mm. ....
   .... *A. parextimum* n. sp.
   Pronotal microsculpture not granulate, submetallic to metallic; microsculpture isodiametric to slightly stretched transverse mesh; elytra with metallic green reflection; legs usually piceous, sometimes testaceous; overall length 6.9-9.3 mm. 10
10. Foretarsi with squamose hairs on ventral surface (males). 11
    Foretarsi glabrous ventrally except for lateral spinose setae (females). 12
11. Discriminant function score (-10.8 - 12.9 prothoracic length - 4.1 maximum elytral width + 7.3 elytral length) less than .213, prothorax with lateral margins sinuate before hind angles (figs. 33e,f,h), overall length 6.9-8.0 mm. (Distributional range, fig. 51). ....
    .... *A. decorum* green-glabrous morph
    Discriminant function score greater than .213, prothorax with lateral margins convex to sinuate before hind angles (figs. 33a-d), overall length 7.1-8.8 mm. (Distributional range, fig. 58). .... *A. elongatulum*
12. Discriminant function score (-11.6 - 10.7 prothoracic length + 1.7 maximum elytral width + 4.9 elytral length) less than .199, prothorax with lateral margins simuate before hind angles (figs. 33e,f,h), overall length 6.9-9.2 mm. (Distributional range, fig. 51). ....
    .... *A. decorum* green-glabrous morph
    Discriminant function score greater than .199, prothorax with lateral margins convex to sinuate before hind angles (figs. 33a-d), overall length 8.7-9.2 mm. (Distributional range, fig. 58). .... *A. elongatulum*

Agonum cyanope (Bates)

(figures 29j; 30a; 31c-e; 35a,d; 36e; 37, 38)

Anchomenus cyanopis Bates, 1882, *Biol. Cent.-Amer. Coleop.* 1(1):94; Bates, 1884, *Biol. Cent.-Amer. Coleop.* 1(4):281.
Agonum cyanopis; Csiki, 1931, *Coleopterorum Catalogus* 115:841.
Agonum cyanope; Liebherr, 1982, *Coleopt. Bull.* 36:153 (male lectotype designated).

Lectotype. male, Mexico, Guanajuato (BMNH)!.

Diagnosis. Darkest species of the group, with metallic purple reflection; many specimens with well-developed granulate isodiametric microsculpture; prothorax quadrate with planar disc, shallow ovoid latero-basal depressions and convex or barely sinuate lateral margins (figs. 31c-e).

Description. *Head.* Slender, eyes moderately projecting, temple moderately expanded behind eye (fig. 30a); antenna with second segment bearing apical ring of 9-10 setae (fig. 29j), third segment pubescent in apical half, segments 4-11 fully pubescent.

Thorax. Pronotum quadrate, lateral margin evenly convex, basal angle poorly developed (figs. 31c-e); lateral depression narrow from front angle to basal seta, narrow margin continuing from basal seta to middle of base; latero-basal depressions ovoid to linear, shallow, with tightly spaced punctulae along the depression; latero-basal depression broadened and weakly indicated forward approximately ½ length of prothorax; surface laterad of basal depressions elevated, smooth; median impression sharp, well-defined, comprised of punctulae in specimens with very granulate microsculpture; anterior transverse impression moderately well-developed and linear to almost absent.

Elytra. Moderately convex and parallel-sided; apical sinuation poorly to moderately developed (fig. 35a); elytral intervals flat to slightly convex, striae smooth, punctulae barely visible at 80X; striae 1-5 strongly joining basal margin; basal margin angulate at shoulder; 4-6 setae in or adjacent to third elytral interval.

Flight Wings. Length 1.2 to 1.3 times elytron length, membranous areas piceous, smoky brown (fig. 36e).

Male Genitalia. Aedeagal shaft straight in apical half with apex slightly upturned (figs. 38e-g), wrinkles in basal curve of aedegus weak to absent (fig. 38f); internal sac of aedeagus as long as straight apical portion of shaft; surface of sac densely covered with spicules, largest on the dorsal surface (everted position).

Female Reproductive Tract. Basal segment of gonostylus with apical fringe of 10 to 12 setae (figs. 38a,b); spermatheca long, 24 to 30 constrictions producing a moniliform configuration (figs. 38c, d); spermathecal gland entering spermathecal duct distant from spermatheca, about 1/3 duct length from bursa copulatrix.

Color. Dorsal surface piceous, with blue to purplish reflection, strongest on head, disc of prothorax and elytra; base and lateral areas of prothorax tending toward blue or blue-green; base of mandible piceous, apex rufo-piceous; maxilla, labium rufo-piceous, palpus dark rufo-piceous; underside of scape rufo-piceous, all other antennal segments piceous, legs uniformly piceous, with knees, external tibial surfaces and tarsi slightly paler; ventral body surface uniformly dark piceous.

Microsculpture. Head with uniform isodiametric microsculpture; pronotum and elytra with more strongly developed isodiametric microsculpture, granulate in specimens from Chihuahua,

Mexico, and Arizona, less granulate in southern parts of range; ventral body surface with isodiametric microsculpture; lateral areas of abdominal sternites with well-developed longitudinal wrinkles.

Length. 7.1-9.2 mm.

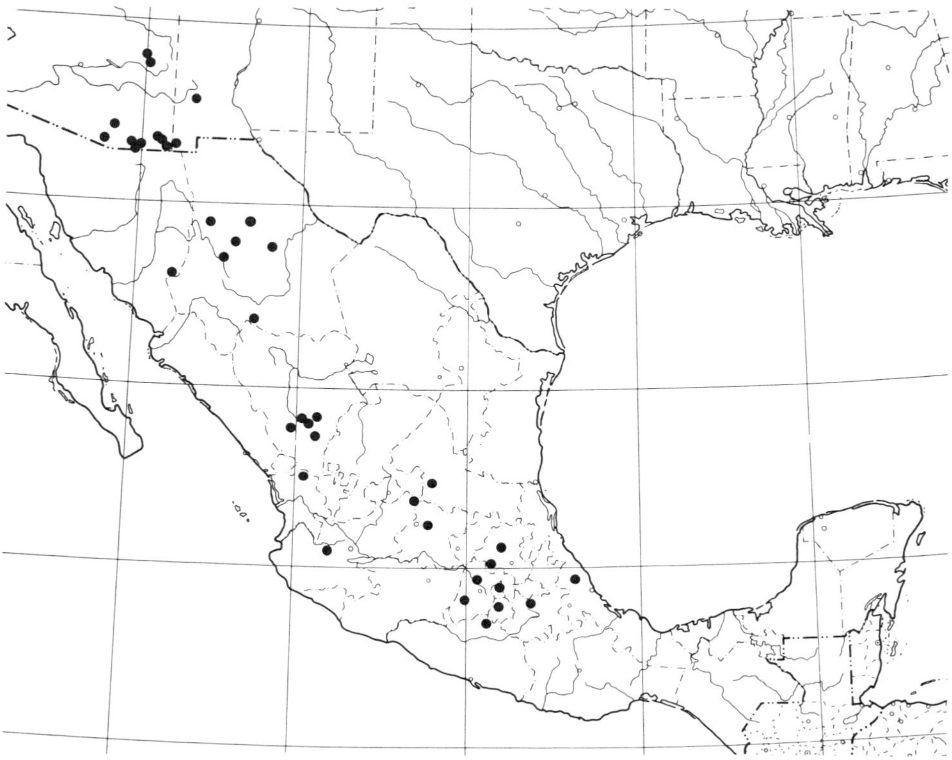

Figure 37. Distributional records of *A. cyanope* (Bates).

Distribution. Arizona, south along the Sierra Madre Occidentale to the Mexican states of Guerrero, Puebla and Vera Cruz (fig. 37).

Distributional Records. 157 specimens: ARIZONA: no other data (2, NMNH; 2 MSUC). COCHISE Co.: 4.7 mi. S Apache, 2-VIII-1967 (1, UCR); Bodie Wash 3 mi. SE Parker Cyn., 23-IX-1970 (1, CAS); Chiricahua Mts., 8-VIII-1933 (2, CAS), 20-IV-1961 (1, AMNH), VII (2, UAZ); Huachuca Mts., Bear Ck., W of Montezuma Pass, 1-VII-1956 (2, CNC); Palominas, San Pedro R., 23-VIII-1960 (3, RTBe); Pridham Cyn., Chiricahua Mts., 30-VIII-1975 (5, SMcC); Rucker Cyn., Chiricahua Mts., 24-III-1970 (1, SMcC), 25-VII-1969 (1, CNC); 26-VIII-1965 (1, UCR), 14-XI-1971 (4, SMcC); N Sonora, Morrison coll. (3, BMNH). NAVAJO Co.: Lakeside, 21-VIII-1948 (2, MCZ); Showlow, VI-1927 (2, CAS), 5-IX-1967 (1, CAS). PIMA Co.: Tucson & vic. 30 mi. S, 24-III-1935 (1, MCZ). SANTA CRUZ Co.: 18-VII-1948 (1, AMNH); Canelo, 14-V-1967 (7, CNC), 18-VII-1948 (3, AMNH), 20-VII-1958 (1, UAZ), 11-

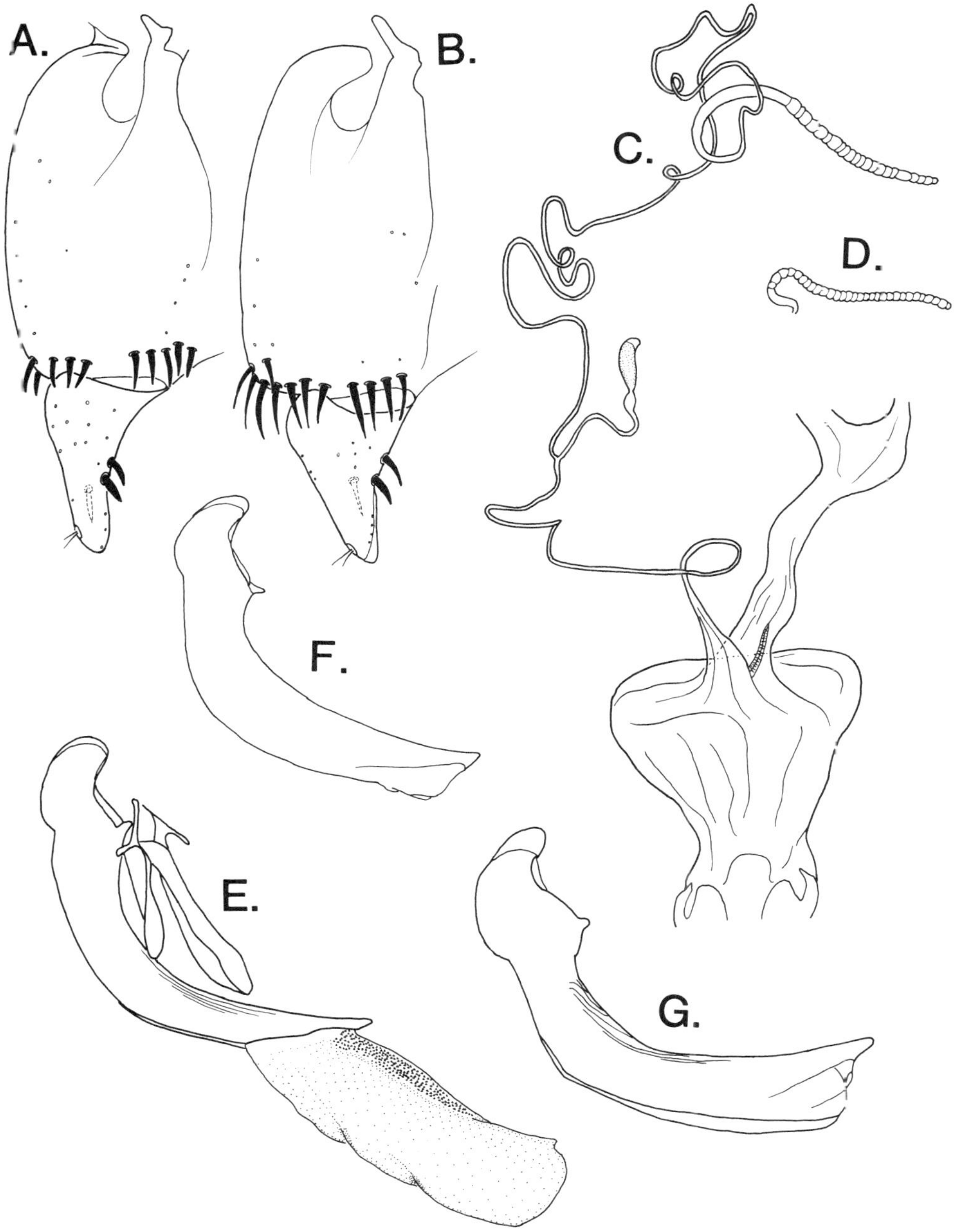

Figure 38. Internal and external reproductive structures of *A. cyanope*. A–B. Left gonostylus, ventral view (dorsal seta of apical segment shown as broken line): A. Arizona; B. MX:Durango, N. Coyotes. C. Female internal reproductive tract, ventral view, MX:Durango, N. Coyotes. Penultimate segments of gonostyli at bottom; spermathecal gland enters spermathecal duct at basal 1/3; spermatheca with regular constrictions. D. Spermatheca, AZ:Cochise Co., SE. Parker Cyn. Lk. E–G. Aedeagus, ventral view (resting position): E. Parameres in resting position, aedeagal internal sac everted, AZ:Canille, F. MX:Durango, N. Coyotes; G. AZ:Canille.

X-1955 (1, UAZ); Canelo Hills, 29-IX (1, UAZ), 21-X-1943 (1, UAZ); Peña Blanca Lk., 26-VII-1961 (1, FDPI), 10-XII-1975 (1, UAZ). NEW MEXICO: Grant Co., Silver City, 1-VII-1932 (1, MCZ); Hidalgo Co., 20 mi. S Rodeo, 28-VIII-1965 (1, UCR). MEXICO. CHIHUAHUA: Bachinavis, 12-VIII-1971 (1, UAZ); Catarinas, 5800′ el., 25-VII-1947 (6, AMNH); Guerrero, Papagochic R., 30-VII-1934 (3, NMNH); Primavera, 5500′-6000′ el., 30-VI-1947 (1, AMNH); San Jose Babicora, 5-VII-1947 (4, AMNH); Santa Barbara, 7000′-8000′ el., 19-IV-1947 (1, AMNH). DURANGO: (4, MCZ); Coyotes, 8 mi. N, 7500′ el., 7-XI-1970 (10, CAS); Durango, (7, BMNH; 1, NMNH), 14-VIII-1947 (1, AMNH); Durango, 15 mi. W, 3-VII-1964 (5, CNC); Durango, 65-70 km SSW, road to La Flor, 2590 m el., 17-IX-1979 (1, CAS); El Salto, 8700′-8800′ el., 20-III-1953 (15, MCZ), El Salto, 10 mi. W, 7-V-1961 (24, CNC), 5-VI-1964 (1, CNC); Otinapa, 8200′ el., 11-VIII-1947 (3, AMNH). DISTRITO FEDERAL: Cuajimalpa, 4-V-1969 (2, NMNH), 19-23-X-1968 (3, NMNH); Mexico City (1, BMNH); Temescaltepec (3, MCZ). GUANAJUATO: Guanajuato (1, BMNH). GUERRERO: Amula, 6000′ el., VIII (1, BMNH). HIDALGO: Guadelupe, 10-VII-1932 (7, MCZ); Tula (1, BMNH). JALISCO: 2 mi. W jct. Hwy. to Ameca, 9-VII-1953 (1, AMNH); Ojuelos de Jalisco, 20-XI-1948 (2, CAS). MEXICO: no other data (1, BMNH); Atlocomulco, 2500m, 1-X-1948 (1, MICH). MORELOS: Cuernavaca (2, BMNH). NAYARIT: Santa Teresa, 2073 m el., 22-25-X-1979 (2, CAS). PUEBLA: Puebla (1, BMNH; 1, CAS). SAN LUIS POTOSI: San Luis Potosi, 24-VI-1953 (4, AMNH). SONORA: Yecora, 7000′ el., 20-22-V-1961 (2, CNC). VERA CRUZ: Jalapa (2, NMNH).

Biology and Habits. *A. cyanope* is a species of the oak-pine floristic zone. In Arizona it can be found in the intermountain valleys, and up to 6,000′ elevation in mesic canyons. In Mexico, recorded localities range from 5,500′ to 8,800′ elevation.

Etymology. The epithet *cyanope* is derived from cyan (Gr. *kyaneos*), dark blue; opisth (Gr. *opisthen*), behind, at the back.

Agonum extimum Liebherr, new species

(figures 1, 29a-h; 30b; 32a,b; 35b,e; 39, 41)

Holotype. female, MEXICO, Coahuila, San Pedro, 31.2 km E, 900 m el., 31-V-1981, J.T. Doyen and J.K. Liebherr collectors (CISC).

Allotype. male with same data (CISC).

Diagnosis. Superficially with same body shape as *A. extensicolle*, but with median basal area of pronotum and latero-basal pronotal impressions impunctate; elytra with dull matte surface, body surface covered with pelage of fine unarticulated microsetae.

Description. *Head.* Frons foreshortened, head broad, with protruding convex eyes, neck broad, temple expanded gradually to meet hind margin of eye (fig. 30b); antenna with scape bearing fine setae irregularly spaced over surface, second segment with apical ring of 11-12 setae as well as numerous setae toward base; segments 3-11 fully pubescent.

Thorax. Pronotum appearing slightly longer than wide, but with maximum width at to slightly before middle averaging 1.12 times median length, (figs. 32a, b); lateral margin sinuate before well-developed hind angle; lateral depression narrow, widened only slightly near basal seta, continuing along base to immediately inside latero-basal depression, basal bead faintly visible to absent on middle of prothorax; latero-basal impression ovoid to linear, continuing shallowly to forward of lateral seta, sometimes faint punctulae in impressions; transverse wrinkles present or absent in impressions as well as on prothoracic disc; median longitudinal impression fine although continuous, deeper toward pronotal base; anterior transverse impressions weak, barely traceable.

Elytra. Long, ovoid, maximum width 2/3 distance to apex, apical sinuation faint close to apex (fig. 35b); elytral intervals slightly to moderately convex; striae well-impressed, faint punctulae visible at 80X; striae 1-5 strongly joining basal margin; basal margin not to moderately angulate at shoulder; 7-16 setae set in punctures in or adjacent to third elytral interval along with irregularly placed macrosetae on or adjacent to the fifth and seventh elytral intervals.

Secondary Setation. In addition to articulated macrosetae associated with punctures on the body surface, surfaces of the head, thorax, elytra, abdominal sternites and legs (fig. 35e) covered with fine setae, a dense pelage; increased pubescence of the antennae correlated with presence of the pelage.

Male Genitalia. Aedeagus with apical half relatively straight, apex slightly upturned (fig. 39d-f); aedeagal sac as long as apical straight portion of aedeagus; spicules covering sac, somewhat more strongly developed on dorsal surface (everted position).

Female Reproductive Tract. Basal segment of gonostylus with apical fringe of 12-17 setae (figs. 39a,b); spermatheca long, 37-40 constrictions producing moniliform configuration (fig. 39c); spermathecal gland entering spermathecal duct distant from spermatheca, about 1/3 duct length from bursa copulatrix.

Color. Dorsal surface of head and prothorax metallic green with bluish reflection; elytra with faint metallic green to blue-green reflection; elytral epipleura rufo-piceous, darker near shoulder, mouthparts and edges of labrum rufo-piceous; center of labrum piceous; third and fourth segments of maxillary palpus and apical segment of labial palp infuscated; antenna with scape rufo-testaceous, the second segment darker, segments 3 to 11 piceous, legs rufo-testaceous; tarsi darker, rufo-piceous, ventral body surface rufo-piceous.

Microsculpture. Head, pronotum, and elytra with uniform isodiametric microsculpture; secondary setation producing punctured appearance; ventral body surface with uniform isodiametric microsculpture, secondary setation producing wrinkles among individual facets, lateral areas of abdominal sternites with well-developed longitudinal wrinkles.

Length. 8.1-10.9 mm.

Distribution. Lowland habitats near Douglas, AZ, east of the Sierra Madre Occidental to Texas (fig. 41). PARATYPES: 238: TEXAS: no other data (1 female, MCZ). CAMERON Co.: Brownsville, 15-III-1972, USDA light trap (1 male, MSUC); Brownsville, 11-16-VI-1933 (9 males, 4 females, MCZ); Brownsville, 4-X-1957 (1 male, 4 females, AMNH). DUVAL Co.: San Diego, 23-26-V (3 males, 1 female, NMNH). HUDSPETH Co.: Dell City, 9 mi. SW, 31-VII-1950 (1 female, AMNH). McLENNAN Co.: Waco, 20-VIII-1968 (1 female, DHKa). SUTTON Co.: Sonora, 6-VIII-1947 (1 female, AMNH). TERRELL Co.: Sheffield, 16 mi. S, Blackstone Ranch, 28-VI - 2-VII-1959 (1 female, UASM). UVALDE Co.: Uvalde, 19-III-1914 (1 male, NMNH). WEBB Co.: Laredo, 2-VII-1947 (1 female, FMNH). NEW MEXICO: HIDALGO Co.: Lordsburg, 2-VII-1956 (1 female, AMNH); Rodeo, 13 mi. NE, 13-VII-1964 (1 male, 1 female, NMNH); Rodeo, 4 mi. N, rte. 80, 23-VII-1963, at light (1 female, FDPI); Rodeo, 2 mi. S, 6-VIII-1959 (2 females, AMNH). LUNA Co.: Deming, 12-VII-1917 (2 females, AMNH; 1 female, CAS), 22-VII (1 male, NMNH), 31-VII-1973, at light (6 males, 3 females, LLLa/PMCh). ARIZONA: COCHISE Co.: Bowie, 28-VII-1970, UV trap (3 males, UAZ); Chiricahua Mts., 17-VII-1958 (1 male, CISC); nr. Double Adobe, 17-VII-1973 (1 male,

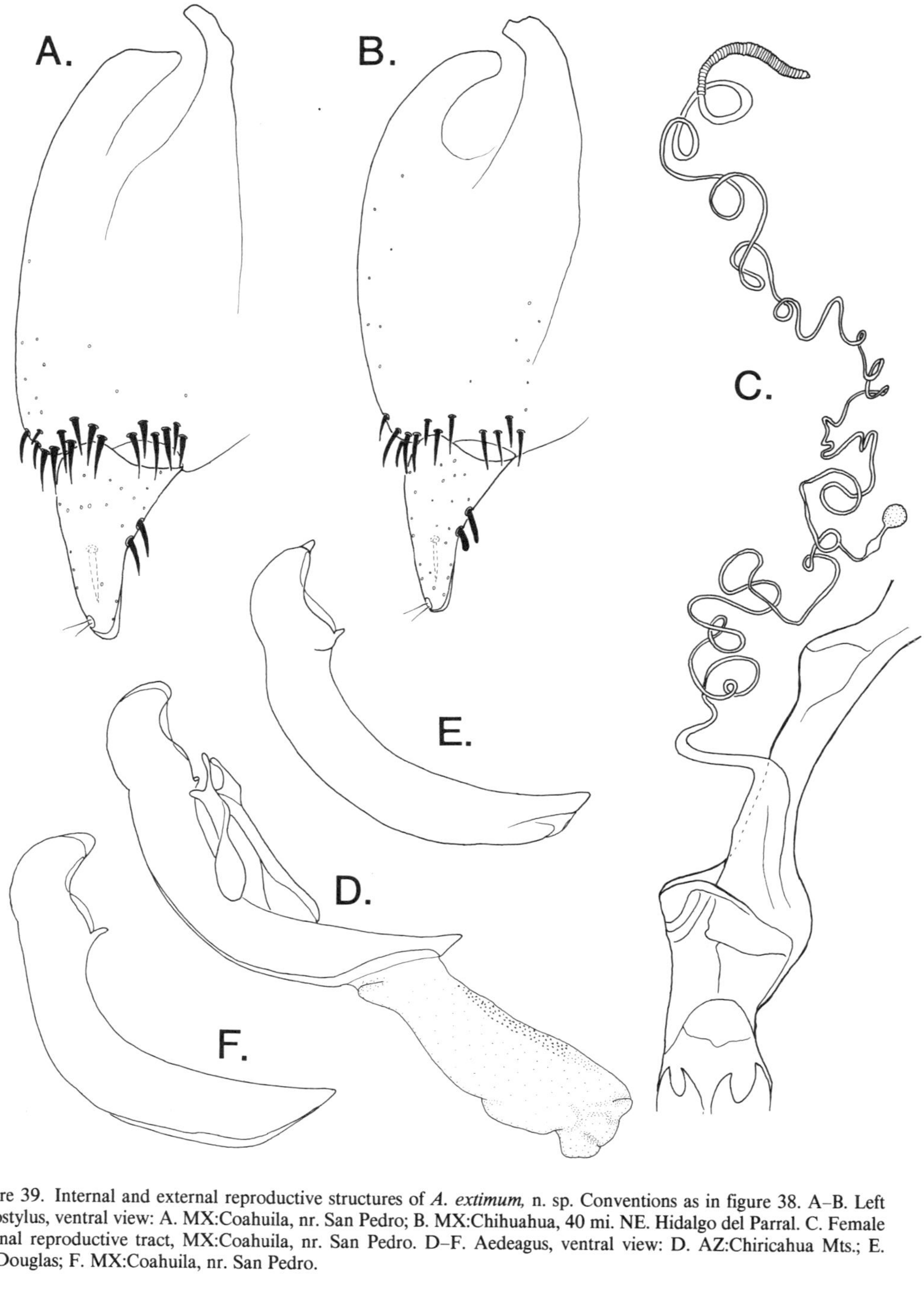

Figure 39. Internal and external reproductive structures of *A. extimum,* n. sp. Conventions as in figure 38. A–B. Left gonostylus, ventral view: A. MX:Coahuila, nr. San Pedro; B. MX:Chihuahua, 40 mi. NE. Hidalgo del Parral. C. Female internal reproductive tract, MX:Coahuila, nr. San Pedro. D–F. Aedeagus, ventral view: D. AZ:Chiricahua Mts.; E. AZ:Douglas; F. MX:Coahuila, nr. San Pedro.

SMcC), 2-VIII-1976 (1 male, SMcC); Douglas, 20-21-VIII-1944 (1 male, 1 female, CISC), 7-IX-1957 (1 female, FMNH), 13-IX-1957 (1 female, FMNH), 18-IX-1957 (1 female, FMNH), 20-IX-1957 (1 male, FMNH); Douglas, at light, 9-VII-1971 (1 female, SMcC), 4-VIII-1971 (1 male, SMcC), 20-VIII-1973 (2 females, AMNH); Gleeson, 7 mi. S, 20-VIII-1977, at light (1 male, SMcC); Guadelupe Cyn., 31-VII-1975, at light (1 male, SMcC); nr. McNeal, at light, 16-VI-1976 (1 female, SMcC), 18-19-VIII-1970 (1 male, 1 female, SMcC); Palominas, San Pedro R., 23-VIII-1960 (1 female, UASM); Portal, 2 mi. E, 15-VII-1955 (1 female, AMNH); Portal, 4700′ el., 29-VII-1964, at light (1 male, UIDA); Portal, 5 mi. W, S.W.R.S., 5400′ el., 22-VII-1956 (1 male, 1 female, AMNH), 1-VIII-1958 (2 males, CISC), 10-VIII-1973 (1 female, LLLa); San Bernardino Ranch, 14-VII-1975, at light (2 males, 3 females, AMNH; 1 male, 3 females, SMcC); Willcox, 6 mi. SE, 28-VII-1952 (1 male, 1 female, AMNH); Willcox, 12-VII-1959, at light (1 female, UASM), 3-VIII-1957 (2 females, UAZ); 18-VIII-1958 (1 female, UAZ); 7-IX-1950 (1 male, AMNH). YAVAPAI Co.: Congress Jct., 20-IX-1940 (1 female, LACM). MEXICO: CHIHUAHUA: Camargo, 20 mi. SW, 4500′ el., 13-VII-1947 (1 male, 1 female, AMNH); 25 mi. SW, 14-VII-1947 (1 male, AMNH), 42 mi. SW, 4900′ el., 15-VII-1947 (1 female, AMNH); Chihuahua, 110 mi. N, 22-VIII-1964 (1 male, UCR); Chihuahua, 17-VII-1969 (1 male, UASM); Hidalgo del Parral, 40 mi. NE, Rte. 45, 5200′ el., UV light, 11-IX-1967 (1 female, UASM); Jimenez, 10 mi. N, 10-11-IX-1950 (1 male, 1 female, AMNH); Parrita, 5 mi. W, Santa Clara Cyn., 1-IX-1956 (1 male, CISC); Samalayuca, 24-VI-1947 (1 female AMNH). COAHUILA: Hermanas, 12 mi. N, 11-VIII-1959 (1 female, UCD); San Pedro, 18.3 mi. S (sic E), 1080 m el., 27-VIII-1974 (5 males, 5 females, UASM); San Pedro, 31.2 km E, 900 m el., 31-V-1981 (50 males, 46 females, JKLi), 31-V-1981 (18 males, 7 females, CISC).

Biology and Habits. *A. extimum* is found in lowland desert habitats near water. The type series was taken in a muddy, drying creek, on the southern shore of Lake Mayran east of San Pedro, Coahuila. The dominant vegetation on the shores of the creek bed is *Acacia*, which shades portions of the banks. The creek bed is composed of very fine grained mud, which had fissured as it had dried. During the heat of the day, beetles were found 10-15 cm below the surface in those fissures. At night, *A. extimum* was plentiful on the surface. Taken along with *A. extimum* were *Megacephala carolina* L., *Omophron americanum* Dej., *Agonum texanum* LeConte, *Polpochila* sp., *Anisodactylus* sp., 3 spp. of *Chlaenius*, 2 spp. of *Brachinus*, and 4 spp. of *Bembidion*. The surface of the mud at the type locality became a scene of coleopteran carnage after sunset. Teneral individuals active on the mud surface fell prey to other fully sclerotized predators. Food was scarce, judged by the appetites of *A. extimum* collected live for larval rearing.

Etymology: From its occurrence in such marginal habitats (L. *extimus*), the most remote.

Agonum parextimum Liebherr, new species

(figures 29i; 30c; 32c,d; 35c,f; 40, 41)

Holotype. female, MEXICO, Sinaloa, Los Mochis, 27-VI-1922, C.T. Dodds (CAS).

Allotype. male with same data (CAS).

Diagnosis. Superficially similar to *A. extensicolle* but with median basal area of pronotum and latero-basal pronotal depressions impunctate; elytra with dull, matte surface; body surface glabrous except for articulated macrosetae.

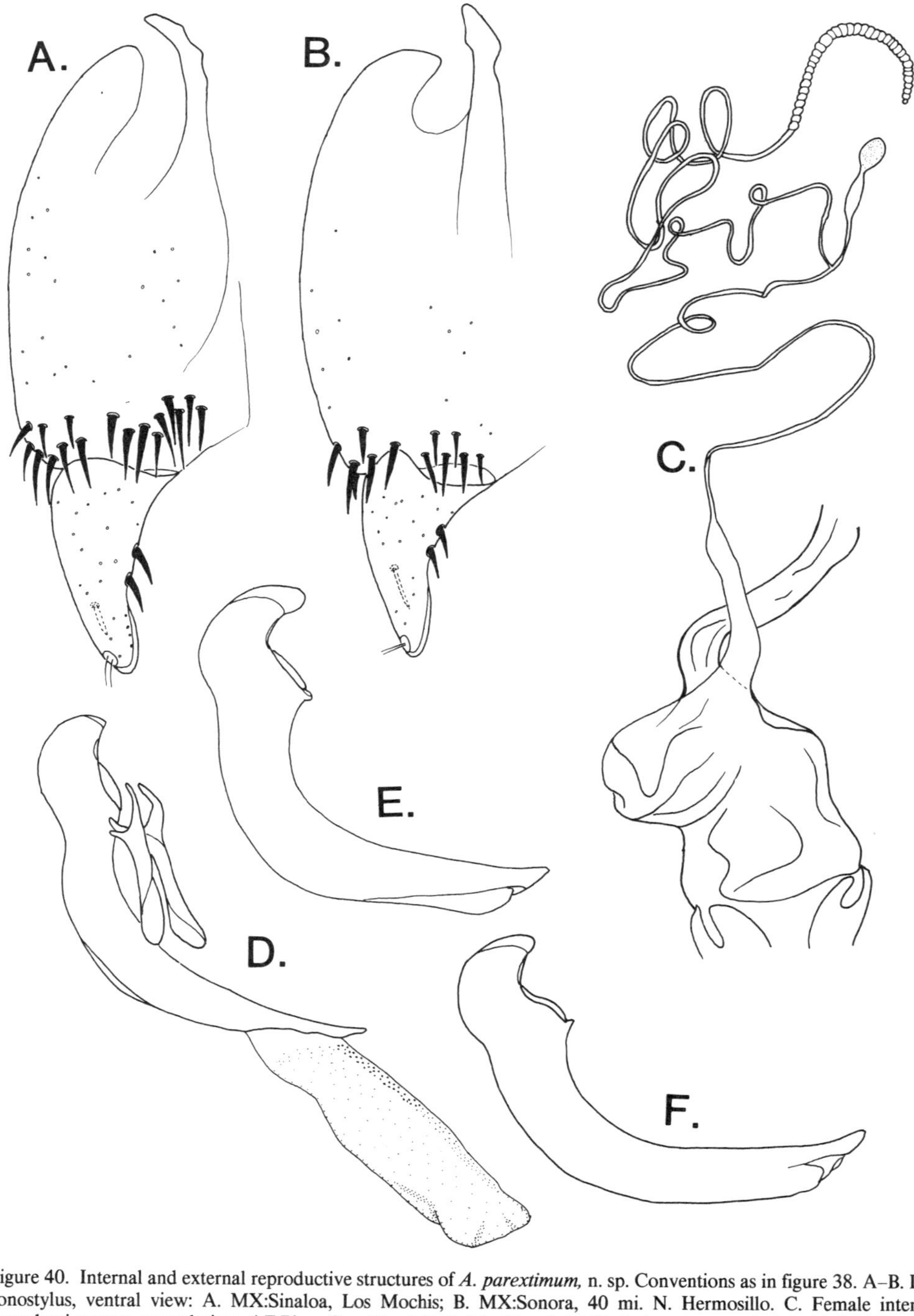

Figure 40. Internal and external reproductive structures of *A. parextimum*, n. sp. Conventions as in figure 38. A–B. Left gonostylus, ventral view: A. MX:Sinaloa, Los Mochis; B. MX:Sonora, 40 mi. N. Hermosillo. C. Female internal reproductive tract, ventral view, AZ:Pima Co., 15 mi. S. Robles Ranch. D–F. Aedeagus, ventral view: D. MX:Sinaloa, Venedio; E. AZ:Pima Co., 27 mi. W. Quijotoa; F. MX:Sonora, 10 mi. N. Santa Ana.

Description. *Head.* Frons foreshortened, head appearing broad due to well-developed protruding eyes; neck broad, temple gradually expanded to meet hind margin of eye (fig. 30c); antenna with second segment bearing apical ring of 11-12 setae (fig. 29i), third segment pubescent in apical half, segments 4-11 fully pubescent.

Thorax. Pronotum appearing slightly longer than wide, but with maximum width before middle averaging 1.15 times median length (figs. 32c,d), lateral margin sinuate before well-developed hind angle, lateral depression narrow, widening slightly inside basal sinuation; basal bead visible immediately inside latero-basal seta, faintly traceable to absent inside latero-basal depression; latero-basal depression ovoid to linear, continuing forward to area of anterior seta, at most with only faint punctulae; transverse wrinkles crossing basal depression slightly elevated, smooth; median longitudinal impression clear, somewhat more well-developed at base of pronotum; anterior transverse impression weakly developed, smooth, barely traceable.

Elytra. Long, slightly ovoid, maximum width 2/3 distance to apex, apical sinuation faint, close to apex (fig. 35c); elytral intervals flat to slightly convex; striae lightly impressed although complete, faint punctulae visible at 80X; striae 1-5 strongly joining basal margin; basal margin slightly angulate at shoulder; 4-6 setae in or adjacent to third elytral interval.

Male Genitalia. Aedeagus with apical half relatively stout, evenly curved (figs. 40d-f); longitudinal wrinkles absent from inside of basal curve of aedeagal shaft; aedeagal sac as long as apical straight portion of aedeagus (fig. 40d), spicules covering sac, somewhat more strongly developed on dorsal surface (everted position).

Female Reproductive Tract. Basal segment of gonostylus with apical fringe of 12-16 setae (fig. 40a,b), spermatheca long, 37-47 constrictions producing moniliform configuration (fig. 40c); spermathecal gland entering spermathecal duct distant from spermatheca, about 1/3 duct length from bursa copulatrix.

Color. Dorsal surface of head, and prothorax metallic green with faint yellowish reflection; disc of prothorax sometimes with brunneus cast; elytra with faint metallic green reflection when viewed under magnification, appearing matte piceous to the eye; elytral epipleura rufo-piceous, darker near shoulder; mouthparts, labrum and clypeal apex rufo-piceous, apical segments of labial and maxillary palps infuscated; antenna with scape rufo-testaceous, segment 2 only slightly darker, segment 3 rufous, segments 4-11 rufo-piceous; legs testaceous, femoral apices slightly infuscated, tarsi slightly darker, rufo-piceous; ventral body surface rufo-piceous.

Microsculpture. Head, pronotum and elytra with uniform isodiametric microsculpture, lack of punctulae producing smooth granulate appearance; ventral body surface with somewhat weaker isodiametric microsculpture, more strongly developed on episterna and lateral areas of abdominal sternites; faint longitudinal wrinkles on lateral areas of abdominal sternites.

Length. 8.0-10.1mm.

Distribution. Lowland habitats near Tucson and Phoenix, AZ, south to Sonora and southern Sinaloa along the Mexican coast (fig. 41). PARATYPES: 61: ARIZONA: PIMA Co.: Quijotoa, 17 mi. W on Rte. 86, 27-VIII-1960 (4 males, 3 females, UASM); Robles Ranch, 15 mi. S, 2900′ el., 3-IX-1950 (1 female, AMNH); Sahuarita, 25-VII-1932 (2 males, MCZ); San Simon, at light, 3-IX-1954 (1 female, LACM); Sasabe, 15-II-1977 (1 male, 2 females, CNC); Tucson (1 female, CAS; 1 female, MCZ), 7-VII-1935, at light (1 male, NMNH), 19-VII-1963 (1 male, 1 female, CAS), 26-VIII (1 female, CAS); Tucson, Desert Museum, 19-VII-1955, light trap (2 males, UAZ); Tucson, St. Xavier Mission, 29-VII-1924 (3 females, CAS); Tucson, 25 mi. W nr. Brawley Wash, 27-VIII-1960 (7 females, UASM). PINAL Co.: Florence (1 female, NMNH); Maricopa, 17-X-1927 (1 male, CAS). MEXICO: SINALOA: Escuinapa, 3-IX-1972, at light (1

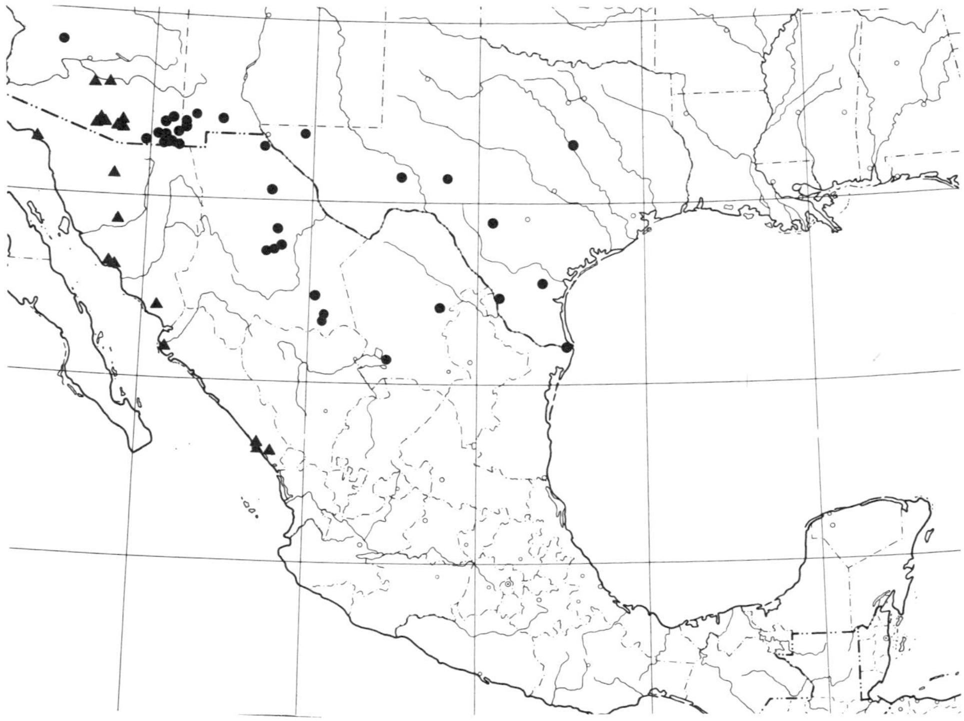

Figure 41. Distributional records of *A. extimum,* n. sp. (circles) and *A. parextimum,* n. sp. (triangles).

female, CDFA); Los Mochis, 27-VI-1922 (4 males, CAS), 1-VII-1922 (1 female, CAS), 20-VII-1922 (1 female, CAS), 25-VII-1922 (1 female, CAS); Mazatlan, 31-VII-1959 (1 male, AMNH); Venedio, 30-VII-1918 (1 male, 1 female, CAS), 31-VII-1918 (1 female, CAS), 10-VIII-1918 (1 male, CAS), 25-VIII-1918 (1 female, CAS), 27-VIII-1918 (1 male, CAS). SONORA: Guaymas, 3-IX-1972, (1 male, CDFA); Guaymas, 18 mi. SE, 24-VII-1954 (1 female, NMNH); Hermosillo, 9-VII-1953 (1 female, CAS); Hermosillo, 40 mi. N, 8-VIII-1960 (3 females, CAS); Navajoa, 13 km. SW, 24-VIII-1954 (1 female, 1 male, NMNH); Rocky Point (Puerto Peñasco), 27-VII-1956 (1 female, AMNH); Santa Ana, 10 mi. N, 3-IV-1958 (1 male, 2 females, CAS).

Biology and Habits. *A. parextimum* resides in lowland habitats similar to those occupied by *A. extimum*; desert water sources subject to desiccation during dry periods.

Etymology. *A. parextimum* is the sister species of *A. extimum.* Its distribution on the west coast of Mexico complements the range of *A. extimum* (fig. 41).

Agonum texanum (LeConte)

(figures 29k; 30d; 31f-h; 34a; 36a; 42; 43; 49a-c)

Anchomenus texanus LeConte, 1878, *Proc. Amer. Philos. Soc.* 17:374.
Anchomenus (Pseudanchus) texanus; Casey, 1920, *Memoirs on the Coleoptera* 9:58.

Agonum (Anchomenus) texanum; Csiki, 1931, *Coleopterorum Catalogus* 115:862.
Agonum (Stictanchus) texanum; Lindroth, 1966, *Opusc. Entomol. Suppl.* 29:629.
Anchomenus megillus Bates, 1891, *Trans. Roy. Entomol. Soc. London* 1891:252 (NEW SYNONYMY).
Agonum (Anchomenus) megillum; Csiki, 1931, *Coleopterorum Catalogus* 115:847.
Agonum megillum; Liebherr, 1982, *Coleopt. Bull.* 36:153 (male lectotype designated).

Types. of *texanum*, holotype female, TEXAS (MCZ)!, type locality designated Clifton, Bosque Co., TX by Lindroth (1966); of *megillum*, lectotype male, MEXICO, Villa Lerdo, (BMNH)!.

Diagnosis. Broadest species of the group, pronotal maximum width 1.25 times median length, lateral margin not or only barely sinuate before basal angle (fig. 31f-h); male aedeagus with straight apex and well-developed lateral sclerites on internal sac (fig. 42e).

Description. *Head.* Broad with moderately projecting eyes, frons relatively foreshortened (fig. 30d). Temple moderately expanded from broad neck; antenna with second segment bearing 6-8 setae in an apical ring (fig. 29k), third segment glabrous except for apical setae, segments 4-11 covered with fine setae as well as apical setae.

Thorax. Pronotum broad, lateral margin not or barely sinuate before hind angle (figs. 31f-h); basal angle obtuse, margin behind basal seta curved outward, lateral bead narrow in anterior 2/3 of pronotum, widening basally to maximum at basal seta, then narrowing until disappearing posterior to latero-basal depression; slight wrinkling and small punctulae adjacent to well-developed latero-basal depression; depression shallower anteriorly, running parallel to lateral margin; slightly raised tubercle present outside latero-basal depression; basal bead variable, from absent (fig. 31h) to easily traceable (fig. 31f); median longitudinal impression clear; anterior transverse impression weak, gently sloped.

Elytra. Elytron broad, more or less flattened on disc, broadest at apical third (fig. 34a); apical sinuation poorly developed; elytral intervals flat to slightly convex, striae smooth except for faint punctulae; elytral striae 1-5 only weakly joining basal margin; basal margin smoothly rounded at shoulder; 4-7 setae in or adjacent to elytral interval 3; setae rarely situated (one side only) in or adjacent to interval 5.

Male Genitalia. Apical half of aedeagal shaft straight (figs. 42e, 49a-c); surface of aedeagus with longitudinal wrinkles inside basal curve; internal sac of aedeagus small in relation to aedeagal length; two lateral sclerites supporting sac when everted (fig. 42e); sclerites visible inside aedeagal shaft when sac uninflated (figs. 49a-c); surface of internal sac with fine spicules.

Female Reproductive Tract. Basal segment of gonostylus with fringe of 11-15 stout setae (figs. 42b-d); spermatheca fusiform with tapering apex, faint cuticular rings on surface (fig. 42a); spermathecal gland entering spermathecal duct at base of spermatheca.

Color. Head metallic green, sometimes with bluish reflection; prothorax royal blue to blue-green in western Arizona populations, metallic green to aeneous on disc from central Arizona to Texas and Chiapas; area laterad of latero-basal depression tending to be greener than center of disc if disc is aeneous; dorsal surface of elytron varying from metallic blue-green in western Arizona populations to aeneous-green in eastern part of range; elytral epipleuron rufo-piceous, somewhat darker at base; mouthparts uniformly rufo-piceous; antenna with scape rufo-piceous, lightest anteriorly and below, segments 2-11 piceous; femora rufo-piceous, infuscated at apex; tibiae rufo-piceous, lighter on outer surface, with external ridge rufo-testaceous; tarsi rufo-piceous; ventral body surface piceous to rufo-piceous.

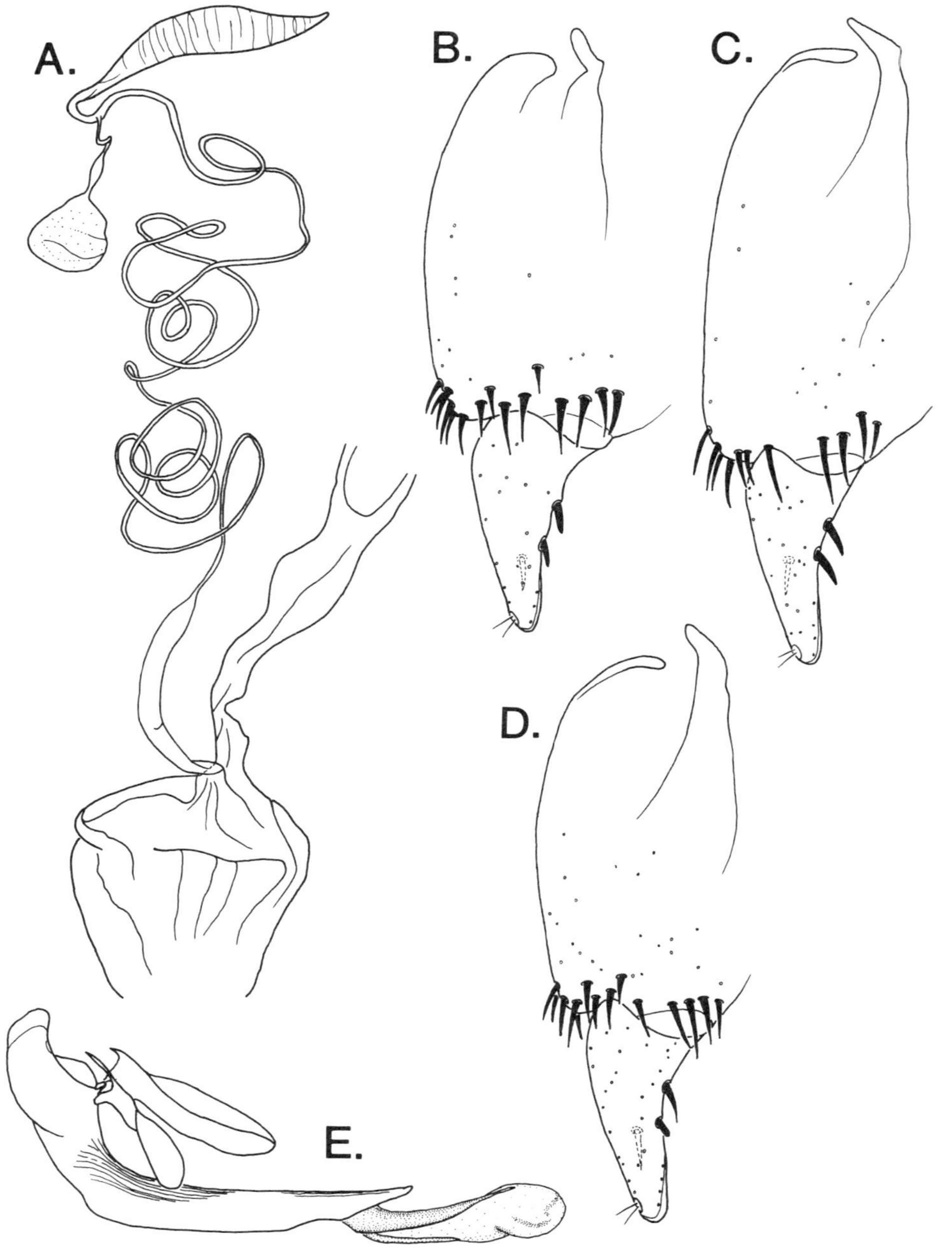

Figure 42. Internal and external reproductive structures of *A. texanum;* conventions as in figure 38. A. Female internal reproductive tract, ventral view, TX:Brownsville. Spermathecal gland enters at base of spermatheca. B–D. Left gonostylus, ventral view: B. MX:Chiapas, 5 km E. Comitan; C. TX:Sinton, Welder Wildlife Found.; D. AZ:Yavapai Co., Camp Verde. E. Aedeagus, ventral view, TX:Brownsville.

Microsculpture. Head with uniform isodiametric microsculpture; pronotum with isodiametric microsculpture on disc, slightly stretched inside lateral depressions with tendency to be granulate outside lateral depressions; elytra with uniform isodiametric microsculpture; ventral body surface with uniform isodiametric microsculpture, lateral areas of abdominal sternites with faint longitudinal wrinkles.

Length. 7.1mm (Chiapas population) - 10.1mm.

Distribution. North of Mexico, *A. texanum* is found in Texas, Oklahoma, Kansas, New Mexico, and Arizona (fig. 43). Kansas records may represent dispersing individuals with permanent residence restricted to regions further south. Kansas records: McPherson Co., McPherson (1, CAS); Trego Co., Cedar Bluff St. Pk., 13-VIII-1966 (1, CISC).

Distributional Records. 787 specimens, records from Mexican localities follow: MEXICO: CHIAPAS, 5 km. E Comitan (air field), 30-V-1972 (21, UASM). CHIHUAHUA: 42 mi. SW Camargo, 4900′ el., 15-VII-1947 (4, AMNH); Chihuahua (2, BMNH); Chihuahua, 17-VII-1969 (2, NMNH); 12 mi. N Chihuahua, 23-VIII-1960 (2, CAS); Delicias, 11-VII-1947, 4150′ el. (1, AMNH); 12 mi. N Escalon, 18-IV-1961 (1, CNC); 40 mi. NE Hidalgo del Parral, 5200′ el., 11-IX-1967 (3, UASM); Cd. Jimenez, 26-VI-1964 (1, NMNH); 10 mi. N Jimenez, 10-11-IX-1950 (7, AMNH); Primavera, 30-VI-1947, 5500′-6000′ el. (1, AMNH); Salaices, 23-VII-1947 (1, AMNH); Samalayuca, 24-VI-1947 (1, AMNH); Sombreretillo, 28-VI-1956 (1, CAS); Valle de los Olivos, 5500′ el., 20-VII-1947 (1, AMNH). COAHUILA: 67 km S Sabinas, Rte. 57, 470 m el., 26-VII-1975 (2, UASM); 18.3 mi. E San Pedro, 1080 m el., 27-VIII-1974 (1, UASM); 19.4 mi. E San Pedro, 900 m el., 31-V-1981 (9, CISC); Torreon, 23-V, 16-VI-1937 (2, CAS). DURANGO: Lerdo, 12-VI-1918 (3, USNM); Villa Lerdo (8, BMNH). SONORA: 17 km SW Moctezuma, 21-22-VII-1980, 944 m el. (1, SMcC). TAMAULIPAS; Lago Almagre, 7.5 mi. W La Pesca, 28-X-1965 (1, UASM); 13 km E Magiscatzin, 28-V-1948 (1, MCZ); Cd. Mante, 100′ el., 18-VI-1949 (1, AMNH); 29 mi. N San Fernando, 300′ el., 24-X-1965 (1, UASM).

Biology and Habits. *A. texanum* is found in lowland habitats near water sources in xeric to mesic *Acacia* scrub habitats. It was taken in leaf litter along Limpia Creek in the Davis Mts., 5,200′ el., 17-IV-1981, along with *A. extensicolle* and *A. decorum*. This capture was made when carabid activity was low, and probably represents overwintering adults. Movement into leaf litter away from the water's edge appears to be an overwintering strategy. *A. texanum* exhibits a great propensity for dispersal by flight.

Infraspecific Geographic Variation. Discriminant analysis is used to study geographic variation within *A. texanum*. For males, 12 of the 13 mensural characters are heterogeneous enough in the various populations to produce a significant F-ratio in the univariate analysis (table 23). Profemur width does not differ significantly in the populations analyzed, and it is removed from the discriminant analysis. The first two discriminant functions account for 72.5% of the variance in the study set, apportioned as 57.1% accounted for by function 1, and 15.4% by function 2. Function 3 accounts for 11.0% of the variance, but the distribution of samples is not meaningful geographically. The function scores for the eight samples analyzed are depicted in figure 44a.

Capture records for *A. texanum* are relatively abundant throughout Texas. Outlying populations occur at the edges of the range of the species. These include populations at Comitan, Chiapas (49), and in western Arizona near Kirkland (57), and the sample from Villa Lerdo, Durango (51). The Chiapas population is aberrant: its individuals are much smaller than those from Texas (Appendix 5). The nearest known population to the one at Comitan is at Jalapa,

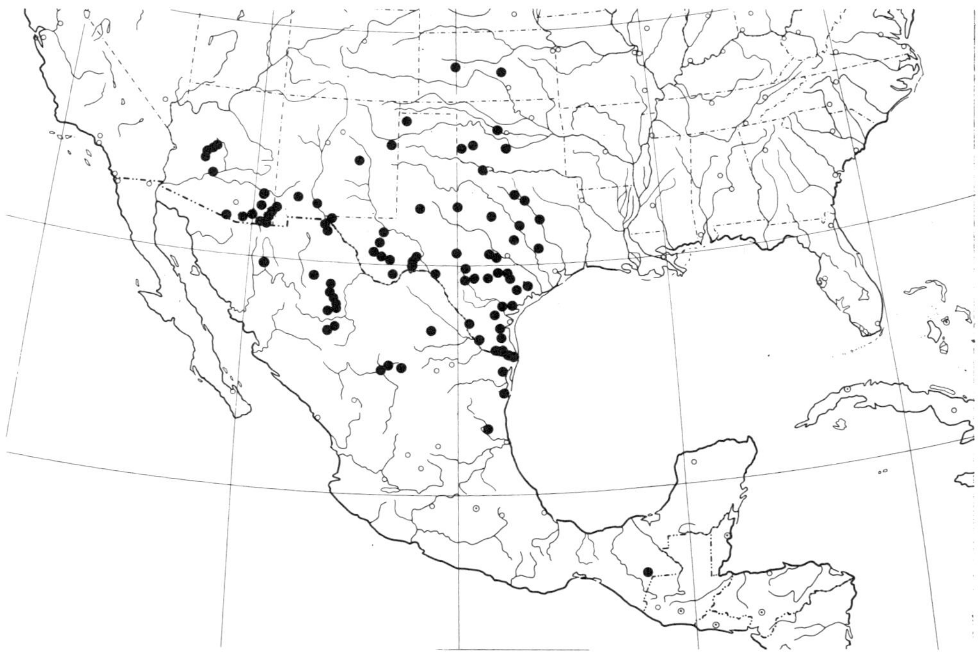

Figure 43. Distributional records of *A. texanum* (LeConte).

Table 23. Wilks' lambda values, univariate F-ratios, and standardized canonical discriminant function coefficients for males of *A. texanum.*

Variable	Wilks'	F*	Coefficients d.f. 1	Coefficients d.f. 2
hw	.564	6.85	1.36	0.23
pl	.714	3.55	0.27	0.07
apw	.727	3.33	0.53	0.15
mpw	.698	3.83	–0.79	–0.60
bpw	.653	4.71	–0.65	–0.24
huw	.719	3.46	–0.13	0.21
mew	.705	3.70	0.65	–0.30
el	.618	5.47	–0.90	1.41
4al	.704	3.72	0.59	–0.29
pfl	.683	4.10	–0.78	–0.08
pfw	.852	1.53*	—	—
3sl	.752	2.93	0.08	0.52
pd	.702	3.75	0.02	–0.04

* All F-ratios significant at P less than .05 except for pfw; pfw excluded from discriminant analysis.

Vera Cruz, a distance of 650 km. The Kirkland, Arizona population is isolated on the eastern border of the Colorado Desert. The population at Villa Lerdo is less isolated, but is a western outlying locality in northern Mexico.

The geographic variation based on biometric data corroborates the outlying nature of populations 49, 51, and 57 (figs. 43, 44a). The Chiapas population is separated from the central populations on function 1 and 2, based on head width and elytral length measurements on function 1, and elytral length on function 2 (table 23). The Chiapas population has smaller than average head width, and short elytral length. The Villa Lerdo population scores negatively on function 1, because of the relatively narrow head (.83mm average) compared with the averages of .84-.93mm for the other populations. The Kirkland population is similar to Villa Lerdo in possession of a narrow head (.84mm).

Table 24. Wilks' lambda values, univariate F-ratios, and standardized canonical discriminant function coefficients for females of *A. texanum.*

Variable	Wilks'	F*	Coefficients d.f. 1	d.f. 2
hw	.763	2.97	–0.72	–0.33
pl	.556	7.63	1.85	0.69
apw	.656	5.02	0.25	0.21
mpw	.670	4.72	–1.06	0.52
bpw	.624	5.77	–0.47	1.47
huw	.611	6.10	0.99	–0.98
mew	.683	4.45	–1.53	0.59
el	.774	2.79	–0.03	–0.45
4al	.704	4.02	0.17	0.33
pfl	.577	7.02	0.82	–1.45
pfw	.630	5.62	–0.12	0.78
3sl	.585	6.78	0.51	0.12
pd	.646	5.24	0.03	–1.46

* All F-ratios significant at P less than .02.

In the female analysis, 61.5% of the variance is accounted for by the first two discriminant functions; 40.9% by function 1, and 20.6% by function 2. The third function accounts for 16.6% of the variance, but as in the male analysis, sample scores do not have any relation to geography. All mensural characters vary significantly among the eight populations analyzed (table 24). The results are much the same as for the males (fig. 44b). The Chiapas population receives a negative d.f. score, partly due to short prothoracic length: 1.54mm average versus averages of 1.70-1.87mm in the other populations. This difference, coupled with a positive coefficient of 1.85 for this character, aids separation. Another character distinguishing the Chiapas population on function 2 is profemur length. The Chiapas population has short legs, profemur length averaging 1.60mm, versus averages of 1.83-1.91mm in other populations. The positive coefficient of .82 produces lower scores for the Chiapas population.

The Kirkland population (57) is somewhat distinct from the Texas populations on function 1. Like the Chiapas population, individuals are smaller than in the Texas populations: prothoracic length averages 1.70mm versus 1.77-1.86mm in Texas populations. The biometric difference

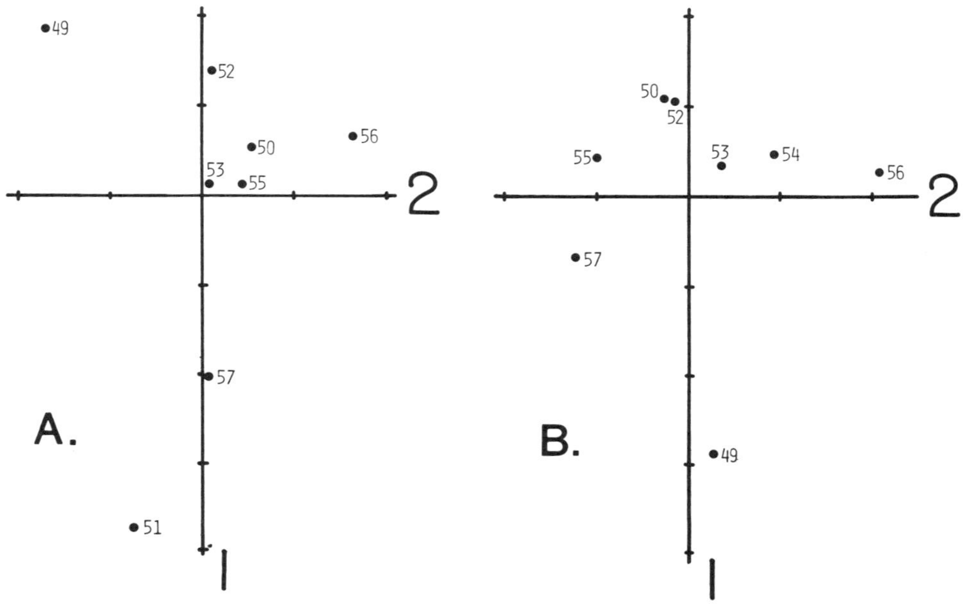

Figure 44. Scatterplots for the first two discriminant functions based on population samples of *A. texanum*. A. Males. B. Females.

between the Kirkland population and the Texas populations is less pronounced in the females than in the males. This trend is seen overall in the *A. extensicolle* group: female body shape is more conservative over the species' ranges than male body shape.

The male and female analyses corroborate the existence of biometrically divergent outlying populations in *A. texanum*. There is little variation within the central part of the range. The peripheral populations are characterized by smaller body size. This implies that geographically marginal populations are under different environmental conditions, causing changes that result in smaller body size.

Agonum extensicolle (Say)

(figures 29l; 30e; 32e-h; 34c; 36b,f-h; 45; 46; 49d-g)

Feronia extensicolle Say, 1823, *Trans. Amer. Philos. Soc.* 2(n.s.):54.

Anchomenus extensicollis; Dejean, 1828, *Spécies général Coléoptères* 3:113; Say, 1834, *Trans. Amer. Philos. Soc.* 4(n.s.):421; Bates, 1882, *Biol. Cent.-Amer. Coleop.* 1(1):93; Bates, 1884, *Biol. Cent.-Amer. Coleop.* 1(4):280; Blatchley, 1910, *Coleoptera of Indiana*, p. 126.

Platynus extensicollis; LeConte, 1854, *Proc. Acad. Nat. Sci. Phila.* 7:45.

Anchomenus (Pseudanchus) extensicollis; Casey, 1920, *Memoirs on the Coleoptera* 9:57.

Agonum (Anchomenus) extensicollis; Csiki, 1931, *Coleopterorum Catalogus* 115:860.

Agonum (Stictanchus) extensicolle; Lindroth, 1966, *Opusc. Entomol. Suppl.* 29:625.

Feronia proximus Harris, 1828, *New England Farmer* 7:132; LeConte, 1854, *Proc. Acad. Nat. Sci. Phila.* 7:45 (synonymy).
Anchomenus obscuratus Chaudoir, 1843, *Bull. Soc. Imp. Nat. Moscou* 16:763; LeConte, 1854, *Proc. Acad. Nat. Sci. Phila.* 7:46 (synonymy).
Anchomenus viridis LeConte, 1848, *Ann. Lyc. Nat. Hist. New York* 4:222; LeConte, 1854, *Proc. Acad. Nat. Sci. Phila.* 7:46 (synonymy).
Anchomenus cyanescens Motschulsky, 1859, *Bull. Soc. Imp. Nat. Moscou* 3:159; Lindroth, 1966, *Opusc. Entomol. Suppl.* 29:625 (synonymy).
Anchomenus (Pseudanchus) cyanescens; Casey, 1920, *Memoirs on the Coleoptera* 9:57.
Anchomenus (Pseudanchus) gaudens Casey, 1920, *Memoirs on the Coleoptera* 9:55; Lindroth, 1966, *Opusc. Entomol. Suppl.* 29:625 (synonymy).
Agonum (Anchomenus) gaudens; Csiki, 1931, *Coleopterorum Catalogus* 115:861.
Anchomenus (Pseudanchus) gaudens clientulus Casey, 1920, *Memoirs on the Coleoptera* 9:55.
Anchomenus (Pseudanchus) vigilans Casey, 1920, *Memoirs on the Coleoptera* 9:56; Lindroth, 1966, *Opusc. Entomol. Suppl.* 29:625 (synonymy).
Agonum (Anchomenus) vigilans; Csiki, 1931, *Coleopterorum Catalogus* 115:862.
Anchomenus elongatulus; LeConte (not Dejean), 1848, *Ann. Lyc. Nat. Hist. New York* 4:222.
Anchomenus (Pseudanchus) elongatulus; Casey (not Dejean), 1920, *Memoirs on the Coleoptera* 9:56.

Types. of *extensicolle*, female neotype, Rumney, N.H., type locality (Lindroth 1966), neotype designation (Lindroth and Freitag, 1969) (MCZ)!; of *proximus*, no type area given, type lost; of *obscuratus*, United States, Amer. sept. (female, possibly type in MNHP)!; of *viridis*, female holotype, Indiana and Ohio River (MCZ)!; *of cyanescens*, female holotype, California (type probably mislabeled, ZMM)!; of *gaudens*, male holotype, Lake Champlain, NY (NMNH)!; of *clientulus*, male holotype, New Jersey (NMNH)!; of *vigilans*, male holotype, North Carolina (NMNH)!.

Diagnosis. Upper body surface metallic blue, metallic green or aeneous; prothoracic lateral margins sinuate before well-developed hind angles, base constricted; latero-basal pronotal depressions and median pronotal base rugose and punctate (figs. 32e-h); legs testaceous to rufo-testaceous.

Description. *Head.* Head moderately elongate, slender, eyes less well-developed, only moderately protruding, temple very gradually expanded from broad neck to hind margin of eye (fig. 30e); antenna with second segment bearing apical ring of 6-8 setae (fig. 29l), third segment glabrous except for apical setae, segments 4-11 pubescent.

Thorax. Pronotum quadrate, with base constricted, lateral margin sinuate before well-developed hind angle (figs. 32e-h); lateral depression narrow, widening in basal 1/3 to become more or less planar with latero-basal depression; crenulations usually present on surface of lateral gutter; basal margin faintly traceable inside basal seta behind latero-basal depressions to completely absent medially; latero-basal depression well-developed, linear anteriorly and widening toward basal margin; tubercle outside latero-basal depression separating it weakly from lateral gutter; tubercle, latero-basal depression, and basal median area of pronotum rugose-punctate; transverse wrinkles often present in latero-basal depressions and on prothoracic disc; median longitudinal impression distinct, breaking up slightly on rugose basal area; anterior transverse impression well-defined, impunctate.

Elytra. Varying in length; if long relative to median prothoracic length, parallel-sided; if shorter relative to prothoracic length, then ovoid with humeri less well-developed; greatest width

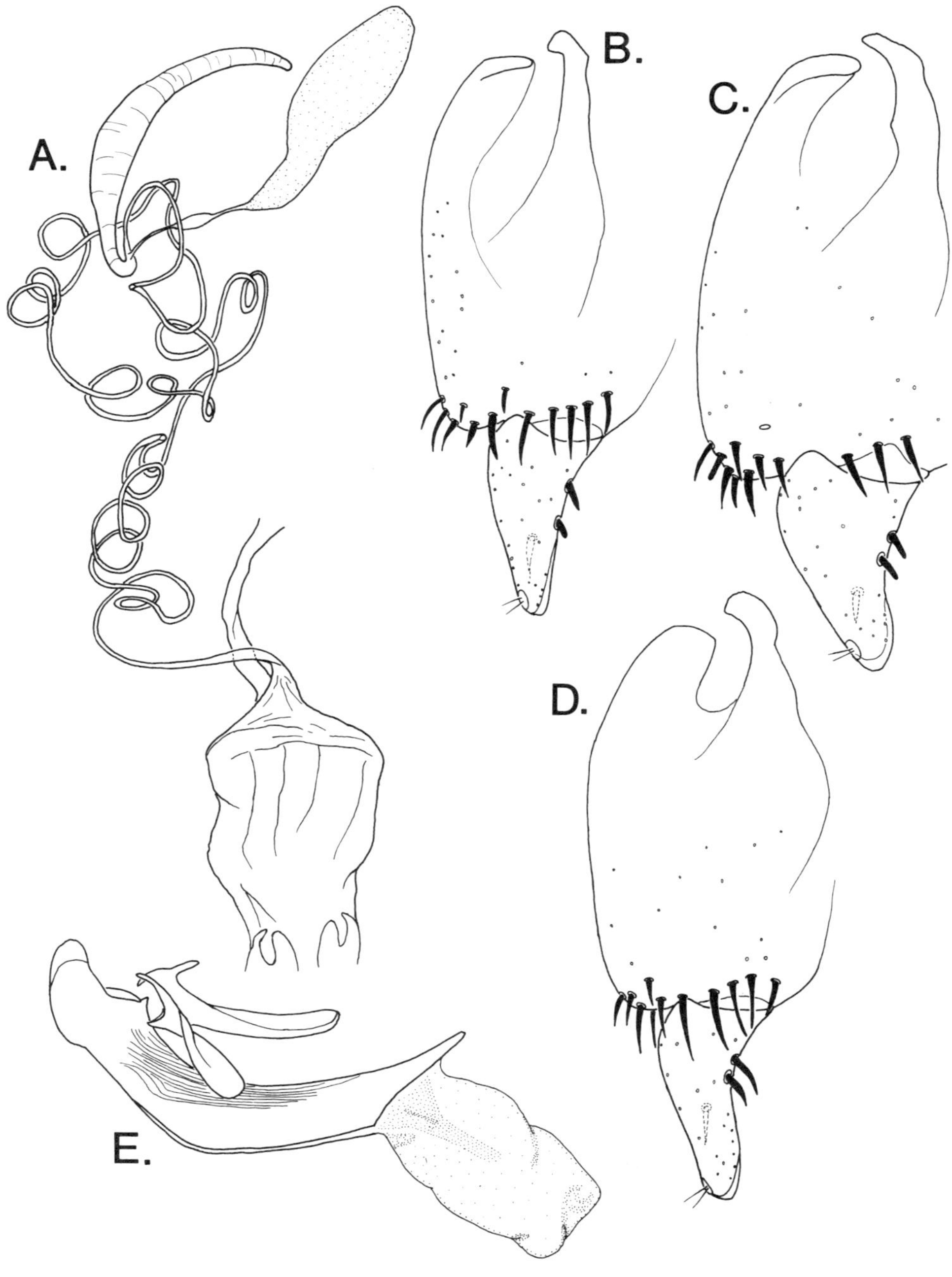

Figure 45. Internal and external reproductive structures of *A. extensicolle*. Conventions as in figure 38. A. Female internal reproductive tract, ventral view, MX:Jalisco, Puente Caquixtle. B–D. Left gonostylus, ventral view: B. CO:Boulder Co., N. Golden; C. NC:Alleghany Co.; D. AZ:Cochise Co., Palominas. E. Aedeagus, ventral view, VT:Burlington.

at apical 1/3; apical sinuation moderately well-developed; interval outside eighth stria relatively broad at apex (fig. 34c); elytral intervals slightly convex to convex; striae well-impressed, with punctulae easily visible at 40X; striae 1-5 strongly joining basal margin; basal margin slightly to moderately angulate at shoulder; 4-7 setae in or adjacent to third elytral interval.

Flight Wings. Normally fully winged with flight wing length to elytral length ratio 1.10-1.35 (fig. 26); rarely flight wings are reduced, with flight wing length to elytral length ratio .86-.91 (fig. 36h). (See dispersal capability analysis for details.)

Male Genitalia. Aedeagus evenly curved, sometimes with apical half slightly straightened, apex slightly upturned (figs. 45e, 49d-g); well-developed longitudinal wrinkles inside basal curve; aedeagal sac as long as apical straight portion of aedeagus, covered with fine spicules; lateral areas of sac with longitudinal areas forming slightly sclerotized bars (fig. 45e).

Female Reproductive Tract. Basal segment of female gonostylus with apical fringe of 11-12 setae (figs. 45b-d); spermatheca long and fusiform, very light cuticular rings on receptaculum; spermathecal gland entering duct at base of spermatheca (fig. 45a).

Color. Extraordinarily variable (see color analysis section for details); head and pronotum metallic green with golden reflection in Quebec and Nova Scotia, becoming metallic green with blue reflection in the eastern and midwestern United States and becoming a metallic royal blue with green reflection from west Texas and Arizona to the southern limit of distribution; elytral color even more variable, ranging from aeneous bronze, to vivid metallic green, to royal blue with purple reflection, over the same geographic limits; elytral epipleura piceous to rufo-piceous at base, paler toward apex; center of labrum piceous with same metallic reflection as head, edges and the rest of mouthparts rufo-piceous; basal segments of labial and maxillary palpi rufo-piceous; apical segments slightly darker; antenna with scape testaceous to rufo-testaceous, segments 2-4 progressively darker, segments 5-11 rufous to rufo-piceous; legs rufo-testaceous to testaceous, coxae slightly infuscated, tarsi hardly darker than femora and tibiae; ventral body dark, piceous with metallic reflection.

Microsculpture. Head with uniform isodiametric microsculpture; pronotum with isodiametric microsculpture along lateral margins and at base, becoming transversely stretched on disc to form poorly developed transverse mesh; elytra appearing granulate due to well-developed uniform isodiametric microsculpture; ventral body surface with well-developed isodiametric to slightly stretched microsculpture, lateral areas of abdominal sternites with well-developed longitudinal wrinkles.

Length. 7.1mm (Mexican specimens) to 10.8mm (West Point, NY specimen).

Distribution. North of Mexico, throughout the eastern United States and southern Canada east of the Rocky Mountains (fig. 46); in the Colorado, Green, and Gila River drainages west of the Continental Divide.

Distributional Records. 4300 specimens, records of collections in Mexico follow: MEXICO: AGUASCALIENTES; 5.3 mi. N Aguascalientes, 6100′ el., 7-IX-1967 (14, UASM); 11 mi. W Aguascalientes, Rte. 70, 5500′ el., 16-I-1966 (1, UASM). CHIHUAHUA: Casa Grandes (1, MCZ); Chihuahua City (1, USNM; 5, BMNH); Madera, 7200′ el., 6-VII-1947 (1, AMNH); 20.1 mi. S Miñaca, 6700′ el., 17-VIII-1966 (1, UASM); nr. Primavera, 5500′-6000′ el., 30-VI-1947 (7, AMNH); Santa Clara, 6500′ el., 2-3-VII-1947 (2, AMNH). DURANGO: 12.2 mi. S El Banco, 5220′ el., 10-IX-1967 (6, UASM); 27.5 mi. S El Banco, 7600′ el., 9-IX-1967 (3, UASM); Graceros, 6000′ el., 6-VII-1961 (10, MSUC); Rio Chico, 16 mi. N Durango, 6500′ el., 11-I-1966 (19, UASM); Rio Chico, 6400′ el., 3-VII-1961 (1, MSUC); Villa Lerdo (1, BMNH); Villa Ocampo, 27-VII-1947 (1, AMNH). HIDALGO: Rio Tula nr. Tasquillo; Rte. 85, 5300′

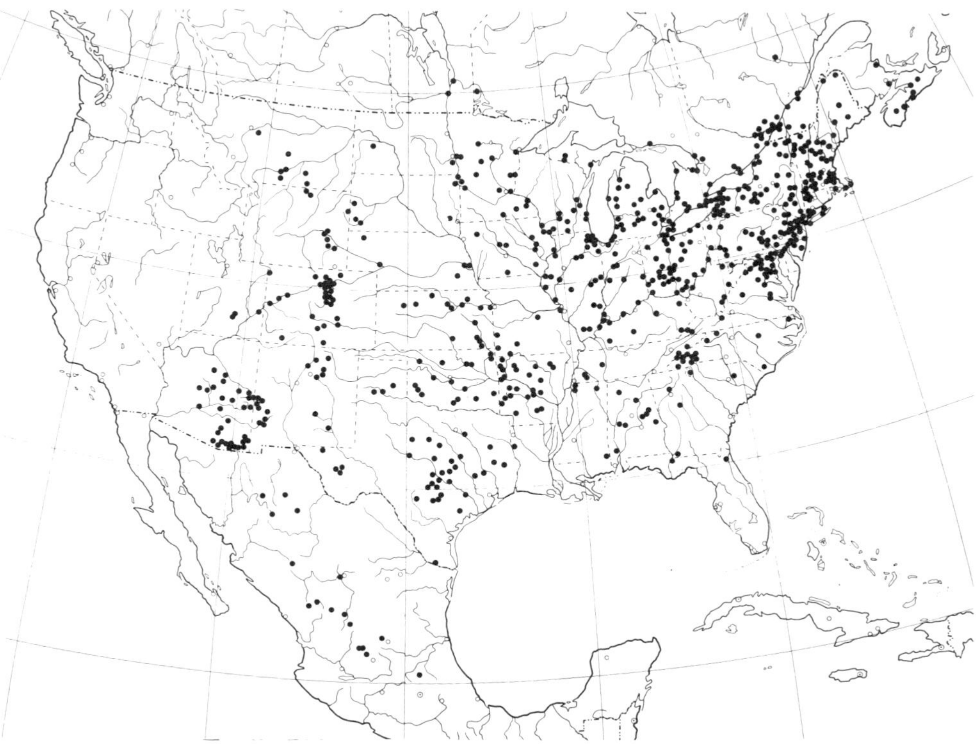

Figure 46. Distributional records of *A. extensicolle* (Say).

el., 13-IX-1965 (1, UASM). JALISCO: Puente Caquixtle, 9.7 mi. E Encarcion de Diaz, 16-I-1966 (2, UASM). SAN LUIS POTOSI: Puente la Parada, 7.5 mi. NW Maxquitic, Rte. 49, 7000′ el., 9-I-1966 (1, UASM). ZACATECAS: Rio Trujillo, 16 mi. NW Fresnillo, 6600′ el., 29-VI-1954 (1, CAS).

Biology and Habits. *A. extensicolle* is found principally on gravelly to rocky banks of smaller streams in sheltered mesic ravines or canyons, e.g., in the Chiricahua Mts., AZ, at 5,000′ elevation and up in shaded areas of boulder-strewn creek banks. It is an early successional streamside hydrophile.

Adults overwinter away from the stream banks in leaf litter or under bark, e.g., on 9-IX-1977, west of Bitely, Newaygo Co., MI, numerous individuals were found aggregated under bark of a downed tree, lying off the ground in a ravine. The log was about 10-15 feet from the water's edge. In Leslie Cyn., Cochise Co., AZ, 25-VIII-1981, 8 individuals were excavated from the last damp spots in the drying creek bed. Adults work downward along fissures in the mud to keep close to moist soil.

Larvae were reared from eggs produced by beetles collected 14-VIII-1979, Sycamore Ck. north of Payson, Gila Co., AZ. Adults were kept in ambient outdoor light, and fed cut *Tenebrio* larvae. Females laid the first eggs 26-27-IX, with larvae hatching until 13-X. These findings suggest reproductive diapause is not inhibited by shortening day length as long as food is available.

Larvae were reared at approximately 22°C, with first instar duration averaging 6.9 days (n=15), second instar 9.3 days (n=14), and third instar 20.4 days (n=10). Only three adults were successfully reared, with the pupal period averaging 8 days. The total period from egg hatch to adult eclosion averaged 44.6 days under these conditions.

A large series of individuals were taken throughout the year at Boulder, CO by Clair Armin. Of the 211 individuals taken between 1-VII and 29-IX-1958, 17 were tenerals: 16-30-VII (3); 1-15-VIII (5); and 16-30-VIII (9). Judging from these results, larvae are present in June and July. At Tucson, AZ, in Rillito Wash, 6 tenerals were taken 6-26-V-1978 (RLDa). This wash is dry except for rainy periods, and the presence of spring emerging adults implies breeding following winter rains, with larvae present in May. Environmental conditions vary greatly over the range of this species. A facultative breeding regime would be the safest strategy for exploitation of streamside habitat. In such a system, local conditions would cause immediate changes in the breeding condition of adult beetles.

Etymology. *A. extensicolle* derives its name from *extensivus* (L.), stretched out; plus *collum* (L.), neck.

Infraspecific Geographic Variation. The electrophoretic analysis used for species circumscription, and the species group level discriminant analysis produced conflicting data regarding circumscription of *A. extensicolle.* The electrophoretic data showed that the Arizona populations possess Pgm-1 locus 1.18 in frequencies from .56-1.0, whereas this allele was not found in the three eastern population samples. The geographic distribution of samples limits the faith that can be placed in this result, but the hypothesis that two species exist must be tested. The species group discriminant analysis does not corroborate the existence of two biometrically distinct entities within populations here called *A. extensicolle.* Further infraspecific analyses are presented that corroborate the species group discriminant analysis. Analysis of biometric characters and an analysis of upper body surface color are presented. Both analyses support recognition of *A. extensicolle* as a variable species of broad geographic distribution.

Biometric Analysis. Discriminant analysis was conducted separately for males and females using population samples 26 to 39 (table 5). For males, 10 of the 13 mensural characters exhibit significant univariate differences among populations (table 25) and are used in the analysis of geographic variation. The male discriminant analysis accounts for 66.8% of total variance in the first two functions, 46.5% accounted for by function 1, and 20.3% by function 2. The centroids based on these functions are plotted in figure 47a for all population samples. The range of individual d.f. scores within each population is included for function 1, providing a view of sample variability and overlap.

The first discriminant function separates the male samples into two rather distinct groups. Populations 27, 29, 30, and 31 receive positive scores on d.f. 1, whereas other population sample scores are approximately zero or negative. The separation of these two groups is achieved principally by the contribution of two characters: humeral width and elytral length. The four positive scoring populations have narrower humeri than the other populations: averages range from 1.09-1.19mm versus 1.21-1.32mm in other populations (Appendix 5). The large negative value, -.93, produces negative scores in the samples exhibiting broader humeri. Elytral length is shorter in the same four populations: average elytral length of 4.85-5.14mm versus 5.15-5.46mm in the other populations. The largest standardized coefficient on d.f. 1 is seen for this character (-1.11) insuring positive scores for individuals with shorter elytra (table 25).

Discriminant function 2 separates populations into two groups: populations 31, 37, 38, and 39 versus the other populations. The three characters contributing most to the discriminant score

Table 25. Wilks' lambda values, univariate F-ratios, and standardized canonical discriminant function coefficients for males of *A. extensicolle.*

Variable*	Wilks'	F*	Coefficients d.f. 1	d.f. 2
hw	.819	2.09	0.68	–0.67
pl	.764	2.92	0.34	–0.05
apw	.805	2.29	0.34	–0.59
mpw	.765	2.91	–0.47	0.74
bpw	.750	3.14	–0.01	–0.48
huw	.529	8.40	–0.93	–0.51
mew	.691	4.23	0.01	0.36
el	.609	6.07	–1.11	–0.01
pfl	.787	2.57	0.58	1.42
pfw	.791	2.50	0.22	–0.55

* 4al, 3sl, and pd do not differ significantly among samples and are not included in discriminant analysis. All other variables differ at P less than .05.

are profemur length, with a coefficient of 1.42; maximum prothoracic width, with a coefficient of .74; and head width, with a coefficient of -.67 (table 25). None of these characters is a very good discriminator, based on respective Wilks' lambda values of .787, .765, and .819; however, several geographic trends can be seen in these attributes. The four populations—Limpia Canyon, Texas (31), Clark County, Virginia (37), Ithaca, New York (38), and Nova Scotia (39)—all have shorter legs on the average than those in most other populations: average profemur length 1.79-1.87mm versus 1.83-2.01mm in the other 10 populations. The Nova Scotia individuals possess the shortest legs, averaging 1.79mm. Maximum prothoracic width does not differ when populations 31, 37, 38, and 39 are compared to the others, and the coefficient for this character simply represents an attempt to make a best fit to the overall data. Head width differs when the two groups are compared, with populations 31, 37, 38, and 39 possessing relatively broader heads: average of .82-.85mm versus .79-.83mm in the other populations. Based on function 2, the most important geographic trends are shortened legs and slightly broadened head relative to the other body measurements in the three eastern population samples (Virginia, New York, and Nova Scotia) and in the Davis Mountains sample. It is of interest that these trends have occurred twice. In the eastern samples these trends are correlated with wide humeri (function 1) whereas in the Davis Mountains sample they are correlated with narrow humeri.

For females, the first two discriminant functions account for 70.8% of the total variance, 49.4% accounted for by function 1 and 21.4% by function 2. Eleven of the 13 mensural characters differ significantly among the samples (table 26); analysis is restricted to these variable characters. Discriminant function 1 separates the samples into the same two distinct groups seen in the male analysis (fig. 47b). Populations 27, 29, 30, and 31 receive high d.f. scores due to heavy contributions by humeral width and elytral length. Both characters are good discriminators based on low Wilks' lambda values (table 26). The four most positive scoring samples possess narrow humeri (average 1.14-1.28mm versus 1.29-1.43mm in other populations) and shorter elytra (average 4.82-5.40mm versus 5.37-6.02mm). The large negative coefficients for these two variables ensure positive scores for populations 27, 29, 30, and 31.

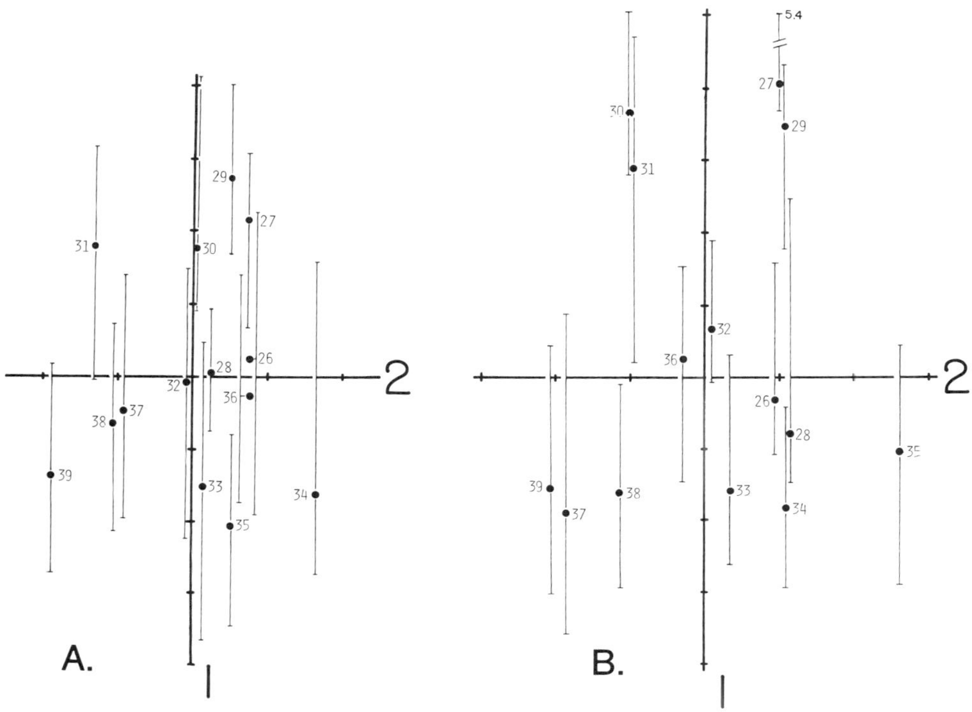

Figure 47. Scatterplots of first versus second discriminant functions based on population samples of *A. extensicolle.* Ranges of individual scores indicated for function 1. A. Males. B. Females.

Table 26. Wilks' lambda values, univariate F-ratios, and standardized canonical discriminant function coefficients for females of *A. extensicolle.*

Variable*	Wilks'	F*	Coefficients d.f. 1	d.f. 2
hw	.719	3.13	0.71	–0.88
pl	.704	3.36	0.35	–0.08
mpw	.741	2.79	–0.33	–0.95
bpw	.749	2.69	0.21	0.49
huw	.534	6.99	–1.26	–0.83
mew	.653	4.26	–0.10	–0.34
el	.523	7.31	–1.63	0.64
4al	.641	4.48	0.04	–0.26
pfl	.719	3.13	0.61	1.52
3sl	.750	2.66	0.02	0.64
pd	.811	1.87	1.22	0.08

* apw and pfw do not differ significantly among samples and are not included in discriminant analysis. All other variables differ at P less than .05.

Based on discriminant function 2, a pattern similar to that seen in males exists for the female samples (fig. 47b). For females, population samples 30, 31, 37, 38, and 39 possess the most negative d.f. scores. As in males, profemur length, maximum prothoracic width, and head width have the largest standardized coefficients. When the raw data is analyzed univariately, the range of values of these characters for populations 30, 31, and 37-39 is not significantly different from the range of values in the other nine populations. Each negatively scoring population differs from the other group of nine population samples in one of the characters: sample 31 in large head width; sample 38 in larger maximum prothoracic width; samples 30, 37, and 39 in smaller profemur length. Combining these trends in the raw data with a head width coefficient of -.88, a maximum prothoracic width coefficient of -.95, and a profemur length coefficient of 1.52 produces negative d.f scores for samples 30, 31, 37, 38, and 39. The northeastern populations (36, 37, 38) are separated from the other seven populations with wide humeri. This is the only geographically contiguous group of populations separated by function 2 in both the male and female analyses. In both sexes, individuals in these populations tend to have broader heads, broader prothoraces, and shorter legs relative to the other body measurements. Such an analysis corroborates Lindroth's intuitive analysis of prothoracic shape for this species (Lindroth 1966, 627, figs. 317a,b).

Could the distinct groupings distinguished by function 1 in both the male and female biometric data correspond to distinct species? Two arguments can be used to counter such a grouping derived from humeral width and elytral length. First, there is significant overlap of individual scores for the two groups on d.f. 1 (figs. 47a,b), caused by variable elytral configuration. In humeral development, differences may signify populational differences in metathoracic development, with narrower humeri signifying reduction of flight capability (Darlington 1936). Comparison of average relative flight wing lengths (table 21) shows that populations 27, 29, 30, and 31 do not possess significantly shorter wings than the other 10 populations (based on Wilcoxon rank-sum test on sample averages, p=.227 for males, and p=.094 for females). It is possible that differential flight capability is correlated with humeral width but not relative wing length. In general, flight reduction in this species group has been accomplished through correlated reduction in relative wing length and metathoracic development (see analysis of dispersal capabilities), but a population level analysis within *A. extensicolle* could produce a more detailed picture.

A second argument against recognition of two entities derives from the geographic distribution of the samples. The positive scoring samples—Aguascalientes (27), Flagstaff (29), Gila River (30), and Limpia Canyon (31)—are all in the southern part of the range of *A. extensicolle*. Other samples from this area possess more well-developed humeri: Rio Chico, Durango (26); Tucson, Arizona (28), and Kerrville, Texas (32). These populations are biometrically similar to those in the remainder of the distributional range. The hypothesis that two broadly overlapping species possess identical states for known qualitative characters but differ in elytral configuration is not tenable. A more tenable hypothesis for the available data is that populational differences in elytral configuration exist within a variable species. The amount of genetic heterogeneity (measured by Wright's F_{ST} index), and the very low percentage of presumed flight captures of *A. extensicolle* are compatible with the interpretation that this species has restricted among-population dispersal.

Color Analysis. The large number of synonyms for *A. extensicolle* can be partially explained by color variation across the range of the species: three general color forms have consistently been recognized. The far western populations in Arizona and Mexico possess a brilliant purplish blue dorsal body surface. This form has been considered *A. cyanescens* Motschulsky (Casey 1920). The populations in central North America from Colorado to eastern Ohio are comprised principally of individuals with a vivid metallic green upper body surface, and are representative of the name *A. viride* LeConte (Lindroth 1966). The populations from the eastern United States, New England, and eastern Canada consist of individuals with a bronzed upper body surface, considered representative of *A. extensicolle.*

In order to study the nature of variation in color, samples of a number of populations have been assessed for head, pronotal, and elytral color. The purpose of this analysis is twofold: (1) to assess whether the color variation occurs clinally over space, or whether the variation occurs in discrete steps; and (2) to ascertain whether the colors of the various body parts vary in a correlated manner, or are independently determined.

Because this analysis is strictly concerned with color transition, relative change over distance is considered. A color scale is presented, with endpoints represented by the most bronzed individuals of the Nova Scotia sample and the most brilliant purplish blue individuals of the Mexican populations. The color of the elytra varies most, so its range is used as the standard. The color transition is coded as a five-state character, one specimen representing each state used in comparison with all analyzed individuals. The five individuals used as color standards are held as vouchers in the Essig Museum of Entomology, University of California, Berkeley.

The standardized specimens were compared to color chips in Smithe (1974), but since such color schemes cannot account for the metallic sheen present in the integument of these beetles, the colors are best described in vernacular terms. To compare specimens with the described colors, they should be viewed at 30X under strong incandescent light, 45° incident to the vertical.

The color states are:

State 0: isodiametric facets of microsculpture aeneous, with golden reflections of incident light;
State 1: facets of microsculpture with golden reflections of incident light and greenish reflections between facets;
State 2: facets reflect brilliant metallic green, and faint golden reflection from between facets;
State 3: facets reflect royal blue, with purple reflections between facets;
State 4: facets reflect purple, with faint blue reflections between facets.

The color of the vertex of the head, pronotum, and elytra is presented for individuals from 20 population samples (table 27). The within-sample average color index for all 20 population samples is presented in figure 48. Elytral color varies the most across the species range, with index averages ranging from 3.86 (Tucson) to 0.00 (Nova Scotia). Pronotal color varies less, with index averages ranging from 3.61 (Tucson) to 0.85 (Nova Scotia). The color of the vertex is much more uniform, with the index averaging from 2.94 (Aguascalientes) to 1.15 (Nova Scotia).

To ascertain whether the transition in color is clinal or results from discrete areas of uniform color being punctuated by areas of abrupt color change, the localities are assigned ordinates on a line following a great circle drawn from the Nova Scotia sample site to Isla Santa Magarita, Baja California Sur (fig. 48). Population samples are placed on the ordination line at a point corresponding to a line passing through the sample perpendicular to the ordination line. This ordination approximates the longest axis of the range of *A. extensicolle*, placing nine populations to the north side of the line and ten to the south. The line also transects the centers of the ranges of the three typological color entities usually recognized within the species. Elytral color index (y) was regressed against geographic distance (x in units of 100 miles) along the great circle

Table 27. Number of individuals in each color category for head, pronotal, and elytral color, and average population value on color scale.

		Vertex color						Pronotal color						Elytral color					
Locality	*n*	0	1	2	3	4	avg.	0	1	2	3	4	avg.	0	1	2	3	4	avg.
Durango, Mexico	25	0	0	9	11	5	2.84	0	0	3	13	9	3.24	0	0	1	12	12	3.44
Aguascalientes, Mexico	17	0	0	3	12	2	2.94	0	0	0	10	7	3.40	0	0	0	6	11	3.65
Tucson, AZ	28	0	0	6	17	5	2.61	0	0	3	5	20	3.61	0	0	1	2	25	3.86
Flagstaff, AZ	18	0	0	7	11	0	2.61	0	0	4	13	1	2.83	0	0	5	11	2	2.83
Gila River, NM	23	0	0	14	9	0	2.39	0	0	5	10	8	3.13	0	0	1	12	10	3.39
Davis Mts., TX	31	0	0	18	13	0	2.42	0	0	12	14	5	2.71	0	0	3	19	9	3.19
Kerrville, TX	26	0	2	16	8	0	2.23	0	0	16	10	0	2.51	0	0	11	13	2	2.65
Boulder, CO	30	0	3	21	6	0	2.10	0	5	19	6	0	2.03	0	2	20	7	1	2.17
Fayetteville, AR	20	0	0	16	4	0	2.20	0	0	12	4	4	2.60	0	0	8	6	6	2.90
Omaha, NB	27	0	2	24	1	0	1.96	0	1	18	8	0	2.26	0	0	9	14	4	2.81
Talladega N.F., AL	28	0	0	22	6	0	2.21	0	0	19	6	1	2.07	0	0	9	15	2	2.46
Moorhead, MN	26	0	5	21	0	0	1.81	1	4	21	0	0	1.77	0	1	23	2	0	2.04
Lk. Winnipeg, Canada	28	3	11	13	1	0	1.43	2	12	12	2	0	1.50	1	5	18	4	0	1.93
Clemson, SC	22	0	0	10	12	0	2.54	0	0	7	15	0	2.68	0	3	7	11	0	2.23
Newaygo, MI	69	0	3	65	1	0	1.97	1	5	60	3	0	1.94	1	8	56	4	0	1.91
Clarke Co., VA	21	0	0	10	10	1	2.57	0	1	9	9	2	2.57	0	14	5	2	0	1.43
Ithaca, NY	45	0	7	26	2	0	1.22	2	11	29	3	0	1.73	26	16	3	0	0	0.49
Ottawa, Canada	27	0	7	20	0	0	1.74	0	5	20	2	0	1.89	19	7	1	0	0	0.33
Windsor Co., VT	26	0	1	25	0	0	1.96	0	2	20	4	0	2.08	11	14	1	0	0	0.61
Nova Scotia, Canada	16	0	5	11	0	0	1.15	3	9	4	0	0	0.85	16	0	0	0	0	0.00

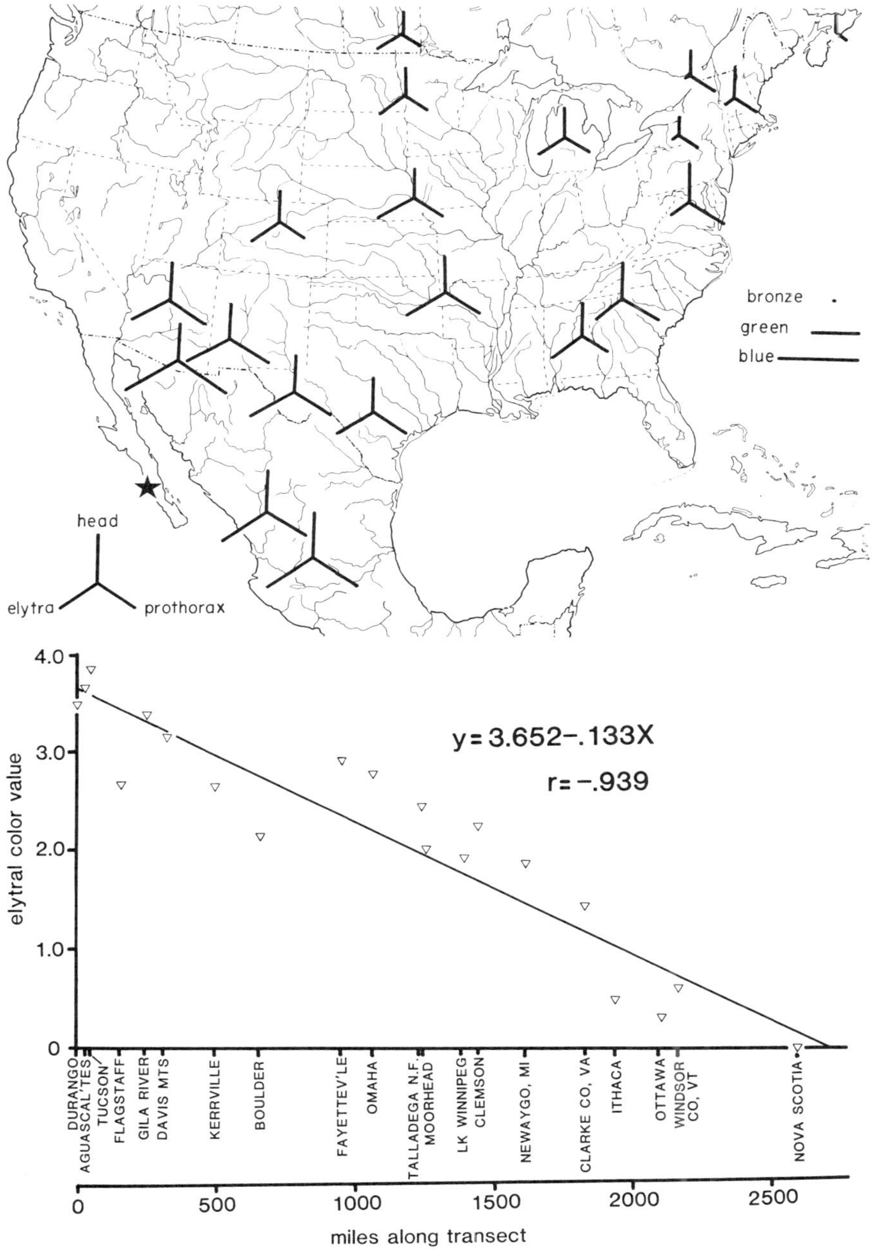

Figure 48. Results of color analysis of *A. extensicolle*. Upper shows color index values presented in table 27. Nova Scotia population represents northern end of transect, southern end denoted by star. Lower shows regression of elytral color value against geographic distance along transect. Unit distance is 100 miles.

transect, with the resultant equation, $y = 3.652 - .133x$, accounting for 88.2% ($r^2 = .8817$) of the variance in color. Deviations from this linear regression occur, and should be expected owing to the great distance of some localities from the line of ordination. Even though the overall change is linear, some major steps in color index value exist. The most noticeable is the change between Newaygo, Michigan and Clark County, Virginia versus Ithaca, New York. The Ithaca and Ottawa samples are much more bronzed than those to the west and south, whereas the Windsor County, Vermont sample lies near the regression line.

Vertex and pronotal color vary in step with elytral color, with the linear regressions for the color indices as follows: for vertex color, $y = 2.647 - .0466x$, $r = -.7494$; for pronotal color, $y = 3.122 - .0698x$, $r = -.8161$. Again, x is measured in units of 100 miles. In all cases, a linear regression explains much of the variance in color: 56.2% for head color , 66.6% for pronotal color, and 88.2% for elytral color. Hence color variation exhibits a clinal transition across the species range. Recognition of this phenomenon is best achieved taxonomically by recognition of *A. extensicolle* as a variable species.

Any attempt to formalize geographic variation in *A. extensicolle* using the subspecific rank should be avoided. Such an action would only serve to cloak the within-species variability in a guise of correlated change of characters. Based on comparison of the discriminant and color analyses, characters change independently in an orderly manner, but not in a correlated manner.

Agonum decorum (Say)

(figures 29m,n; 31a; 32i; 33e-i; 34d; 36c; 49i,j; 50; 51-54)

Feronia decora Say, 1823, *Trans. Amer. Philos. Soc.* 2(n.s.):53.
Anchomenus decorus; Dejean, 1828, *Spécies général Coléoptères* 3:115; Say, 1834, *Trans. Amer. Philos. Soc.* 4(n.s.):421; LeConte, 1848, *Ann. Lyc. Nat. Hist. New York* 4:223; Bates, 1884, *Biol. Cent.-Amer. Coleop* 1(4):280; Blatchley, 1910, *Coleoptera of Indiana*, p. 126.
Platynus decorus; Leconte, 1854, *Proc. Acad. Nat. Sci. Phila.* 7:46.
Anchomenus (Pseudanchus) decorus; Casey, 1920, *Memoirs on the Coleoptera* 9:68.
Agonum (Anchomenus) decorum; Csiki, 1931, *Coleopterorum Catalogus* 115:860.
Agonum (Stictanchus) decorum; Lindroth, 1966, *Opusc. Entomol. Suppl.* 29:629; Liebherr, 1983, *Ann. Entomol. Soc. Amer.* 76:349-358.
Anchomenus californicus Dejean, 1828, *Spécies général Coléoptères* 3:127; Mannerheim, 1843, *Bull. Soc. Imp. Nat. Moscou* 16:197 (NEW SYNONYMY).
Platynus californicus; LeConte, 1854, *Proc. Acad. Nat. Sci. Phila.* 7:47.
Anchomenus (Pseudanchus) californicus; Casey, 1920, *Memoirs on the Coleoptera* 9:61.
Agonum (Anchomenus) californicum; Csiki, 1931, *Coleopterorum Catalogus* 115:859.
Agonum (Stictanchus) californicum; Gray and Hatch, 1941, *Univ. Wash. Publ. Biol.* 10:17; Lindroth, 1966, *Opusc. Entomol. Suppl.* 29:627.
Anchomenus thoracicus Dejean, 1828, *Spécies général Coléoptères* 3:114 (NEW SYNONYMY).
Agonum (Stictanchus) thoracicum; Lindroth, 1966, *Opusc. Entomol. Suppl.* 29:629.
Anchomenus obscurus LeConte, 1848, *Ann. Lyc. Nat. Hist. New York* 4:223; LeConte, 1854, *Proc. Acad. Nat. Sci. Phila.* 7:46 (synonymy).
Platynus simplex LeConte, 1854, *Proc. Acad. Nat. Sci. Phila.* 7:46 (NEW SYNONYMY).
Anchomenus charmis Bates, 1884, *Biol. Cent.-Amer. Coleop.* 1(4):280 (NEW SYNONYMY).

Agonum (Anchomenus) charmis; Csiki, 1931, *Coleopterorum Catalogus* 115:862.
Platynus hornii Hausen, 1890, *Can. Rec. Science* 4:235; Lindroth, 1966, *Opusc. Entomol. Suppl.* 29:629 (synonymy).
Platynus (Anchomenus) testaceonotus Hausen, 1891, *Natur. Can.* 20:162.
Agonum (Anchomenus) decorum testaceonotum; Csiki, 1931, *Coleopterorum Catalogus* 115:860 (synonymy).
Anchomenus (Pseudanchus) solutus Casey, 1920, *Memoirs on the Coleoptera* 9:60 (NEW SYNONYMY).
Agonum (Anchomenus) solutum; Csiki, 1931, *Coleopterorum Catalogus* 115:862.
Anchomenus (Pseudanchus) impictus Casey, 1920, *Memoirs on the Coleoptera* 9:60 (NEW SYNONYMY).
Agonum (Anchomenus) impictum; Csiki, 1931, *Coleopterorum Catalogus* 115:861.
Anchomenus (Pseudanchus) irruptus Casey, 1920, *Memoirs on the Coleoptera* 9:60 (NEW SYNONYMY).
Agonum (Anchomenus) irruptum; Csiki, 1931, *Coleopterorum Catalogus* 115:861.
Anchomenus (Pseudanchus) vinnulus Casey, 1920, *Memoirs on the Coleoptera* 9:61 (NEW SYNONYMY).
Agonum (Anchomenus) vinnulum; Csiki, 1931, *Coleopterorum Catalogus* 115:862.
Anchomenus (Pseudanchus) luxatus Casey, 1920, *Memoirs on the Coleoptera* 9:67; Lindroth, 1966, *Opusc. Entomol. Suppl.* 29:629 (synonymy).
Agonum (Anchomenus) luxatum; Csiki, 1931, *Coleopterorum Catalogus* 115:861.
Anchomenus (Pseudanchus) decorus arenarius Casey, 1920, *Memoirs on the Coleoptera* 9:68; Lindroth, 1966, *Opusc. Entomol. Suppl.* 29:629 (synonymy).
Agonum (Anchomenus) decorum arenarium; Csiki, 1931, *Coleopterorum Catalogus* 115:860.
Anchomenus (Pseudanchus) tepidus Casey, 1920, *Memoirs on the Coleoptera* 9:68; Lindroth, 1966, *Opusc. Entomol. Suppl.* 29:629 (synonymy).
Agonum (Anchomenus) tepidum; Csiki, 1931, *Coleopterorum Catalogus* 115:862.
Anchomenus uteanus Casey, 1924, *Memoirs on the Coleoptera* 11:81 (NEW SYNONYMY).
Agonum (Anchomenus) uteanum; Csiki, 1931, *Coleopterorum Catalogus* 115:862.
Anchomenus decorus syracusensis Hatch, 1926, *J. New York Entomol. Soc.* 34:247.
Agonum decorum syracusense; Csiki, 1931, *Coleopterorum Catalogus* 115:860.
Agonum (Anchomenus) extensicolle cubanum Darlington, 1934, *Psyche* 41:97 (NEW SYNONYMY).

Types. of *decorum*, male neotype, type locality Arlington, MA (Lindroth 1966), neotype designation (Lindroth and Freitag 1969) (MCZ)!; of *californicum*, female holotype, California (MNHP, examined by Lindroth [1955]); of *thoracicum*, female holotype, type locality Utah (Lindroth 1966) (MNHP, examined by Lindroth [1955]); of *obscurum*, female holotype, Lake Onondaga, NY (MCZ)!; of *simplex*, male holotype, Colorado River, CA (MCZ)!; of *charme*, female holotype, Mexico City (MNHP)!; of *hornii*, holotype, Canada (MCZ)!; of *testaceonotum*, Ste. Rose, Quebec (type apparently lost); of *solutum*, female holotype, Reno, NV (NMNH)!; of *impictum*, male holotype, San Joaquin Co., CA (NMNH)!; of *irruptum*, male holotype, Priest River, ID (NMNH)!; of *vinnulum*, male holotype, Battle Mountain, NV (NMNH)!; of *luxatum*, male and female syntypes, Utah (NMNH)!; of *arenarium*, male holotype, Galveston, TX (NMNH)!; of *tepidum*, female holotype, Tucson, AZ (NMNH)!; of *uteanum*, male and female syntypes, Callao, UT (NMNH)!; of *syracusense*, male holotype, Onondaga Co., NY (Hatch Coll., ORSU); of *cubanum*, male holotype, Soledad, Cienfuegos, Cuba (MCZ)!.

Diagnosis. A polymorphic species consisting of 4 morphs: (1) with rufo-testaceous prothorax, testaceous legs, and upper body surface covered with a pelage of fine setae; (2) with rufo-testaceous prothorax, testaceous legs, and body surface glabrous except for fixed macrosetae; (3) with piceous to metallic green prothorax, piceous legs, and glabrous body surface except for fixed macrosetae; (4) with piceous to metallic green prothorax, piceous legs, and upper body surface covered with a pelage of fine setae. Pronotum quadrate, maximum width slightly greater than median length, lateral margin sinuate before rounded hind angles, latero-basal depressions sparsely punctate, basal median area impunctate. The sister species, *A. elongatulum*, is very similar morphologically, although *A. decorum* is smaller on the average, with lateral pronotal margin more sinuate before hind angles.

Description. *Head.* Slender, moderately elongate with moderately protruding convex eyes; temple expanding moderately abruptly to meet hind margin of eye (fig. 31a); antenna with second segment bearing apical ring of 6-8 setae in glabrous morphs, 11-12 setae in hirsute morphs (figs. 29m,n); third segment usually glabrous or with a limited number of setae at apex plus an apical ring of larger setae in glabrous morphs, fully pubescent in hirsute morphs; segments 4-11 pubescent.

Thorax. Pronotum quadrate with base constricted; lateral margin sinuate before rounded to well-developed hind angle (figs. 32i, 33e-i); lateral depression narrow, widening at basal 1/3 to slope into latero-basal depression; basal margin curved outward behind basal seta, basal bead an indistinct raised border laterally, visible to poorly developed medially; latero-basal depression finely impressed, linear, becoming slightly wider and shallower a short distance forward, then becoming a shallow broad impression extending as far forward as lateral seta; area outside basal impression flat to slightly convex, sometimes faintly punctate (fig. 33e); median basal area impunctate; median longitudinal impression clearly impressed, sometimes associated with transverse wrinkles (figs. 33h,i); anterior transverse impression variable, median impression visible (figs. 33e,g) to almost absent (fig. 33h).

Elytra. Parallel-sided with well-developed humeri, widest between apical 1/2 and 1/3, apical sinuation moderately well-developed (fig. 34d); elytral intervals flat to slightly convex; striae finely impressed, punctulae visible in striae at 80X; striae 1-5 strongly joining basal margin; basal margin at most slightly angulate at shoulder, usually rounded; 4-8 setae set in punctures in or adjacent to third elytral interval.

Secondary Setation. Hirsute morph characterized by pelage of fine setae on dorsal surface of head, prothorax, and elytra; increased antennal pubescence associated with dorsal pelage; ventral body surface and elytral epipleura essentially glabrous except for fixed macrosetae, single setae, or groups of fine setae occasionally present; legs as in glabrous morph.

Male Genitalia. Aedeagus with shaft evenly arcuate, tip upcurved, flattened area inside curve of shaft in larger individuals (figs. 49i,j; 50d-g), longitudinal wrinkles inside curve of shaft, extending a variable distance toward ventral membrane; aedeagal sac as long as apical portion of shaft (fig. 50d), covered with fine spicules; lateral areas of sac slightly more sclerotized, indicated by deeper yellow in cleared specimens.

Female Reproductive Tract. Basal segment of female gonostylus with apical fringe of 11-17 setae (figs. 50a,b); spermatheca long and fusiform; spermathecal gland entering spermathecal duct proximad of basal curve of spermatheca (fig. 50c).

Color. Polymorphic. *Red morph*: dorsum of head and gena metallic green with bluish reflection; gula rufo-piceous; mandibles rufo-piceous; maxilla, labium, and palps rufo-testaceous, apical segments of palps slightly darker; antenna with scape rufo-testaceous, segments 2 and 3 progressively darker, segments 4-11 rufo-piceous; pronotum rufous to rufo-testaceous; scutellum and basal margin of elytra rufo-testaceous; elytral disc dark, piceous with metallic reflection varying from metallic green to aeneous; elytral epipleura rufo-testaceous with infuscated cast; mesosternum and mesepisternum rufous; mesepimeron, metepisternum, and metepimeron piceous; metasternum and metacoxae rufo-piceous; abdominal sternites rufo-piceous to piceous; coxa of fore- and meso-legs, femora, tibiae, and tarsi tarsi somewhat infuscated, especially toward apices. *Dark or green morph*: dorsum of head, pronotum, and elytra metallic green with a blue to aeneous reflection; mouthparts rufo-piceous, apical segments of palps darker, antennae with scape rufo-piceous, segments 2-11 piceous; prosternum and proepisternum rufo-piceous, meso- and metasternum the same or slightly darker; elytral epipleura rufo-piceous with slight metallic reflection; legs with coxae and trochanters lightest, infuscated rufo-testaceous; femora more heavily infuscated; tibiae and tarsi darker, rufo-piceous.

Microsculpture. Head and pronotum with uniform isodiametric microsculpture, sometimes stretched on center of vertex and pronotum forming a poorly developed transverse mesh; elytra with uniform isodiametric microsculpture strongly developed, creating a granulate appearance; ventral body surface with isodiametric microsculpture, slightly stretched in some areas; lateral areas of abdominal sternites with moderately developed longitudinal wrinkles.

Length. 6.9-9.3mm.

Distribution. Throughout North America from southern Canada to the Isthmus of Tehuantepec, and on the Greater Antilles. Green-glabrous morph restricted to west of the continental divide from Vancouver to Sinaloa, in southern Mexico, and on Cuba, Jamaica, Hispaniola and Grand Cayman Island (fig. 51). Recorded from polymorphic populations in Utah, Colorado, Arizona and Texas. Green-hirsute morph found in two disjunct areas—from Lake Champlain to the vicinity of Ottawa, and throughout the southwest in polymorphic populations (fig. 52). Red-glabrous morph known only from the Lake Bonneville drainage of Utah, in Death Valley, California, and in polymorphic populations in Arizona and Texas (fig. 53). Red-hirsute morph generally distributed east of the Rocky Mountains and Sierra Madre Occidental, from Durango to Manitoba, to Florida and southern Quebec (fig. 54).

Distributional Records. 4,629 specimens. Data on polymorphic populations are presented elsewhere (Liebherr 1983a). Mexican localities follow. All specimens green-glabrous morph except where noted: MEXICO: State unknown: Ayutla (1, MCZ). BAJA CAL. NORTE: 30-III-1947 (1, FMNH); Ojos Negros, 11-V-1951 (1, CNC); Bajada, 3 mi. E Ojos Negros, 10-IX-1961, at light (1, CAS). CHIHUAHUA: 5.5 mi. W Madera, 2270′ el., 30-VIII-1974 (1 red-hirsute, UASM); DISTRITO FEDERALE: Guadelupe (1, MCZ); Mexico (1, BMNH), 12-V-1949 (1, MSU). DURANGO: Durango, 27-28-VI-1964 (1 red-hirsute, NMNH); Graceros, 7-VII-1961, 6000′ el. (1, MSUC); Ventanas, 2000′ el. (2, BMNH). HIDALGO: Guadelupe, 10-VII-1932 (11, MCZ). JALISCO: Tecolotlan, 18-IX-1975 (2, UCD). MEXICO: Atlocomulco, 2500 m el., 1-X-1948 (2, MICH); Toluca (1, BMNH). MORELOS: Cuernavaca (2, MCZ), VI (1, MCZ), XI (1, MCZ). OAXACA: Oaxaca (1, BMNH). PUEBLA: Puebla (1, BMNH). SINALOA: 54 mi. S Culiacan, 540′ el., 23-IV-1969, U.V. light (1, UCR); Escuinapa, 3-IX-1973, lights (2, CDFA); La Noria (2, BMNH); Mazatlan, Mazatlan R., 10-I-1889 (1, MCZ); Mazatlan, 17-23-VII-1963, at black light (9, NMNH); Presidio (Mazatlan) (2, BMNH); Rosario, 20-III-1918 (11, CAS). TABASCO: San Juan Bautista (Villahermosa) (2, BMNH). VERA CRUZ: Cotaxtla exp. sta., 24-VI-1958, light trap (1, CNC); Jalapa (7, BMNH; 1, MCZ).

Biology and Habits. *A. decorum* is found in marshy areas with a well-developed organic mat of vegetation or detritus. In the southeastern United States, it is usually found near areas shaded

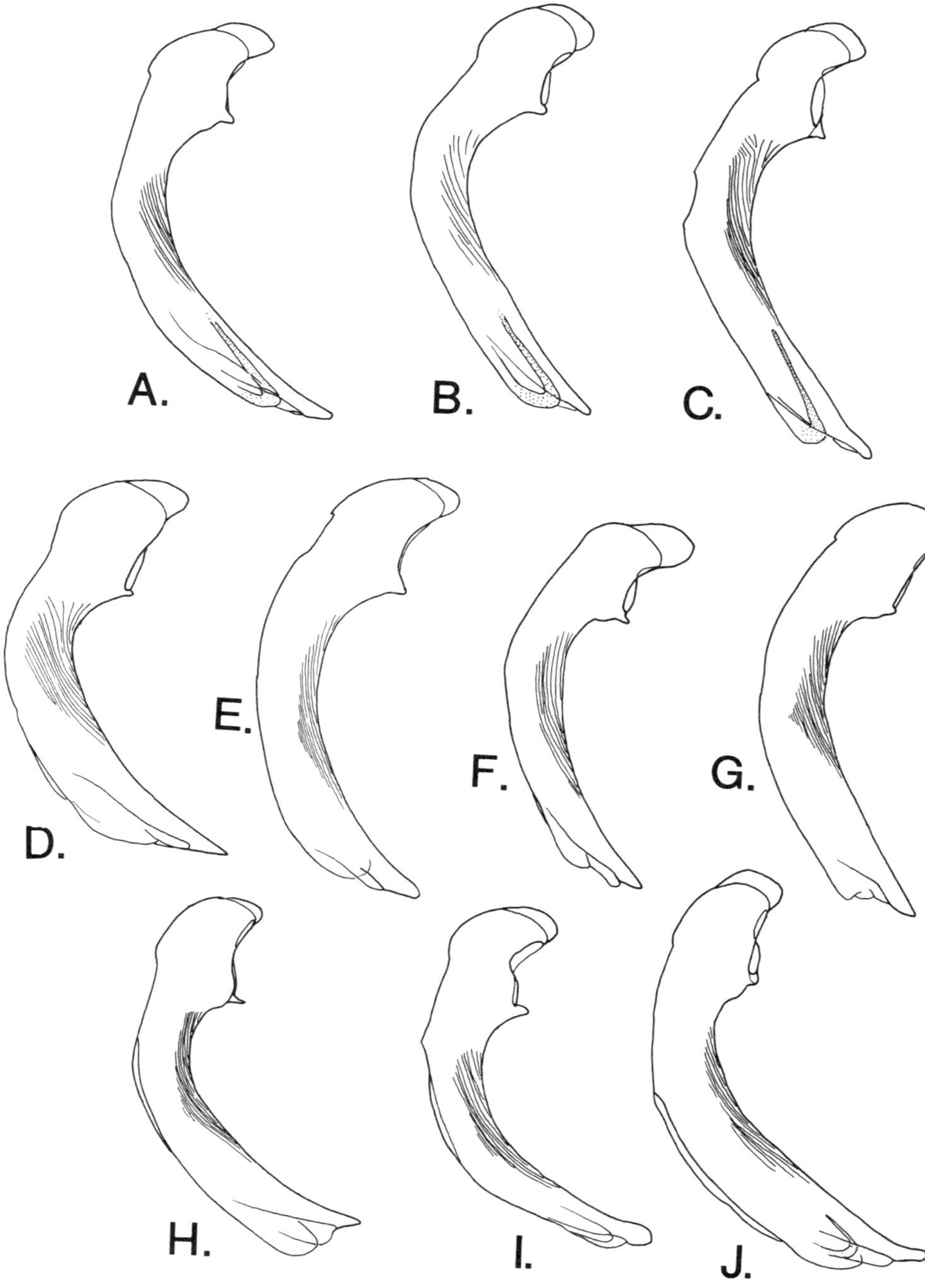

Figure 49. Aedeagus, ventral view. A–C. *A. texanum*: A. MX:Chiapas, 5 km E. Comitan; B. TX:Sinton, Welder Wildlife Found.; C. AZ:Cochise Co., Portal. D–G. *A. extensicolle:* D. MX:Aguascalientes, Aguascalientes; E. MT:Yellowstone N. P.; F. NC:Black Mts.; G. NJ:Morris Plains. H. *A. elongatulum,* FL:Brooksville. I–J. *A. decorum:* I. MX:Jalisco, Tecolotlan; J. FL:Marion Co., Juniper Spgs.

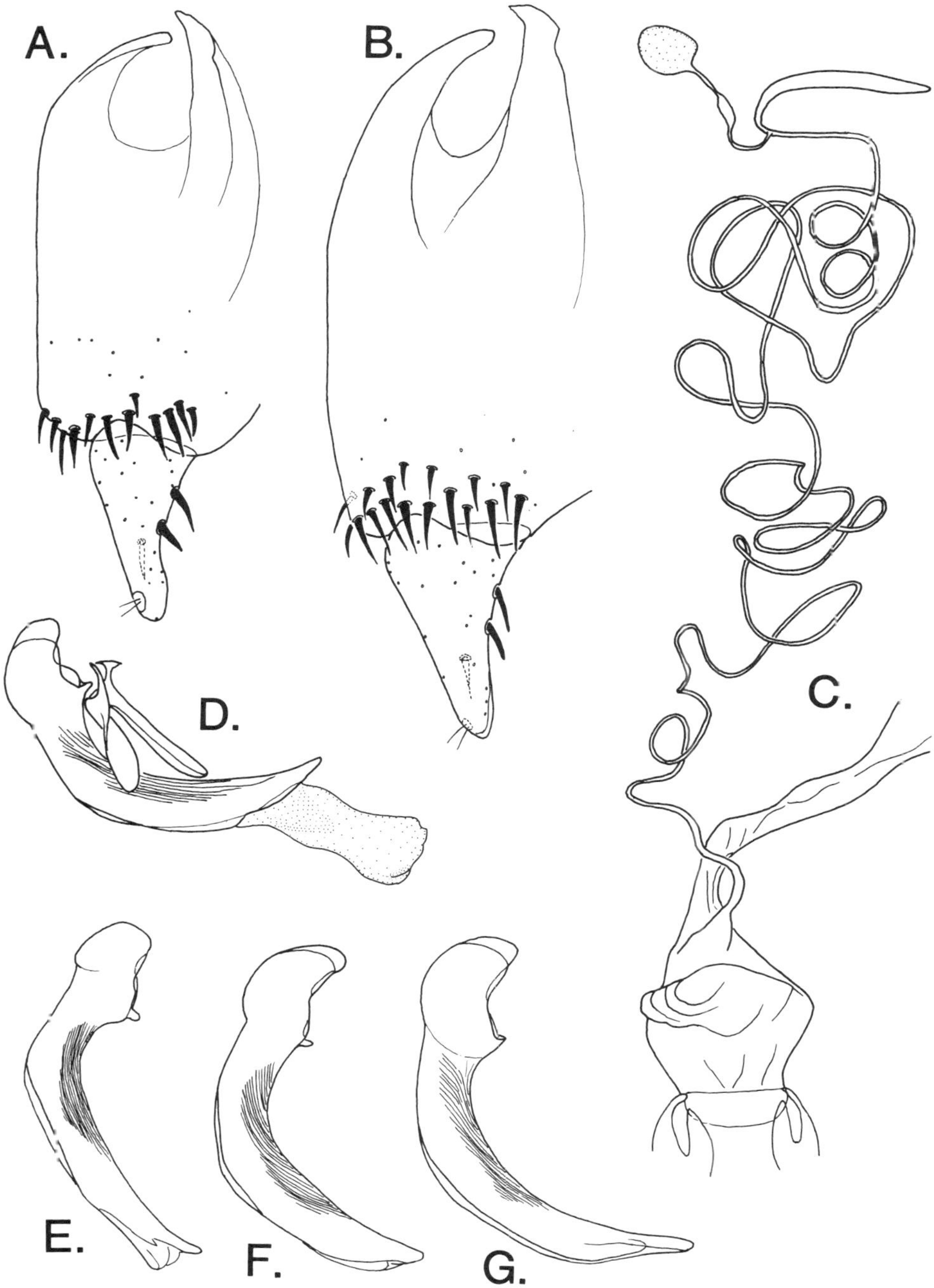

Figure 50. Internal and external reproductive structures of *A. decorum*. Conventions as in figure 38. A–B. Left gonostylus, ventral view: A. CA:Sta. Clara Co., Lk. Lagunitas; B. Cuba:Cienfuegos, Soledad. C. Female internal reproductive tract, ventral view, CA:Sta. Clara Co., Lake Lagunitas. D–G. Aedeagus, ventral view: D. CA:Sta. Clara Co., Lk. Lagunitas; E. AZ:Gila Co., Payson Cyn.; F. NV:Elko Co., Humboldt R.; G. AZ:Sta. Cruz Co., Peña Blanca.

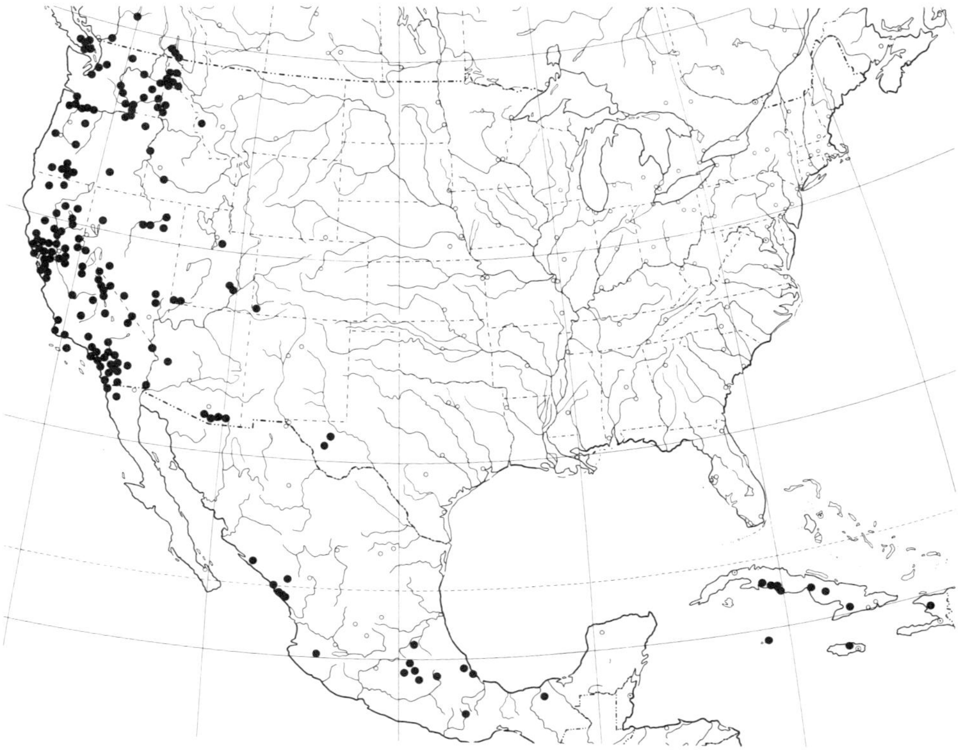

Figure 51. Distributional records of green-glabrous form of *A. decorum* (Say).

by vegetation. In the Great Basin and interior-draining valleys of eastern California, it occurs around springs, caused by the fault-block thrusting occurring with orogenic activity. The upward thrusting exposes aquifers, creating relatively stable though limited marshy habitats. Most of the habitats in which *A. decorum* has been collected are cool, fresh-water springs; however, this species is not very fastidious regarding water quality. At Carlin, Nevada it was collected in a roadside ditch near thermal springs that vented water hotter than 49°C. Beetles were numerous in the marshy habitats surrounding Crystal Springs, near Hiko, Lincoln County, Nevada, which vents water at 27°C. The pH of the water also has little effect on beetle presence, with pH of habitats containing abundant *A. decorum* ranging from 7.5-8.0 (Crystal Springs) to 10.0 (Little Lake, Inyo County, California). *A. decorum* is found in habitats over a broad altitudinal range in the Great Basin. In California it is found in springs from -125 feet elevation (Saratoga Springs, Death Valley, San Bernardino County) to 5,200 feet (Buckhorn Springs, Deep Springs Valley, Inyo County). At the Franklin River, east of the Ruby Mountains, Elko County, Nevada, elevation 6,200 feet, it was taken in riparian habitat associated with *Agonum subsericeum* and *Agonum brevicolle*. Somewhat higher at Comins Lake, White Pine Co., NV, elevation 6,600 feet, only *A. subsericeum* and *A. brevicolle* were taken, though the habitat appeared suitable for *A. decorum*.

Overwintering occurs away from the water's edge, as in *A. extensicolle*. Buckhorn Springs,

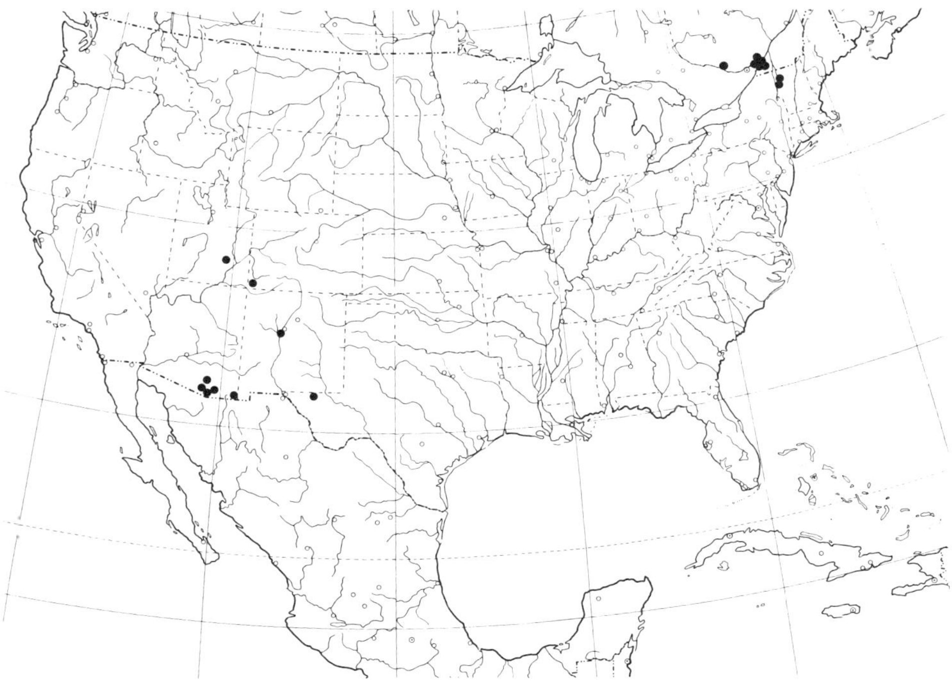

Figure 52. Distributional records of green-hirsute form of *A. decorum* (Say).

Deep Springs Valley, Inyo County, California was visited 15-XII-1979. The spring issues at the bottom of a talus slope. Near the water's edge there were no *A. decorum*, but approximately 200 individuals were taken just above the wet soil near one of the spring outlets, in a soil mixture of granitic gravel and sandy loam. The total area excavated was about 30 X 30 X 30 cm, with suitable microhabitat for overwintering extending beyond the area excavated.

Numerous larvae were reared for study of the infraspecific polymorphisms in *A. decorum* (Liebherr 1983a). The duration from egg hatch to adult emergence averaged 31-34.5 days in the various F_1 broods cited in that paper, under 16:8 L:D and approximately 20-22°C, fed *Tenebrio* chunks every two days. Eggs take up to six days to hatch. The first larval instar typically lasts 4-7 days depending on whether the larva feeds promptly. Second instar duration lasts 3-10 days, and the third instar 7-13 days. Much of the variation in duration is based on larval feeding, with voracious larvae developing more quickly. Pupal stage duration is also variable, lasting 3-8 days. Based on 117 successfully reared adults, males do not develop any faster than females.

The F_1 adults were set up in pairs for F_2 crosses while kept under a 16:8 L:D regime and fed cut *Tenebrio* larvae. The first fertile eggs from the adults appeared 5-6 weeks after the F_1 adults emerged, indicating that there is no obligatory reproductive diapause if larvae and adults are held under long day conditions.

Mortality under rearing conditions is concentrated in the first instar. Egg mortality was not estimated because of the inability to ascertain fertilization. Of 155 eggs hatched, 24 larvae (15.5%) died during the first instar. Only 7 second instar larvae (5.3%), 3 third instar (2.4%), and

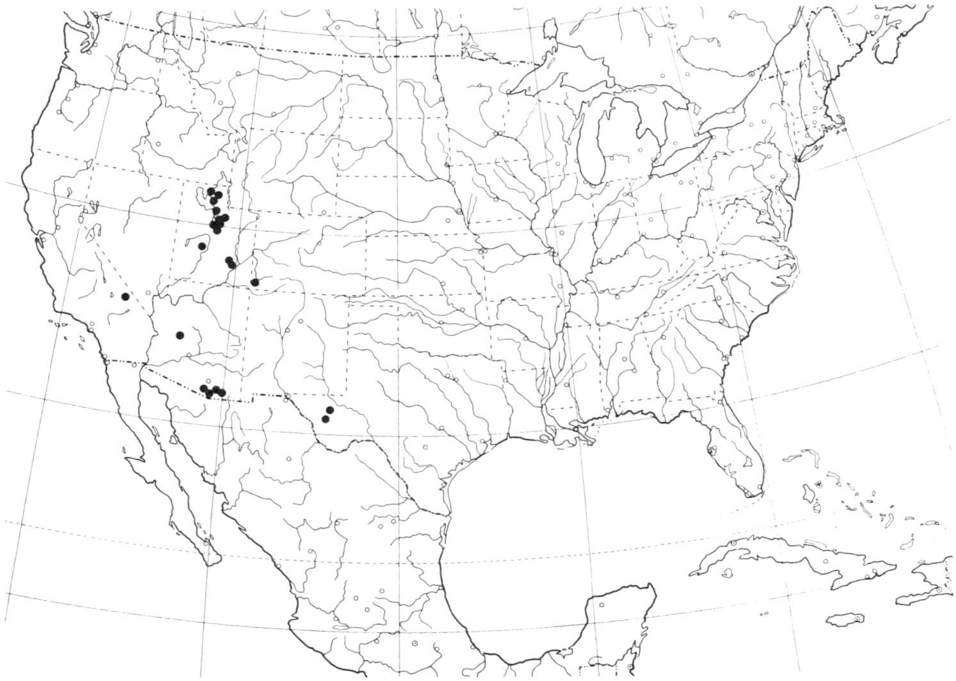

Figure 53. Distributional records of red-glabrous form of *A. decorum* (Say).

one pupa (0.8%) died. The high mortality for first instars is probably due to their inability to tolerate moisture under rearing conditions, resulting in drowning.

Etymology. *A. decorum* can be considered the elegant or decorated *Agonum* (L. *decorus*).

Infraspecific Geographic Variation. The conspicuous polymorphisms of color and setation noted in the description are discussed in Liebherr (1983a). Each polymorphism is governed by one diallelic locus. Based on F_1 and F_2 rearings made with individuals from polymorphic populations in Arizona, it is known that the red state of the color polymorphism is determined by an allele dominant to the green state. For the setational polymorphism, the hirsute state is dominant to glabrous. Polymorphic populations are found throughout southeastern Arizona to west Texas, where the eastern red-hirsute form introgresses with the western green-glabrous form.

The conspicuous polymorphisms indicate major population sets delimited by major geographic barriers. Gene flow, as well as recolonization of glaciated territory since the Wisconsin glaciation, has spread these alleles homogeneously over large regions. The biometric analysis of geographic variation affords another view of the variation within *A. decorum*. Much of the biometric variation follows latitudinal gradients, transcending the color or setational polymorphisms. Whereas the genes for color and setation can serve as markers for gene flow (Liebherr 1983a), the geographic trends in biometric variation most likely illustrate developmental effects due to climate.

The discriminant analysis of *A. decorum* males is based on 18 population samples. Each of

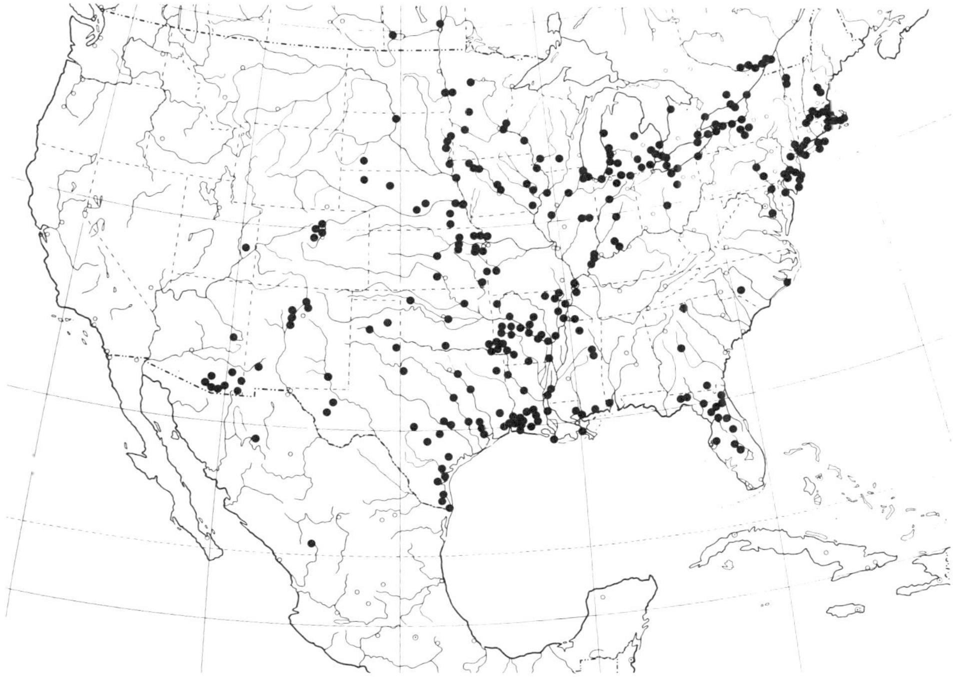

Figure 54. Distributional records of red-hirsute form of *A. decorum* (Say).

these is composed of individuals from a single locality, with the exception of population 15 from Mexico, which is pooled material from the *Biologia Centrali-Americana* (Bates 1882, 1884) (table 5). All 13 characters measured vary significantly among populations based on the univariate F-test (table 28). The first three discriminant functions account for 75.7% of the total variance in the data set, 37.1% being accounted for by function 1, 24.8% by function 2, and 13.8% by function 3. There is significant discrimination of population samples on all three functions (fig. 55). Function 1 produces positive scores in northern and western populations: Utah Lake, UT (6), San Diego, CA (7), Carlin, NV (8) and Chase Lake, WA (11). The populations from more southern localities receive large negative scores: Oktibbeha Co., MS (12); Kenedy Co., TX (13); Cuba (16); Florida (1); Jamaica (18). The characters exhibiting latitudinally correlated clinal variation include head width, fourth antennomere length, profemur length, and profemur width (table 28). The western and northern populations tend to have broader heads relative to other body measurements, with the coefficient of .88 increasing their d.f. score. Fourth antennomere length is shortest in the northern and western populations. Averages range from .43-.52mm in populations 6, 7, 8, and 11 versus .46-.57mm in the other populations. Profemur length tends to be shorter in samples 6, 7, 8, and 11 (average 1.49-1.55mm) compared to the others (average 1.47-1.82mm), although overlap in values does occur. Profemur width is also smaller in the four positive-scoring localities: averages range from .49-.54mm versus .48-.59mm in the other less positive scoring samples.

Discriminant function 2 ordinates samples principally on elytral length, which has a

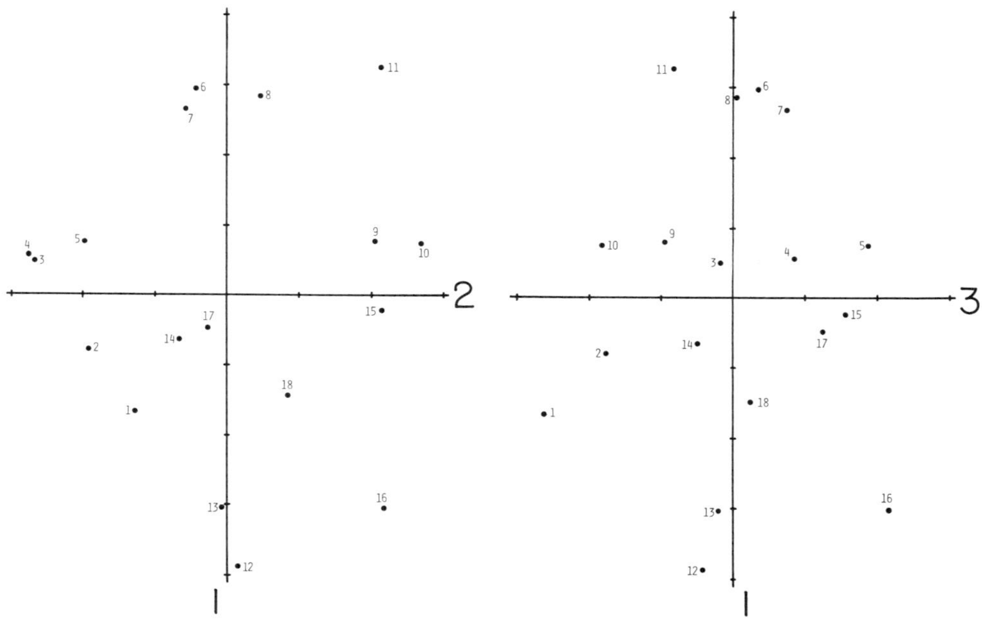

Figure 55. Scatterplots of first versus second, and first versus third discriminant functions based on samples of male *A. decorum.*

Table 28. Wilks' lambda values, univariate F-ratios, and standardized canonical discriminant function coefficients for males of *A. decorum.*

Variable	Wilks'	F*	Coefficients d.f. 1	d.f. 2	d.f. 3
hw	.344	18.05	0.88	–0.80	–0.42
pl	.526	8.53	–0.65	0.27	–0.43
apw	.667	4.74	0.12	0.27	–0.07
mpw	.626	5.66	–0.20	0.50	–0.36
bpw	.639	5.34	0.54	0.28	–0.30
huw	.497	9.59	0.05	0.52	0.88
mew	.551	7.71	0.01	0.26	0.25
el	.438	12.15	0.40	–1.18	0.66
4al	.382	15.30	–0.58	–0.47	–0.17
pfl	.412	13.53	–0.24	0.26	0.35
pfw	.474	10.49	–0.60	–0.12	–0.31
3sl	.623	5.73	0.08	–0.06	0.34
pd	.541	8.05	0.04	–0.14	0.01

* All characters vary significantly at P less than .001.

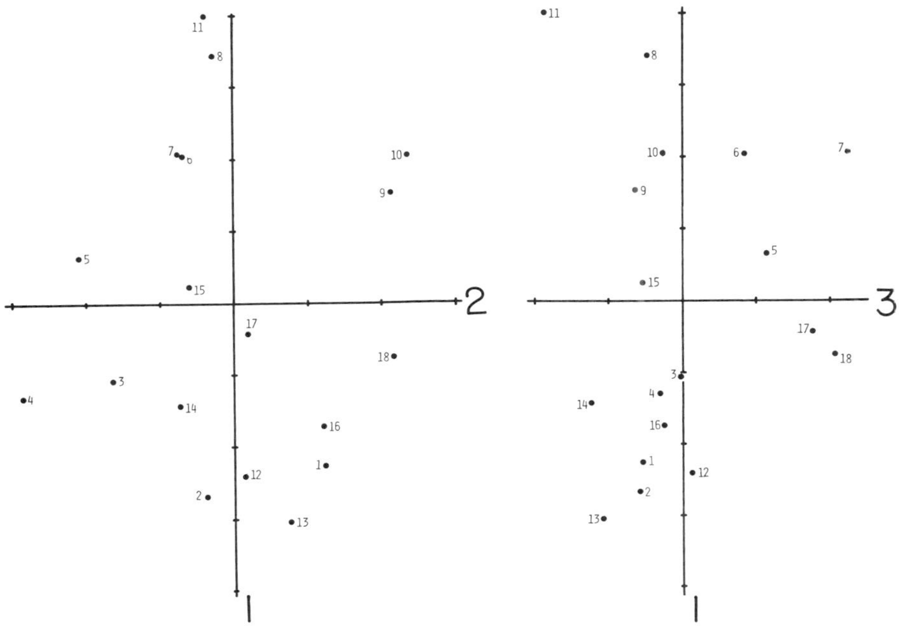

Figure 56. Scatterplots of first versus second, and first versus third discriminant functions based on samples of female *A. decorum*.

coefficient of -1.18. Populations 2-5 have elytral lengths averaging 5.05-5.25mm, whereas populations 9-11 have shorter elytra—averages range from 4.34-4.61mm. Other characters contribute significantly to the score, obscuring the univariate trend in elytral length, but the overall trend is for smaller individuals to score positively on function 2, whereas larger individuals score negatively.

The lack of grouping by geographic locality based on discriminant analysis proves almost as interesting as the concordance of d.f. score with geography. The samples from Diaz Lake, CA (9) and Lake Lagunitas, CA (10) do not receive scores similar to populations 6-8 and 11 on function 1, but they do receive high scores on function 2, like population 11. Populations 6-8 can be characterized as having short appendages, with moderate elytral lengths. Populations 9 and 10 have longer appendages and very short elytra, whereas the Chase Lake, WA population (11) is characterized both by short appendages and short elytra.

The three Antillean populations—Cuba (16), Haiti (17), and Jamaica (18)—are not that similar biometrically. They fall somewhat close to the Mexican sample (15) when all three functions are considered.

The 18 samples of female *A. decorum* prove as variable biometrically as the males. All 13 characters vary significantly among the populations based on univariate F-tests (table 29). The first three discriminant functions (fig. 56) account for 74.1% of the variance in the study set: 42.1% is accounted for by function 1; 20.3% by function 2; 11.7% by function 3.

Function 1 is most clearly interpreted in terms of fourth antennomere length (table 29). Populations 6-11 have shorter antennal segments (averages of .41-.51mm) than those of the

Table 29. Wilks' lambda values, univariate F-ratios, and standardized canonical discriminant function coefficients for females of *A. decorum.*

Variable	Wilks'	F*	Coefficients d.f. 1	d.f. 2	d.f. 3
hw	.562	6.97	0.35	–0.19	0.41
pl	.496	9.09	–0.36	0.86	–0.81
apw	.607	5.79	0.42	0.40	–0.07
mpw	.519	8.28	–0.04	0.45	0.12
bpw	.545	7.48	0.37	0.46	0.64
huw	.547	7.42	0.90	0.02	0.43
mew	.570	6.76	–0.44	–0.72	–0.23
el	.514	8.46	–0.08	–0.92	–0.30
4al	.355	16.21	–0.58	–0.05	0.40
pfl	.384	14.34	–0.50	0.27	1.09
pfw	.488	9.39	–0.29	0.38	–0.41
3sl	.554	7.18	–0.01	–0.37	0.30
pd	.557	7.10	0.02	–0.68	–0.59

* All characters vary significantly at P less than .001.

other populations (.52-.57mm). The trend toward shorter appendage lengths in these same samples is seen in profemur length. With the exception of the San Diego sample (7), in which profemur length averages 1.80mm, samples 6 and 8-11 have average profemur length of from 1.50-1.64mm. The other, more southern and eastern samples have profemur length averages of 1.67-1.86mm. Thus, function 1 appears to represent appendicular length in females just as it does in males.

Function 2 is somewhat refractory to analysis of the importance of various characters in producing d.f. scores. Those characters with large coefficients also have relatively large Wilks' lambda values, indicating large within-group variability (table 29). Populations 9 and 10 (Diaz Lake and Lake Lagunitas, CA) are distinguished from the other Great Basin and Pacific Coast samples. This can be partly explained by their shorter average elytral lengths—4.81mm and 4.78mm versus averages of 4.82-5.45mm in samples 6-8 and 11. As in the male analysis, western populations are distinguished by elytral length in function 2, with smaller individuals scoring higher due to the coefficient of -.92.

As in the males, function 3 has no characters contributing in great measure to the function scores. Inspection of sample scores on all three functions allows assessment of population variability in the various geographic regions. As in the male analysis, the Antillean samples (16-18) are heterogeneous biometrically. The samples from the southern United States—Archbold Bio. Sta., FL (1), Ware Co., GA (2), Oktibbeha Co., MS (12) and Houston, TX (13)—are quite similar biometrically. All possess long appendages; for example, fourth antennomere length averages range from .55-.56mm. Beside long antennae, the average individuals possess very narrow humeri: average humeral width ranges from 1.15 -1.18mm. Other samples possess wider elytra, with average humeral width ranging from 1.16-1.43mm.

The geographic trends common to males and females are decreasing appendicular length from south to north and decreasing elytral length from east to west and northwest. The Washington state and coastal northern California samples are among the smallest beetles in the analysis, based on shorter elytral length. These samples, plus the red-glabrous samples from Utah Lake, UT and the green-glabrous samples from Carlin, NV and San Diego, CA all possess antennae with relatively shorter articles. Shorter antennal length transcends the red-green polymorphism, and appears to be largely correlated with cooler climate.

Agonum elongatulum (Dejean)

(figures 29o; 31b; 33a-d; 34b; 36d; 49h; 57, 58)

Anchomenus elongatulus Dejean, 1828, *Species* général Coléoptères 3:112.
Platynus floridanus LeConte, 1878, *Proc. Amer. Philos. Soc.* 17:374 (NEW SYNONYMY).
Agonum (Anchomenus) floridanum; Csiki, 1931, *Coleopterorum Catalogus* 115:861.
Anchomenus (Pseudanchus) gravidulus Casey, 1920, *Memoirs on the Coleoptera* 9:59 (NEW SYNONYMY).
Agonum (Anchomenus) gravidulus; Csiki, 1931, *Coleopterorum Catalogus* 115:861.
Anchomenus (Pseudanchus) collisus Casey, 1920, *Memoirs on the Coleoptera* 9:59 (NEW SYNONYMY).
Agonum (Anchomenus) collisus; Csiki, 1931, *Coleopterorum Catalogus* 115:859.

Types. of *elongatulum*, male and female syntypes, Amer. sept., (MNHP) male!, type locality restricted to Winter Park, FL by Lindroth (1966); of *floridanum*, female holotype, Capron, FL (MCZ)!; of *gravidulum*, female holotype, Indian River Haulover, FL (NMNH)!; of *collisum*, male holotype, Marion Co., FL (NMNH)!.

Diagnosis. Very similar to the green-glabrous morph of *Agonum decorum*; on the average slightly larger, pronotum broader at base, lateral margin less sinuate; elevated tubercle usually present laterad of latero-basal pronotal depression.

Description. *Head.* Slender, moderately elongate with moderately protruding convex eyes; temple expanded moderately abruptly to meet hind margin of eye (fig. 31b); antenna with second segment bearing apical ring of 6-8 setae (fig. 29o); third segment usually glabrous except for apical ring of setae, occasionally with fine secondary setae in apical half; segments 4-11 pubescent.

Thorax. Pronotum quadrate with base moderately constricted; lateral margin not or barely sinuate before hind angle (figs. 33a-d); lateral depression narrow, widening at basal 1/3 to 1/4 to meet latero-basal depression; basal margin curved outward inside basal seta, basal bead traceable inside latero-basal depression; latero-basal depression linear, moderately impressed, with transverse wrinkles on inner face, widening behind, becoming shallower forward, continuing shallowly to a line connecting lateral setae; area outside latero-basal depression often raised to form a flat round tubercle, the surface sometimes finely punctate (figs. 33b,d); median basal area smooth; median longitudinal impression smooth, lightly impressed in apical half, somewhat more impressed behind middle, often with transverse wrinkles, anterior transverse impression linear, moderately impressed to absent.

Elytra. Parallel-sided to slightly ovoid, humeri well-developed, lateral margin slightly expanded behind middle, apical sinuation moderately well-developed (fig. 34b); elytral intervals flat to slightly convex; striae finely impressed with punctulae visible at 80X; striae 1-5 strongly joining basal margin; basal angle slightly to moderately angulate at shoulder; 4-8 setae in or adjacent to third elytral interval.

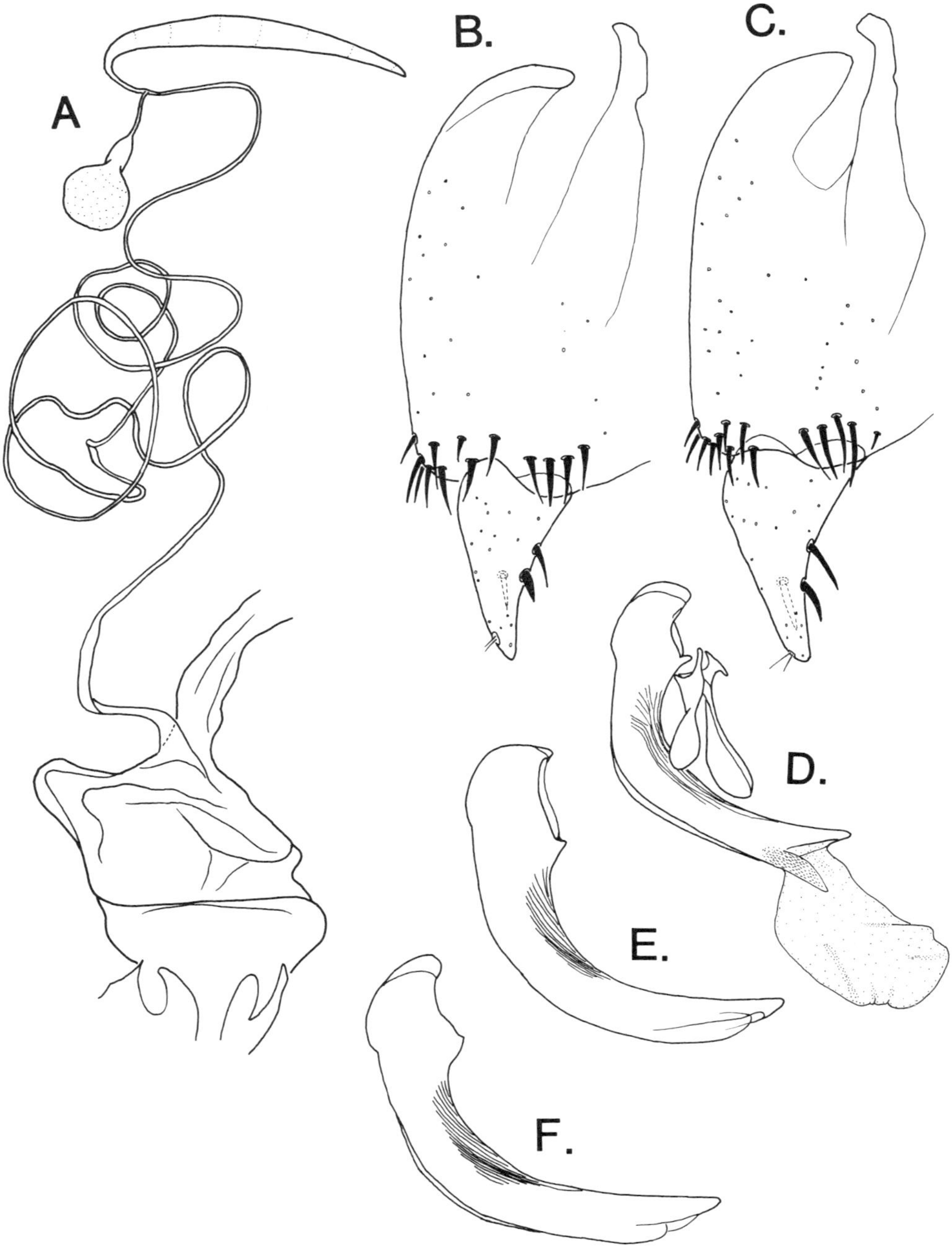

Figure 57. Internal and external reproductive structures of *A. elongatulum.* Conventions as in figure 38. A. Female internal reproductive tract, ventral view, FL:Myakka R. St. Pk. B–C. Left gonostylus, ventral view: B. B.W.I.:South Bimini Isl.; C. FL:Brooksville. D–F. Aedeagus, ventral view: D. FL:Highlands Co., Archbold Biol. Sta.; E. GA:Ware Co., NE. Fargo; F. FL:Myakka R. St. Pk.

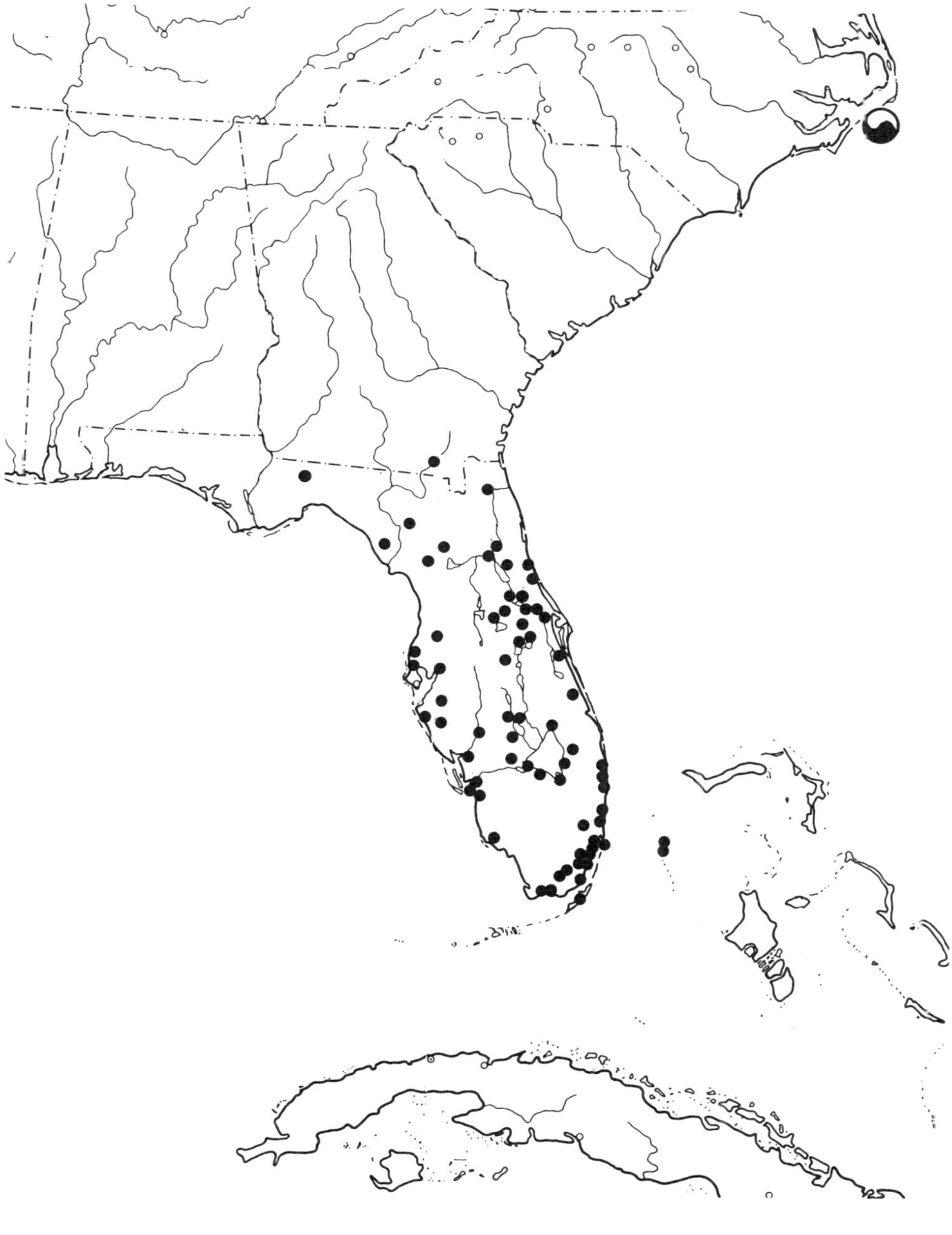

Figure 58. Distributional records of *A. elongatulum* (Dejean). Half-filled circle represents NC:Ocracoke Island record, representative of putative *A. decorum* × *A. elongatulum* hybrids.

Male Genitalia. Aedeagus indistinguishable from that of *A. decorum*; shaft evenly arcuate, tip upcurved, flattened area in curve of shaft in larger individuals (figs. 49h, 57d-f); longitudinal wrinkles well-developed inside basal curve of shaft; aedeagal sac as long as apical portion of shaft (fig. 57d), covered with fine spicules; lateral areas of sac with lightly sclerotized bars indicated by a deeper yellow in cleared specimens.

Female Reproductive Tract. Basal segment of female gonostylus with apical fringe of 13-14 setae (figs. 57b,c); spermatheca long and fusiform; spermathecal gland entering spermathecal duct proximad of basal curve of spermatheca.

Color. Head and pronotum metallic green with bluish or aeneous reflection; elytra metallic green with aeneous reflection; elytral epipleura piceous basally, slightly paler apically; mouthparts rufo-piceous; antenna with scape infuscated, rufo-piceous, segments 2-11 piceous; ventral body surface dark, pronotal epipleura and center of abdominal sternites rufo-piceous, remaining areas darker; legs with femora, tibiae and tarsi infuscated rufo-piceous; coxae slightly lighter. Very rarely a red morph occurs; prothorax ferruginous, but unlike *A. decorum*, mesosternum and mesepisternum piceous. Other body parts colored differently from regular dark form include: antennal scape rufous to rufo-piceous; coxae of pro- and meso-legs rufo-testaceous, femora of all legs rufo-testaceous; tibiae with rufo-testaceous base and infuscated apex.

Microsculpture. Head with uniform isodiametric microsculpture; pronotum with uniform isodiametric microsculpture laterally and basally, becoming slightly stretched in center of disc; elytra with well-developed, uniform isodiametric microsculpture appearing slightly granulate; ventral body surface with well-developed isodiametric microsculpture slightly stretched in some areas; lateral areas of abdominal sternites with moderately developed longitudinal wrinkles.

Length. 7.1-9.0mm.

Distribution. Peninsular Florida, generally south of the Suwanee River (fig. 58); Bimini Islands. A series of putative *A. decorum* X *A. elongatulum* hybrids known from Okracoke Island, NC is discussed in the section on delimitation of these two species.

Biology and Habits. *A. elongatulum* is found in a variety of habitats near water. At Pahokee St. Pk., Palm Beach Co., FL, adults hid during the day under algal mats on the rocky-sandy shore of Lake Okeechobee. At Myakka River St. Pk. *A. elongatulum* was found in abundance on soil in a riparian marsh that had been disked for control of water hyacinth. Both localities had no shaded areas.

A. decorum is rare in Florida, whereas *A. elongatulum* is plentiful. This is correlated with the former's preference for marshes with a well-developed mat of vegetation, and at least somewhat reduced solar insolation, a habitat less common in Florida than the open marshy expanses between hammocks in which *A. elongatulum* occurs.

No larvae of *A. elongatulum* have been reared, but a third instar *Agonum* larva was taken with adult *A. elongatulum* at Pahokee St. Pk., 29-IV-1981. This would indicate *A. elongatulum* breeds in the spring; but life cycles are facultative in this group, and mating might occur whenever conditions permit.

Etymology. Latin *elongatus*; prolonged. This epithet better fits *A. extensicolle*, no doubt aiding in the confusion regarding proper application of the two names.

PHYLOGENY OF AGONUM EXTENSICOLLE GROUP

To test hypotheses concerning phylogenetic relationships of taxa, the only approach is to hypothesize affinities, using different, independently generated data sets rather than direct experimental procedure. If different data sets have been obtained using methods allowing observation of true phyletic relationships, the congruence of those data sets will corroborate the independently generated hypotheses. Testing for taxonomic congruence can be viewed as the experimental approach by which phylogenetic relationships are estimated.

The use of electrophoretic data for estimation of relationships invites comparison with estimates based on morphological characters. Mickevich and Johnson (1976) compared estimates of phylogeny based on electrophoretic and morphological data for five species of *Menidia* (Telostei: Atherinidae). The electrophoretic alleles, coded as present or absent, constituted one basis for phylogenetic estimation. The morphological data set consisted of measurements of six characters. This is better termed a morphometric or biometric data set. Agreement between the electrophoretic and biometric data was maximal when both types of data were analyzed using the minimum length Wagner method (Farris 1970). Other phenetic methods produced assessments of affinities that did not share as much information.

Turner (1974) reports different results in a study of desert pupfish, *Cyprinodon* species (Teleostei: Cyprinidae). Electrophoretic allelic data, analyzed using allelic frequencies, and biometric data produce conflicting estimates of phylogenetic affinities. Extreme environmental conditions have no doubt exerted strong selection pressure on these species. One response to these conditions has been the use of a sophisticated mating hierarchy by certain species (Soltz and Naiman 1978). Such a system reduces effective population size, making the occurrence of genetic drift more probable. The lack of congruence for the different data sets used to study these species should not be totally unexpected.

Berlocher and Bush (1982) compare classifications of *Rhagoletis* species (Diptera: Tephritidae) based on electrophoretic allelic frequency data and morphological data (Bush 1966), with a phylogeny based on *Rhagoletis* host plant classification. They report little congruence between the electrophoretic and morphological data. If one uses the component analysis of Nelson (1979), four of the ten components present in the morphological classification of Bush (1966) are shared with the shortest tree based on electrophoretic data (Berlocher and Bush 1982, fig. 5). Three of the four preserved components define species complexes, whereas the fourth is defined by placement of the genus in the study least closely related to *Rhagoletis* and closely related genera. There is also lack of congruence between host plant classification and electrophoretic classification. Bush (1966) reported similar findings for the morphological versus host plant classifications.

Through a study of the *A. extensicolle* group, three types of data have been generated which can be used to estimate affinities: electrophoretic data based on soluble proteins and enzymes; morphological data derived from qualitative morphological characters; and biometric data based on morphometric characters.

If electrophoretic data from soluble enzymes is used to estimate affinities, shared alleles imply ancestry, and the relative magnitude of shared alleles can be used to group taxa in a hierarchical manner. The consideration of electrophoretic data as phylogenetic has been loosely defined. If one assumes that genetic material evolves in steady relation to time, meaning that alleles change due to mutation at a constant rate, the percentage of shared alleles is a good indicator of phylogenetic relationships. Such an analysis can be confounded by nonconstant evolutionary rates within and among lineages, the insensitivity of the technique to certain substitutions (Lewontin and Hubby 1966), and the imprecision of allelic scoring due to experimental error or protocol (Ayala 1982).

Morphological data are presented that allow grouping of taxa on the basis of shared, derived character states: the synapomorphous characters of Hennig (1966). Characters are homologous attributes which are assumed to evolve from a nonderived or primitive state to a derived or advanced state. As in electrophoretic analysis, the question of homology is critical, but with morphological characters the assumption of uniform change through time is ignored. Morphological characters usually are assumed to manifest a genetic apparatus more complex than the coding for a single enzyme. This precludes simply looking for shared character states. Shared, derived character states must be used to group taxa, with the distribution of the derived states determining the level at which the character can be used to group taxa. Characters that evolve at different rates will be of use in grouping taxa of different rank. The performance of this type of analysis requires determination of derived and primitive character states. The general belief that out-group comparison provides the best means for this estimation (Maslin 1952; Watrous and Wheeler 1981) requires assumption of an out-group or groups to determine the generalized state of a character. Limited knowledge of what the sister group might be limits the validity of polarity assessment in this type of analysis.

Biometric data are used for discriminant analysis of inter- and infra-specific variation. The 13 measurements account for lengths and widths of functionally differing regions of the body. The underlying genetic basis for body shape probably is polygenic, with changes in developmental sequences and rates causing changes in form (Goldschmidt 1940; Lande 1980). The results show that major changes in appendicular length, humeral width, and body size can be correlated with environmental or climatological factors. This indicates the extent to which these polygenically determined attributes may be influenced by differential selection, and reduces the reliability of these attributes for estimation of phylogenetic relationships. Similar pressures may induce similar body shape configuration in parallel, or by genetic mechanisms that are not homologous. Such long-term instability is not desirable as a basis for phylogenetic hypotheses.

TYPES OF DATA AND ANALYSES

Electrophoretic Data Sets. In addition to the large data set of 32 population samples used in the circumscription of species, an exemplar data set is used to estimate genetic distances within the *A. extensicolle* group. Live individuals were obtained for five of the seven species in the group. Samples were obtained of: *A. extimum* (6 individuals, 31-V-1981, Mexico: Coahuila, nr. San Pedro); *A. texanum* (5 individuals, 31-V-1981, Mexico: Coahuila, nr. San Pedro); *A. extensicolle* (6 individuals, 24-V-1981, Vermont: Chittenden Co., Shelburne); *A. elongatulum* (6 individuals, 1-V-1981, Florida: Sarasota Co., Myakka R. St. Pk.); *A. decorum* red-hirsute form (6 individuals, 24-V-1981, Vermont: Chittenden Co., Shelburne); *A. decorum* green-glabrous form (6 individuals, 7-IV-1981, California: Marin Co., Nicasio Rsvr.). As for the larger data set, 17 electrophoretic loci coded by 15 genes are included (Appendix 7). Protocols are given elsewhere (Liebherr 1983a).

The electrophoretic data are analyzed using the genetic distance coefficient of Rogers (1972) and the Manhattan distance based on presence/absence data. The allelic-frequency-based index of Rogers is used, even though it tends to produce underestimates of distances for loci having multiple allelic substitutions (Wright 1978). Because the maximum Rogers' genetic distance of 1.0 occurs between the three pair-wise combinations of a doubly substituted locus, Rogers' index fails the triangle inequality criterion of a metric coefficient (Sneath and Sokal 1973). Such failure causes distortion of genetic distance that are based on quickly evolving loci.

There are two major reasons for using Rogers' index in the present study. The species within the *A. extensicolle* group are relatively closely related, with species pairs separated in the large data set by a Rogers' distance of as little as .09 (fig. 14). For species this closely related, discarding frequency data would remove much of the information available for estimating genetic distance. The sharing of rare alleles by two taxa would become more important in estimating genetic distances. Such rare alleles might be of interest phylogenetically, but sampling them adequately requires multiple population samples within a species and relatively large sample sizes within populations. Large samples were not available for all species in this study, necessitating reliance on a frequency-based distance estimate. Even if large samples are used, presence/absence coding tends to group samples by what they do not share, rather than by what they share. Loss of a rare allele in one sampled population is a likely event, and should not be given undue emphasis in estimating populational affinities.

A second reason for using Rogers' index lies in the data themselves. Within the study set, 8 of the 17 loci are monomorphic or have the same most common allele in all taxa. Three of the loci exhibit single substitutions across the study set. For three of the other six loci, three different alleles are most common in the various species. For this type of allelic distribution at a locus, the estimates of genetic distance would best fit a network with an unresolved trichotomy at its center. In a study set containing six samples, shared alleles would still unite several of the taxa. For the final three loci, which have four or five most common alleles occurring within the six samples, the data for electrophoretic distance are nearly useless. But even here, allelic frequencies at Pgm-1 and 6-Pgd help unite the two samples of *A. decorum*. Even though multiple allelic substitutions occur among the species of the *A. extensicolle* group, the use of a linear index such as Rogers' was deemed essential for resolution of the most closely related species.

In a second analysis, alleles are coded as present or absent, with Manhattan distances between taxa used for computation. In this procedure, all alleles that are shared by all taxa, and those unique to only one taxon, are removed from the analysis. The allelic character matrix can be derived from Appendix 7, and is not presented. This analysis is included to determine whether such a presence/absence coding could be effectively used with such small samples. An attempt was made to code the various alleles in the shared allele and transformation series analyses of Mickevich and Mitter (1981), but the large number of alleles for several loci precluded use of these methods. The independent allele method used here, coding the presence of each allele as a derived character state, follows that of Mickevich and Johnson (1976).

Morphological Data Set. By examination of external and internal structures, 20 characters were found to be uniform enough within taxa and variable enough across taxa to permit differential diagnosis. Use of such characters for phylogenetic analysis followed the method of Hennig (1966) and determination of character state polarity was based on out-group comparison. Other species of *Agonum* and *Platynus*, as well as members of other tribes of Carabidae, were used to determine the ancestral and derived states for characters observed in the species of the *A. extensicolle* group. States generally distributed throughout these taxa are regarded as primitive. If a state is restricted to a portion of taxa in the group, and absent outside the group, it is considered derived within the group.

The character states and the reasons for assigning a polarity are:

1. *lateral saccal sclerite I.* The internal sac of the aedeagus may possess a lightly scletorized chitinous area on both lateral areas (everted position) (figs. 45e, 50d, 57d). This state is not observed uniformly in *Agonum* species, and is present or absent in various carabids. Presence is assumed to be derived, although this polarity is weakly supported. The consequences of reversing the polarity of this character are discussed below.

2. *lateral saccal sclerite II.* The lateral sclerotized areas of the sac are well-developed, fingerlike sclerites (fig. 42e) that hold the sac erect when they are everted. This state was observed only in *A. texanum* and is assumed to be derived.

3. *internal sac with dorsal spicules.* In several species the internal sac of the aedeagus possesses larger spicules on the dorsum (everted position) than on the lateral areas (figs. 38e, 39d, 40d). This differential spicule configuration is not observed in the other *Agonum* dissected and is assumed to be derived.

4. *aedeagal shaft with longitudinal wrinkles.* In several *Agonum* species, the aedeagus bears longitudinal striations on the middle portion of the shaft (figs. 38e, 42e, 45e, 50d, 57d). These are not observed in *Platynus* and other exemplar carabids in the Bembidiini, Patrobini, and Pterostichini, and are assumed to be derived.

5. *spermathecal duct position.* Among carabids examined in which the spermatheca is associated with a distinct gland, the gland enters at the base of the spermatheca. This is seen throughout *Agonum*, *Platynus* and related genera (figs. 42a, 45a, 50a, 57a). In several species of this group, the gland enters the spermathecal duct about 1/2 to 1/3 the length of the duct from the bursa (figs. 38c, 39c, 40c). This configuration is considered derived.

6. *spermathecal shape.* In most Carabidae the spermatheca is a spindle-shaped, straight tube (figs. 42a, 45a, 50a, 57a). Such an undifferentiated duct with apical spermatheca occurs in *Cicindela* (Freitag 1979) and *Carabus* (Iuga 1966). In several *Agonum* the spermatheca is regularly constricted, taking a moniliform configuration (figs. 38c, 39c, 40c) that is considered derived.

7. *number of moniliform beads in spermatheca.* It is assumed that increased development of moniliform constrictions is the derived configuration for spermathecal shape. *A. cyanope* possesses 25-30 beads (figs. 38c,d) whereas *A. extimum* and *A. parextimum* possess 37-47 beads (figs. 39c, 40c). The latter state is assumed to be more derived.

8. *apical half of antennomere 3 setose.* Throughout *Agonum*, *Platynus*, and many other carabid genera, the third antennomere is glabrous except for an apical ring of setae. The apical half of antennomere 3 is setose in a limited number of *Agonum* species, and this state is considered derived.

9. *wing polymorphism.* Generalized carabids possess fully developed flight wings, and reduction is considered derived. *A. extensicolle* is the only species with a discrete reduced state for wing development (figs. 36g,h). This polymorphism is assumed derived.
10. *pilosity over entire body.* Most Carabidae possess glabrous integument, with the exception of articulated macrosetae *A. extimum* possesses a pelage on the dorsum and ventrum composed of very small unarticulated microsetae. This state is assumed derived.
11. *pilosity restricted to dorsum of body.* Several of the morphs of *A. decorum* possess a pelage of unarticulated microsetae on the dorsal body surface. Fine setae are absent to very scarce ventrally. The induction of such a condition on *A. decorum* is probably caused by a hormonal mechanism similar to the causal mechanism for the pelage in *A. extimum*, but the two characters differ qualitatively. For this reason, assessment of homology is uncertain. In coding these two characters, one could take a reductionist stance, coding regions or sclerites of the body as pilose or glabrous, and considering the states homologous for some sclerites and nonhomologous for others. More likely, the pelage is an integrated feature, induced in full condition by a switch in hormonal control of development. This viewpoint assumes an underlying mechanism for expression of the character. It also recognizes that assessment of homology between characters 10 and 11 is uncertain and reduces the likelihood of unwarranted homoplasy in a cladistic analysis. The price for such a reduction is an increase in autapomorphic, and therefore phyletically uninformative characters.
12. *prothoracic shape.* The prothorax in Patrobini, Bembidiini, and Trechini as well as in many *Pterostichus*, *Agonum* and *Platynus* is cordiform. The lateral margins are sinuate inwardly before the hind angles. In *A. cyanope* and *A. texanum* the basal lateral areas of the prothorax are broadened (figs. 31c-h), making the prothoracic shape noncordate. A broad, noncordate prothorax is assumed derived.
13. *wing pigmentation.* The flight wings of most carabids examined are largely transparent, clear membrane, which may be piceous near veins and in association with the wing-folding mechanism. In *A. cyanope* much of the wing membrane is pigmented a smoky piceous color (fig. 36e). This state is assumed to be derived.
14. *leg color.* Most carabid legs are concolorous with the body. In this species group a number of species have pale testaceous legs: the derived state. For polymorphic species, presence of pale legs in at least one morph implies possession of the derived state.
15. *color of prothorax polymorphic.* Most carabids have concolorous body regions. In *A. elongatulum*, rare individuals have a red to reddish-green prothorax, a state distinct from the normal metallic green prothorax. This polymorphism is considered an autapomorphy in this species.
16. *color of pro- and mesothorax polymorphic.* In *A. decorum*, the red morph possesses ferruginous prothorax, mesosternum, and mesepisternum. As in the presence of the pelage in *A. extimum* versus *A. decorum*, the color polymorphisms in *A. elongatulum* and *A. decorum* may have a similar hormonal basis. If pigmentation proceeds in *Agonum* as it does in scale pattern formation in Lepidoptera (Goldschmidt 1940), timing of developmental events will determine the color pattern. The mechanisms producing the color patterns in the red morphs of *A. elongatulum* and *A. decorum* may be similar, but they are not homologous. As in the coding of characters 10 and 11, a conservative approach to assessment of homology is assumed for coloration; this strategy reduces unwarranted homoplasy at the expense of reducing possible synapomorphies.
17. *number of dorsal punctures on or near interval 3 of elytra.* Generally there are 2, 3, or 4

dorsal elytral setae in carabids (i.e. Bembidiini, Trechini, Patrobini). In the *A. extensicolle* group most species have 4 to 8 setae. *A. extimum* possesses 7-16 dorsal setae (fig. 35b), essentially a doubling of the number commonly seen. The doubled number of dorsal setae is assumed to be derived.

18. *basal punctation of pronotum.* The pronotal base of most *Agonum* is smooth medially, although the latero-basal pronotal depressions may be punctate. The pronotum of *A. extensicolle* has punctures medially as well as in the lateral depressions (figs. 32e-h), and this state is considered derived.

19. *microsculpture on center of pronotum.* Generally microsculpture throughout the Carabidae is of the granulate isodiametric type (Lindroth 1974b). Species occurring in marshy areas tend to have a more transverse microsculpture. *A. extensicolle*, *A. elongatulum* and *A. decorum* all have moderately transverse meshes in the microsculpture on the disc of the pronotum. This state is assumed to be derived.

20. *elytral striae weak before basal margin.* In most *Agonum*, the elytral striae are well-defined even near the basal margin of the elytra. In *A. texanum* the elytral striae are weaker at their base, often becoming evanescent before their grooves join the depression of the basal margin (fig. 34a). This state is considered derived.

The above assessment of character polarities is compatible with the unit character coding of Farris, Kluge, and Eckhardt (1970). In this data set, characters 1 and 2, and 6 and 7 represent unit character coding of multistate characters. The derived state for character 2 or 7 is a restriction of the derived state for character 1 or 6. Such coding causes no problems as long as the final cladogram is inspected to make certain that the transformation of the more restrictive derived state does not occur basal to the change in the less derived state. The complete matrix of taxa by character states is presented in table 32.

Biometric Data Sets. The biometric data, consisting of 13 measurements over 56 population samples of both sexes, are used to generate estimates of interspecific phenetic distances. The centroids (figs. 18, 19) are used as exemplars for the species. Euclidean distances, for each sex, between pairs of centroids derived from the first three discriminant functions can be considered to summarize body shape differences. These distances are purely phenetic constructs. The degree to which affinities based on the phenetic distances are correlated with results of the electrophoretic and morphological data sets determines the value of biometric data for estimating phylogeny.

Methods of Comparison. The purpose of this study is to compare estimates of affinities based on the various data sets; two methods of comparison are presented. In the first, the electrophoretic frequency, electrophoretic presence/absence, and male and female biometric data sets are summarized as interspecific distance matrices (tables 30, 31, 34) and submitted to the Wagner 78 algorithm of Farris (1970) run on the CDC-6400 computer, University of California, Berkeley. The distance and mid option are specified, rooting the resultant network at its midpoint. The taxon by morphological character matrix (table 32) can be converted to a Manhattan distance matrix (table 33) with the minimum-length Wagner tree computed by hand using the method of Sneath and Sokal (1973). This results in a network rooted at a hypothetical ancestor with primitive states for all characters.

Component analysis (Nelson 1979) is used to compare the five rooted networks. The networks are compared to determine what phyletic information is shared, based on the ratio of number of shared components to number of possible shared components. Each component is a monophyletic group of any rank. For each rooted network with *n* terminal taxa, there are *n* - 2 possible components, assuming bifurcations at each node of the network.

Nelson (1979) analyzed Michener's (1977) data on allodapine bees (Hymenoptera: Apoidea) by looking for shared components among phenograms and cladograms presented by Michener. Nelson's method searches for compatible components that are either inclusive, exclusive, or replicated. Inclusive components are those that are composed of different sets of taxa but that have some taxa in common. The shared taxa must possess identical phyletic relationships in the two networks. Exclusive components are groups present on only one of the networks: the groups hold no members in common. Replicated components are those that are identical. If one compares two branching networks, one can search for any of the above types of components in the two networks and combine them compatibly to form a consensus tree. This tree will contain all the components present in one or the other network that do not conflict. Such nonconflicting component analysis, in which information shared between two networks is complemented by nonconflicting elements unique to one or the other network, is here termed relaxed component analysis. Relaxed component analysis is analogous to combining cladograms under assumption two of Platnick (1981, 224).

One may be more rigorous in comparing branching networks. In this comparison only components shared by two networks would be included in a consensus network; this would correspond to including only replicated components in the consensus network. This approach is termed strict component analysis, and corresponds to assumption one of Platnick (1981).

For this study, comparison of branching networks is first done under the strict assumption of component analysis. The assumptions are changed to the relaxed mode only for the consensus network judged best using the strict assumption.

The second manner in which the various data sets are compared is based on the correlation between the distance matrices of the five data sets. Each distance matrix is standardized (Cain and Harrison 1958) by scaling each distance—dividing it by the largest distance in the data set. Pearson's product-moment correlation coefficients can be determined for all pairwise comparisons of data matrices. Whenever one matrix contains taxa not present in the other, the unpaired values are ignored. To enable comparison of the electrophoretic data sets with all others, the pairwise distances between the red-hirsute and green-glabrous morphs versus the other taxa have been averaged. The pairwise comparison of distance matrices is presented by clustering the various correlation coefficient values using the UPGMA algorithm (Sneath and Sokal 1973) and by a minimum length, or Prim network. The resultant phenogram permits assessment of the phenetic similarity between the various distance matrices. Such an assessment of distance matrix similarity is unhindered by the compromises necessary in construction of a minimal-length Wagner network. Such compromises involve obtaining a best fit for all intertaxon distances and defining a root for component analysis. Placement of a root on a network implies directionality. If two taxa separated by small taxonomic distances are situated on opposite sides of an arbitrarily defined root, their affinity will not be reflected in component analysis. If taxonomic distance matrices are compared directly, their affinities will be taken into account in the comparison.

RESULTS AND DISCUSSION

Results of Data Sets. The allelic frequencies for the electrophoretic data set (Appendix 7) can be used to generate Rogers' genetic distance estimates (table 30). As expected, the intraspecific pair, red-hirsute and green-glabrous *A. decorum*, are most similar genetically. Both samples of *A. decorum* are equidistant from *A. elongatulum*. All of the interspecific genetic

Table 30. Electrophoretic distances between taxa in exemplar data set based on Nei's genetic distance above diagonal (Nei 1978) and Rogers' genetic distance below diagonal (Rogers 1972). Standardized Rogers' distances in parentheses.

Taxon	2 EXTM	4 TEXA	5 EXTN	6 ELON	7a DECg	7b DECr
2 EXTM	—	.322	.276	.413	.356	.304
4 TEXA	.318(.86)	—	.323	.304	.276	.261
5 EXTN	.307(.83)	.332(.89)	—	.192	.222	.164
6 ELON	.371(1.0)	.313(.84)	.263(.71)	—	.098	.097
7a DECg	.354(.96)	.320(.86)	.280(.75)	.207(.56)	—	.037
7b DECr	.322(.87)	.291(.78)	.241(.65)	.197(.53)	.142(.38)	—

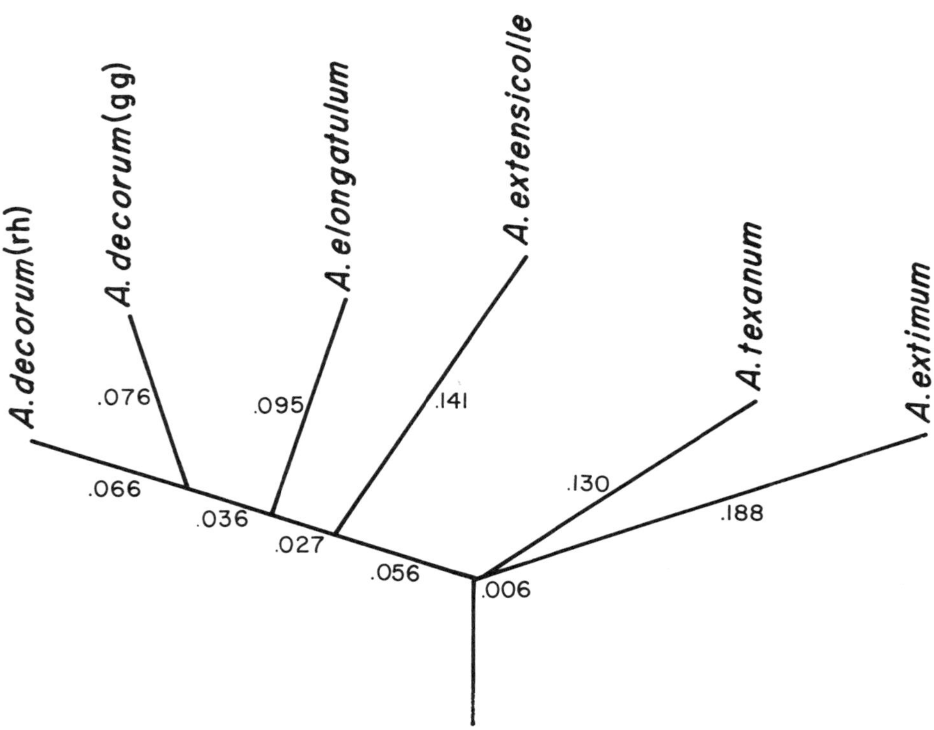

Figure 59. Minimally connected Wagner network (Farris 1970) based on Rogers' genetic distance. Root is placed at midpoint of longest between-sample distance.

distances exhibited in the 32-population, minimally linked network (fig. 14) are repeated by the smaller data set. The minimal-length Wagner network rooted at the midpoint of the longest intersample distance on the network is presented in figure 59. The similar genetic distances seen between *A. extimum* and *A. texanum* versus the other four samples causes a lack of resolution at the midpoint of the network. This structure is caused by repeated allelic substitution at certain loci in this group, and supports Farris' (1981) concerns over the indiscriminate use of genetic distance indices at inappropriate taxonomic ranks. This analysis approaches the limit of electrophoretic resolution of phyletic relationships for some of the loci picked for analysis.

Table 31. Electrophoretic distances between taxa in exemplar data set based on presence/absence data (below diagonal) and standardized to a maximum of 1.0 (above diagonal).

Taxon	2 EXTM	4 TEXA	5 EXTN	6 ELON	7a DECg	7b DECr
2 EXTM	—	.88	.61	.97	1.0	.85
4 TEXA	35	—	.94	.82	.97	.88
5 EXTN	20	35	—	.79	.88	.85
6 ELON	32	33	32	—	.52	.55
7a DECg	37	38	37	17	—	.64
7b DECr	32	33	32	20	25	—

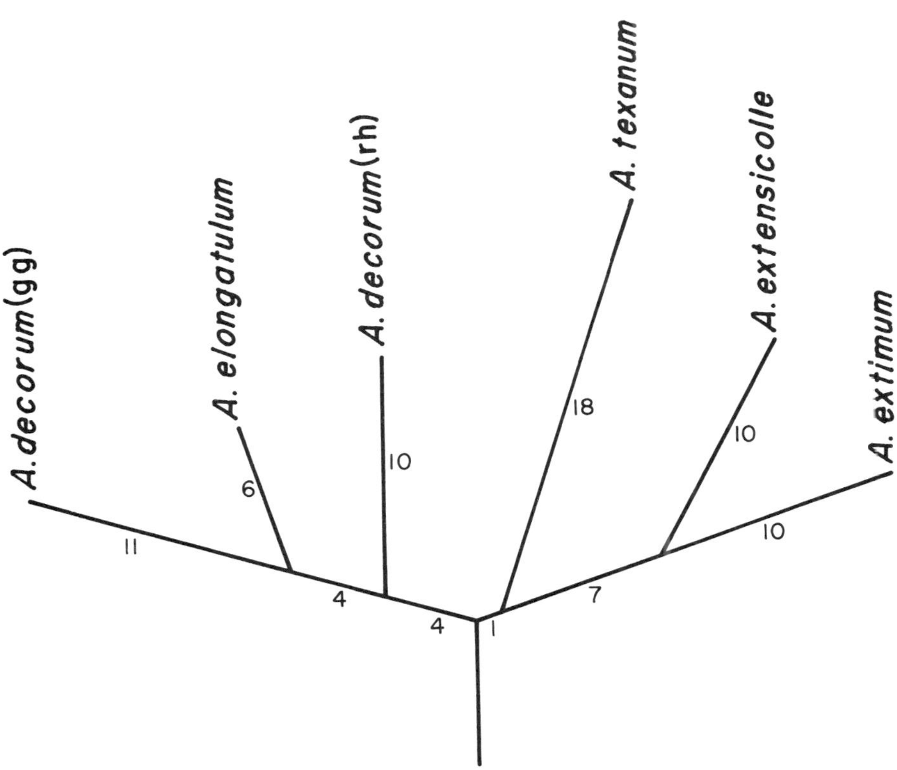

Figure 60. Minimally connected Wagner network (Farris 1970) based on presence/absence of alleles. Root is placed at midpoint of longest between-sample distance.

The genetic distances based on the presence/absence electrophoretic allele coding (table 31) are generally similar to those based on the frequency data, but two major differences are evident, especially when comparing the standardized distances (tables 30, 31). One difference caused by coding for presence or absence involves the distances within samples of *A. decorum* and *A.*

Table 32. Character state matrix for qualitative morphological data set. 0 = primitive, 1 = derived character states.

	Characters																			
Species	1	2	3	4	5	6	7	8	9	10	11	12	13	14	15	16	17	18	19	20
A. cyanope	0	0	1	1	1	1	0	1	0	0	0	1	1	0	0	0	0	0	0	0
A. extimum	0	0	1	0	1	1	1	1	0	1	0	0	0	1	0	0	1	0	0	0
A. parextimum	0	0	1	0	1	1	1	1	0	0	0	0	0	1	0	0	0	0	0	0
A. texanum	1	1	0	1	0	0	0	0	0	0	0	1	0	0	0	0	0	0	0	1
A. extensicolle	1	0	0	1	0	0	0	0	1	0	0	0	0	1	0	0	0	1	1	0
A. elongatulum	1	0	0	1	0	0	0	1	0	0	0	0	0	1	1	0	0	0	1	0
A. decorum	1	0	0	1	0	0	0	1	0	0	1	0	0	1	0	1	0	0	1	0

Table 33. Manhattan distance matrix based on qualitative morphological characters (below diagonal); standardized distances are above diagonal.

	1	2	3	4	5	6	7
Species	CYAN	EXTM	PARX	TEXA	EXTN	ELON	DECO
1 CYAN	—	.50	.33	.58	.92	.83	.75
2 EXTM	6	—	.17	1.0	1.0	.83	.92
3 PARX	4	2	—	.92	.83	.67	.75
4 TEXA	7	12	11	—	.58	.58	.67
5 EXTN	11	12	10	7	—	.33	.42
6 ELON	9	10	8	7	4	—	.25
7 DECO	10	11	9	8	5	3	—

elongatulum. In the presence/absence data, the distance between the morphs is greater than the interspecific distances. This places the red-hirsute morph of *A. decorum* as sister to *A. elongatulum* and green-glabrous *A. decorum* (fig. 60). Increased importance of rare alleles in this analysis, due to limited sampling of rare alleles because of small sample size, causes this deviation.

A second difference between the two methods of analysis involves placement of *A. extensicolle*. The placement of *A. extensicolle* as sister to *A. extimum* is caused in part by shared alleles at the Mpi and 6-Pgd loci (Appendix 7). The two species differ drastically in allelic frequency at these loci, but this information is lost by the presence/absence coding of data. The high frequency of allele .50 for Acph in both of these species supports placement of *A. extensicolle* and *A. extimum* as sisters. Other loci, such as Pgm-2 and Acon-1, better support *A. texanum* and *A. extimum* as sisters.

The rooted Wagner network derived from the morphological data is presented in figure 61. Characters were unweighted in the analysis, but several are clearly better indicators of

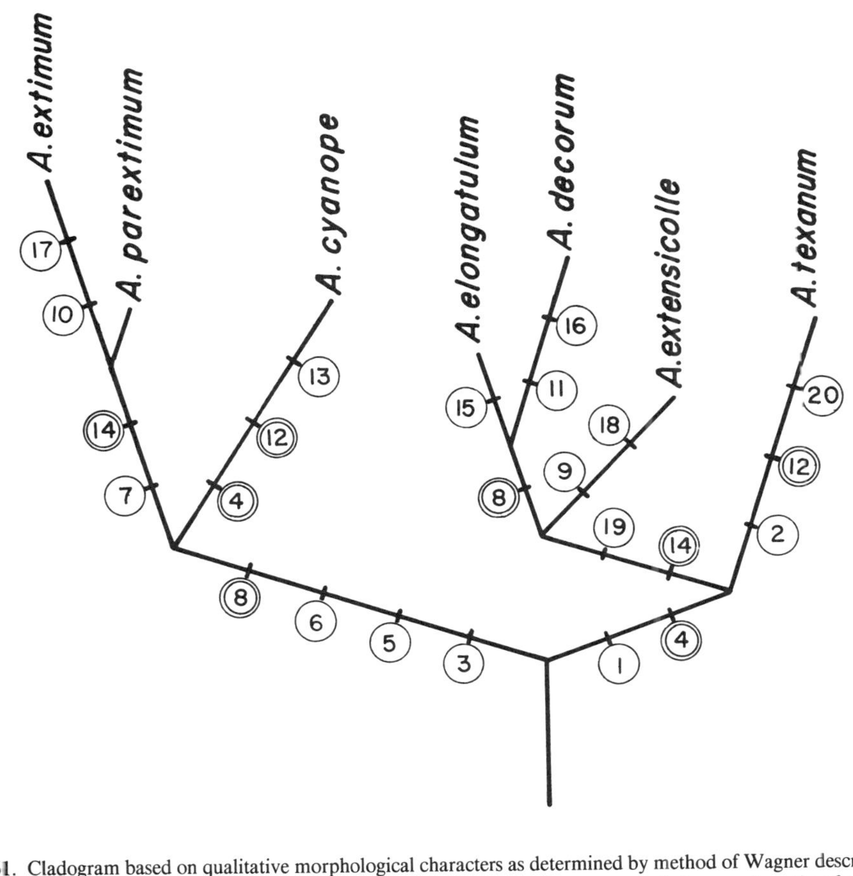

Figure 61. Cladogram based on qualitative morphological characters as determined by method of Wagner described by Sneath and Sokal (1973). Doubly circled characters change states more than once on cladogram. Lengths of stems on network proportional to number of character state changes.

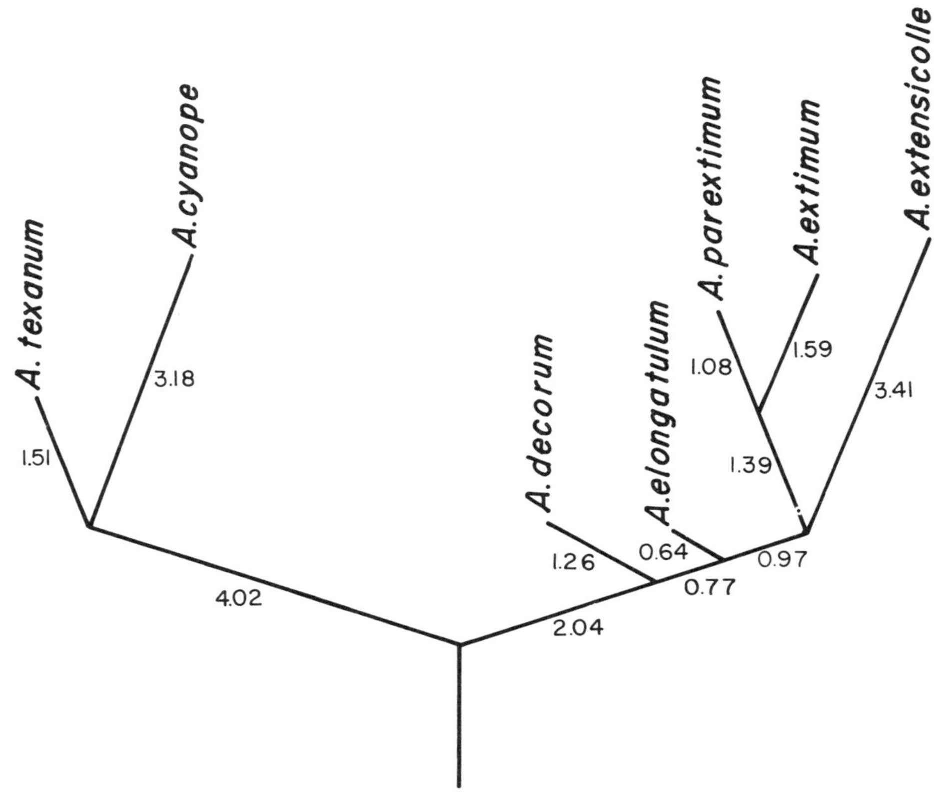

Figure 62. Minimally connected Wagner network (Farris 1970) based on intercentroid Euclidean distances derived from discriminant analysis of males of the *A. extensicolle* species group. Network is rooted at midpoint of longest intersample distance.

evolutionary affinities than others. The characters most likely to indicate true phylogenetic relationships (based on a criterion of complex character states), and distributed over relatively few taxa, are those of the female internal reproductive tract (5, 6, 7). These characters represent much of the structure in the *A. cyanope* lineage within this group. Other characters for which polarities can reasonably be assumed include transverse microsculpture (19), and a number of autapomorphies (2, 9, 10, 11, 13, 15, 16, 17, 18, 20). This cladogram indicates that rounded prothoracic shape (12) has originated independently in *A. texanum* and *A. cyanope*, leg color (14) has become testaceous in two lineages, and antennomere 3 has become apically setose in two lineages (8). For one character of the male genitalia, more complete out-group comparison may reverse the assumed polarity. The longitudinal wrinkles of the aedeagal shaft (4) occurs in other *Agonum*, such as species of the *Agonum melanarium* group. Whether this character state is readily gained and lost is unknown. The variability within *A. cyanope* (figs. 38e-g) implies such a possibility. A change in polarity would eliminate a parallelism in this character, substituting for it one loss. This would make a more parsimonious cladogram, but it would not change the structure.

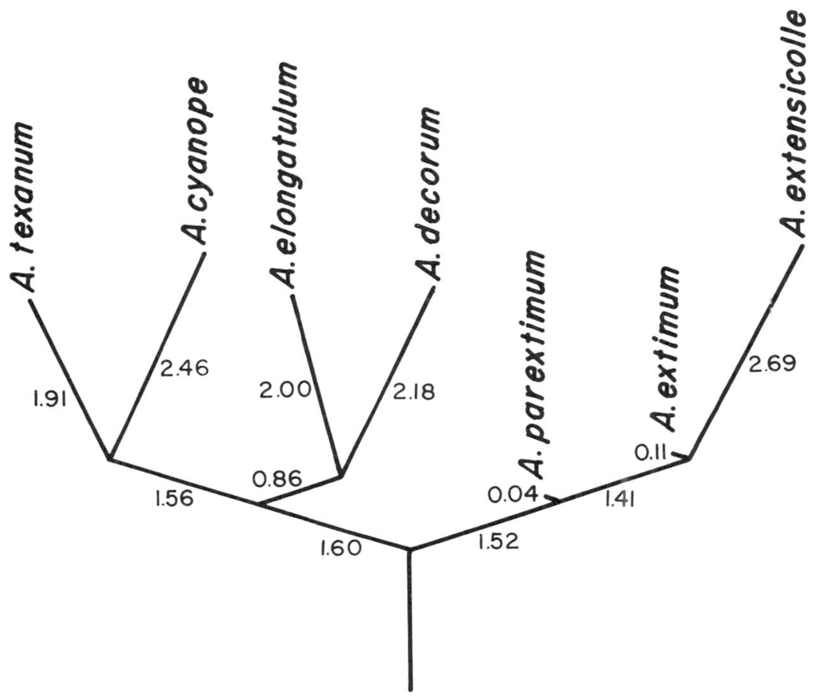

Figure 63. Minimally connected Wagner network (Farris 1970) based on intercentroid Euclidean distances derived from discriminant analysis of females of the *A. extensicolle* species group. Network is rooted at midpoint of longest intersample distance.

The derived nature of Character 1 (lightly sclerotized lateral regions of the internal sac) also could be incorrectly interpreted. These sclerotized areas are somewhat variable within species, and occur sporadically within *Agonum*, and in other carabid genera (e.g. *Scaphinotus*). Reversal of the transformation series would weaken resolution of the *A. extensicolle* lineage from the common ancestor. As with character 4, the structure of the cladogram would remain unchanged.

The Euclidian distances derived from the male and female species group discriminant analyses (figs. 62, 63) are included in table 34. In general, for both sexes the greatest relative distances occur in comparing *A. texanum* and *A. cyanope* to *A. extensicolle. A. extimum* and *A. parextimum* are both relatively similar biometrically to *A. extensicolle*. The distances between *A. decorum* and *A. elongatulum*, versus the other taxa are moderate, indicating the central position of these species in the discriminant analysis scores. The male (fig. 62) and female (fig. 63) minimum-length Wagner networks show similar relationships of the two broad species (*A. cyanope* and *A. texanum*) to the three species with strongly cordate prothoraces (*A. extensicolle, A. extimum*, and *A. parextimum*). The phenetic distances between *A. decorum* and *A.*

Table 34. Euclidean distances between species centroids derived from discriminant analysis of biometric data set. Distances derived from female data above diagonal, distances from male data below diagonal; standardized distances in parentheses.

Species	1 CYAN	2 EXTM	3 PARX	4 TEXA	5 EXTN	6 ELON	7 DECO
1 CYAN	—	12.43(.84)	10.10(.68)	4.70(.32)	14.84(1.0)	10.65(.72)	9.69(.65)
2 EXTM	8.20(.82)	—	2.67(.18)	11.85(.80)	4.40(.27)	4.58(.31)	5.94(.40)
3 PARX	7.19(.72)	1.56(.16)	—	9.41(.63)	5.87(.39)	4.07(.27)	5.46(.37)
4 TEXA	4.38(.44)	8.12(.81)	6.64(.66)	—	13.07(.88)	8.97(.60)	8.83(.60)
5 EXTN	9.15(.91)	2.80(.28)	4.14(.41)	10.01(1.0)	—	4.41(.30)	6.40(.43)
6 ELON	6.89(.69)	6.94(.69)	5.88(.59)	4.60(.46)	8.27(.83)	—	2.66(.18)
7 DECO	5.79(.58)	6.65(.65)	6.20(.62)	6.52(.65)	6.61(.66)	4.19(.42)	—

Table 35. Matrix of correlation coefficients between matrices used to estimate affinities of species.

Classification	(a)	(b)	(c)	(d)	(e)
(a) electrophoretic frequency	—				
(b) electrophoretic presence/absence	.764	—			
(c) qualitative morphological	.737	.308	—		
(d) male biometric	.334	.703	.181	—	
(e) female biometric	.442	.587	.310	.720	—

elongatulum and the other species differ between the two sexes. Comparing the male and female distance values for the 11 standardized distances containing either or both of these two species, one obtains an r = .374. The overall correlation of male and female biometric data is much higher; r = .720 (table 35), indicating that placement of *A. decorum* and *A. elongatulum* is the major cause of disagreement between the two biometric data sets.

Component Analysis of Data Sets. The rooted branching networks corresponding to the five data sets are presented in figure 64. In this representation, distance along each stem is not considered: only the branching pattern is of interest for comparison. Each component is numbered, and if present on more than one network the identifying number is the same. To assess the shared information, pairwise consensus trees can be constructed (fig. 65) that present the replicated components in the pairs of trees. The possible number of components in a bifurcating network is *n* - 2, where *n* is the number of terminal taxa. This number constitutes the denominator of the ratio to the right of each network. The higher the ratio of actual to possible replicated components, the more phyletic information is shared.

In general, the biometric analyses (d and e) do not correspond to the electrophoretic or morphological analyses. Comparisons of a and b with d show no shared information, whereas comparisons of c with d, and c with e, preserve only one component. Component 10 represents *A. extimum* and its sister *A. parextimum.* This component is shared in the consensus network between the morphological (c) and male (d) biometric data. For females, the consensus network of biometric and morphological data contains only component 2; this represents *A. decorum* and *A. elongatulum.* This pair is similar morphologically, electrophoretically, and biometrically, and both taxa are represented in all data sets, causing recurrence of component 2 on many consensus networks.

Among the best ratios of shared to possible shared components are those for consensus networks for electrophoretic frequency versus morphological data, and electrophoretic presence/absence versus female biometric data. The former (a, c) contains components 1 and 2. The basal trichotomy of *A. texanum, A. extimum* and the *A. extensicolle* lineage in the electrophoretic frequency data set (fig. 59) precludes full resolution of this network. The network combining electrophoretic presence/absence and female biometric data contains components 2 and 7. Component 2 is present six times in the ten comparisons, for reasons stated above. Component 7 represents *A. extimum* and *A. extensicolle.* These two species are separated on the morphological cladogram by 12 changes in character state, and are not closely related. Their similarity in the electrophoretic presence/absence data results from the loss of frequency information by the coding process and on the limited sampling of rarer alleles, owing to the small sample size. Their similarity in the female biometric data set reflects shared slender form with basally constricted, or cordate, prothoraces. This similarity is of limited value for establishment of phyletic affinities.

If the type of components used in the consensus network is changed from only replicated components to all nonconflicting components, only one pair of analyses shows a marked increase in the ratio of actual to possible shared components. The comparisons between analyses c, d, and e do not change because they already share all taxa in common. The actual to possible components ratios for electrophoretic presence/absence data compared with the other data do not increase under the relaxed assumption. Only in comparing the electrophoretic frequency and morphological data sets does an increase in the ratio of actual to possible components result from the relaxed assumption (fig. 66). The only conflict between the electrophoretic frequency and morphological data sets involves placement of *A. texanum.* In analysis a, *A. texanum* is the

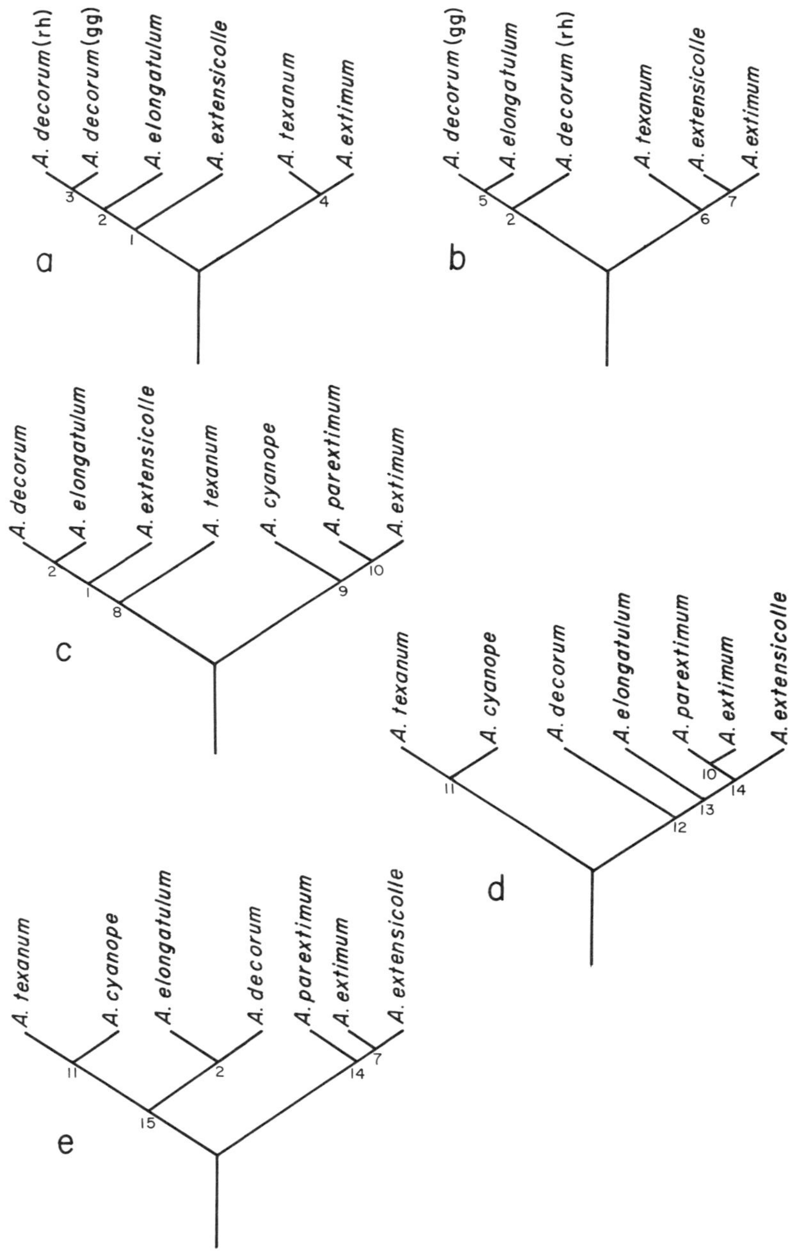

Figure 64. Rooted networks based on the various data sets used to estimate affinities of species in the *A. extensicolle* group. Components are numbered at their basal nodes, with replicated components possessing the same identifying number. A. Network based on electrophoretic frequency data. B. Network based on electrophoretic presence/absence data. C. Network based on qualitative morphological data. D. Network based on intercentroid distances of male discriminant analysis. E. Network based on intercentroid distances of female discriminant analysis.

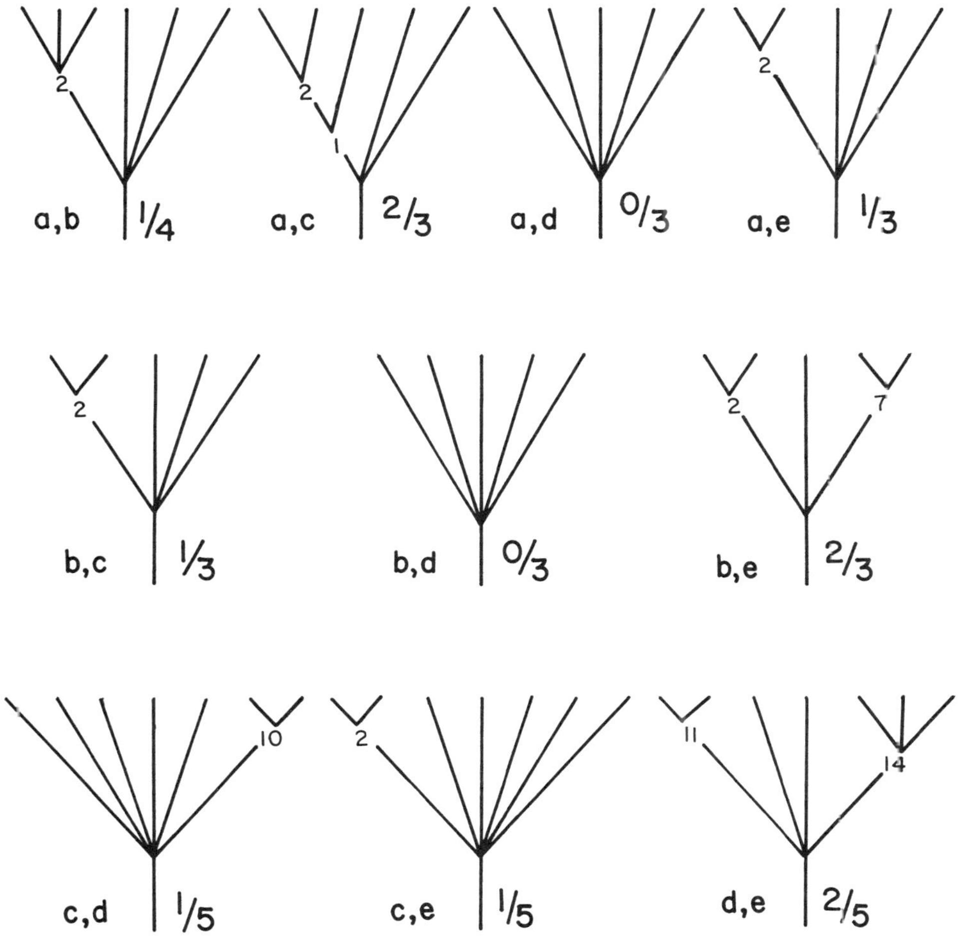

Figure 65. The ten possible pairwise consensus trees determined using the strict assumption, i.e., replicated components. Components and letters as in figure 64. Ratio to right of network is number of components represented on consensus tree to maximum possible number of shared components.

sister to *A. extimum*, although the root of the network is only .006 Rogers' genetic distance from the stem leading to *A. texanum* (fig. 59). The network is essentially rooted at a basal trichotomy, but it was considered a basal bifurcation (fig. 64a). In the morphological cladogram, *A. texanum* is actually a sister to the rest of the *A. extensicolle* lineage (fig. 61). Placement of *A. texanum* on alternate sides of the root in the two networks necessitates placing it on the root under the relaxed assumption.

The rooted consensus network based on electrophoretic and morphological data made under the relaxed assumption (fig. 66) can be used to illustrate three hypotheses. 1) *A. parextimum* and *A. extimum* are very similar genetically. This hypothesis could be easily tested if live material for both species were available. 2) *A. cyanope* is the sister of the *A. extimum* - *A. parextimum* species pair. This hypothesis is based on four synapomorphies, one involving a character that is

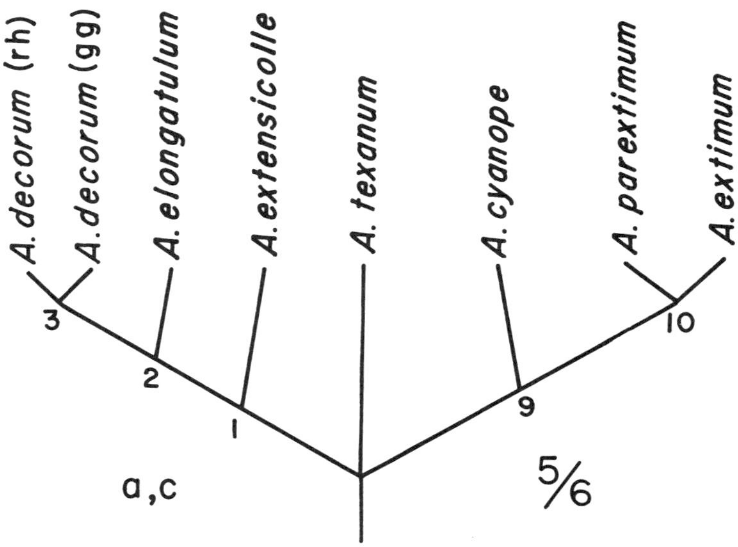

Figure 66. Consensus tree obtained under the relaxed assumption using electrophoretic frequency and qualitative morphological characters. Components listed are all those that are nonconflicting. Ratio is number of nonconflicting components to maximum possible number of such components.

homoplasious on the shortest cladogram (fig. 61). Even though *A. cyanope* looks very different from the *A. extimum* complex, it should be similar genetically. 3) *A. texanum* is the sister of the rest of the *A. extensicolle* lineage. This hypothesis is based on two synapomorphies in the morphological data set; one a possibly homoplasious character, and the other a character of uncertain polarity.

Comparison of Distance Data. The five data sets can be compared based on their shared distance relationships between taxa. If the correlation between distance matrices is used as the means of comparison, the intermediate step of ordination on a minimum length network can be eliminated.

All possible pairwise product-moment correlation coefficients based on the standardized distance matrices are presented in table 35. The results of this analysis are presented as a phenogram, based on the U.P.G.M.A. algorithm (fig. 67a). The two electrophoretic analyses share the most distance information, with the male and female biometric data sets almost as similar. The morphological distances are much more similar to the distances in the electrophoretic analysis than to the distances of the biometric data sets (table 35). The average *r* between morphological and electrophoretic matrices is .522, whereas the average *r* between the morphological and the two biometric data sets is .246.

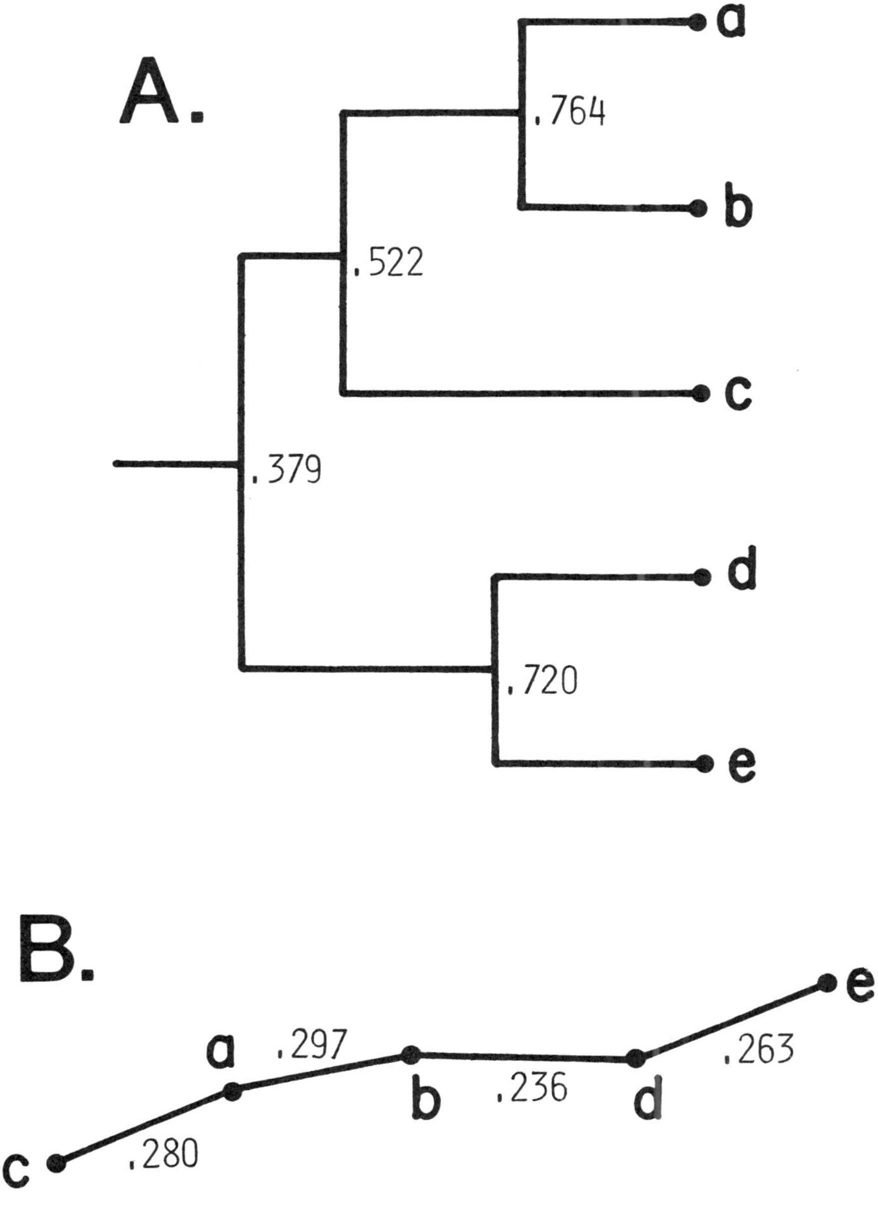

Figure 67. A. UPGMA phenogram based on matrix of between-network correlation coefficients (table 35). Letters correspond to networks described in figure 64. B. Minimally connected Prim network based on between-network correlation coefficients. Correlation coefficients from table 35 are subtracted from 1.0 to obtain distances between networks.

Even though the phenogram clusters the two electrophoretic data sets and then the morphological data set, this relationship is partly due to the UPGMA algorithm. Matrices a and b are very similar, as are a and c, but b and c are notably different (coefficient = .308). An unrooted minimal-length network gives another representation (fig. 67b). In this comparison the male and female biometric data sets (d and e) fall at one end of the network, with the morphological data set (b) at the other end, closest to the electrophoretic frequency data set (a). As in the component analysis, these results support joint use of the electrophoretic frequency and morphological data sets because they produce similar estimates of interspecies affinities, while being independently generated.

SUMMARY

Electrophoretic data, analyzed using a genetic distance index, are compatible with the results achieved using morphological characters. The lack of resolution at the root of the electrophoretically based network is partly due to repeated substitution of alleles at the most actively evolving loci.

Body shape is an inadequate indicator of phylogeny, as evidenced by the incongruence between it and other data sets and between male and female biometric data. Cladistic analysis, using character polarities determined by out-group comparison, appears to be a technique applicable to the broadest array of taxa and taxonomic ranks. Electrophoretic data may corroborate hypotheses based on such data. If the two types of data conflict, the cause may be the type of analysis used for the electrophoretic data. Studies with large sample sizes and a larger number of taxa are required to test methods of analysis such as those of Mickevich and Mitter (1981).

BIOGEOGRAPHIC ANALYSIS

If we can hypothesize phylogenetic relationships between taxa with confidence, we can attempt to correlate that phylogeny with past geological events. If such a correlation can be made, we have developed a hypothesis that attempts to explain how certain organisms have arisen, and why they occur where they do today. Testing of individual biogeographic hypotheses entails the comparison of several hypotheses formulated for different groups of organisms. If common events have created common distributional patterns, we will see regular relationships between taxa and the areas they inhabit. The confounding phenomenon in biogeographic analysis is dispersal of taxa. Individuals expand the taxon's range and the point of origin of the taxon is obscured.

In this analysis, the phylogeny hypothesized is viewed in light of distributions of the species. A common pattern of sister group relationships can be observed, giving a clue to external mechanisms causing speciation. Sister group patterns are then contrasted with those of other groups of organisms to see if common causes could be operating to generate diversity. Timing of past events is estimated using an electrophoretic clock. Finally, a possible scenario is presented which attempts to explain the diversity of the *A. extensicolle* species group.

Phylogenetic and Chorological Patterns. The consensus cladogram (fig. 66) and the morphological cladogram (fig. 61) support recognition of *A. cyanope*, *A. extimum*, and *A. parextimum* as a distinct clade. The phyletic relationships and distributions of these species are presented in figure 68. *A. extimum* and *A. parextimum* are both species of lowland habitats. Their complementary distributions suggest that populations in both ranges were once continuous and conspecific across the lowlands of southern New Mexico and Arizona. This area is the only region in which mesic lowland habitat has connected the Chihuahuan and Sonoran deserts since the major Sierra Madrean orogeny (Morafka 1977). Morafka termed this area the Cochise filter barrier. Past climatic fluctuations during episodes of Pleistocene glaciation have alternately changed this region from prairie and riparian grassland to xeric Madro-Tertiary vegetation. The occurrence of these two species today in lowland areas to the east and west of the Cochise filter barrier suggests that changes in climate caused disjunction of the range of a common ancestor, facilitating allopatric speciation.

The sister species of *A. extimum* and *A. parextimum* is *A. cyanope*, which occurs in upland pine-oak woodland at elevations above 5,000′ in Mexico, and between 3,600′ and 6,000′ in Arizona. If it is assumed that reduction of complex structures is much more likely than their origin, it can be concluded that *A. cyanope* arose from an ancestor with full metathoracic development. By the same criterion, it is unlikely an ancestor like *A. cyanope*, with reduced

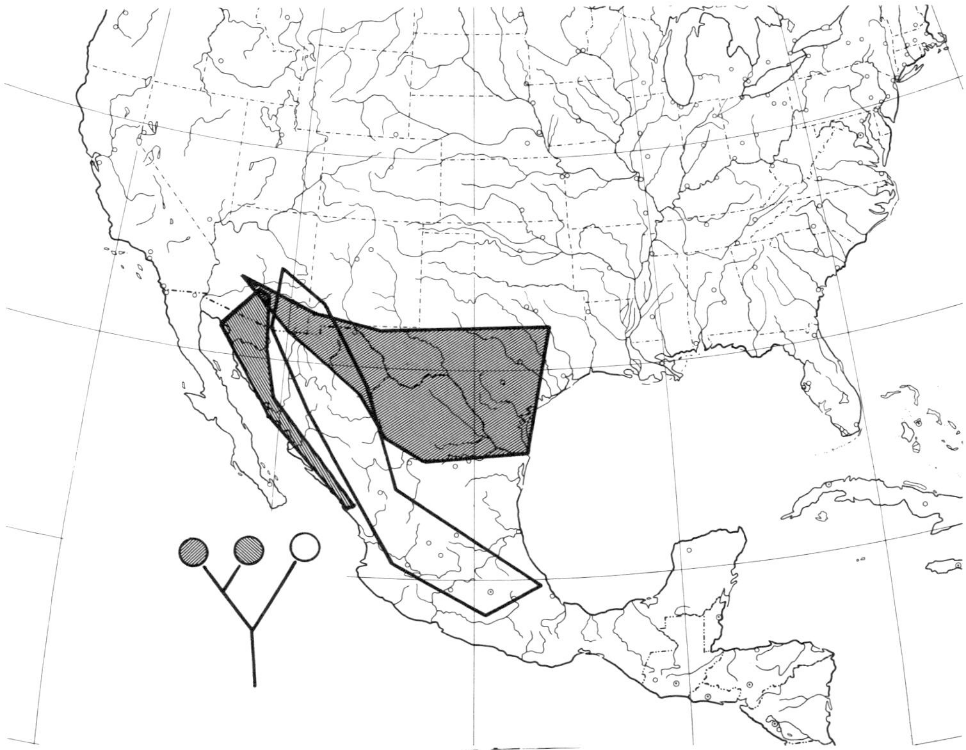

Figure 68. Species ranges in the *A. cyanope* lineage determined by connecting outlying distributional records, accompanied by hypothesis of phylogenetic relationships. Open region represents *A. cyanope*. Shading from lower left to upper right represents *A. extimum*, n. sp.; from upper left to lower right represents *A. parextimum*, n. sp.

metathoracic development, would give rise to species like *A. extimum* and *A. parextimum*. Isolation of peripheral populations of the common ancestor of this lineage in mesic areas could have precipitated speciation of *A. cyanope*. A possible explanation for this event involves north-south vicariance across the Arizona lowlands. During mesic periods, desert riparian populations would have been widely distributed from the Mogollon Rim down the west coast of Mexico and east into the Chihuahuan desert region. Warmer, drier climatic regimes would restrict Sonoran and Chihuahuan lowland riparian habitats, and isolate the more northern populations in more mesic upland habitats. *A. cyanope* could have arisen in mesic canyons north of the drying riparian grasslands of lowland Arizona. Subsequent climatic oscillations could permit tracking of suitable Madrean evergreen woodland throughout the Sierra Madre Occidental by *A. cyanope*.

The other lineage in the *A. extensicolle* group consists of *A. decorum*, *A. elongatulum*, *A. extensicolle*, and *A. texanum*. The affinities of *A. texanum* are not fully resolved, this species being placed at the base of the relaxed consensus network (fig. 66). In this discussion, *A. texanum* can be considered a member of the *A. extensicolle* lineage, as shown in the morphological cladogram (fig. 61). It can also be assumed to be equally related to both lineages, as shown on the relaxed consensus network.

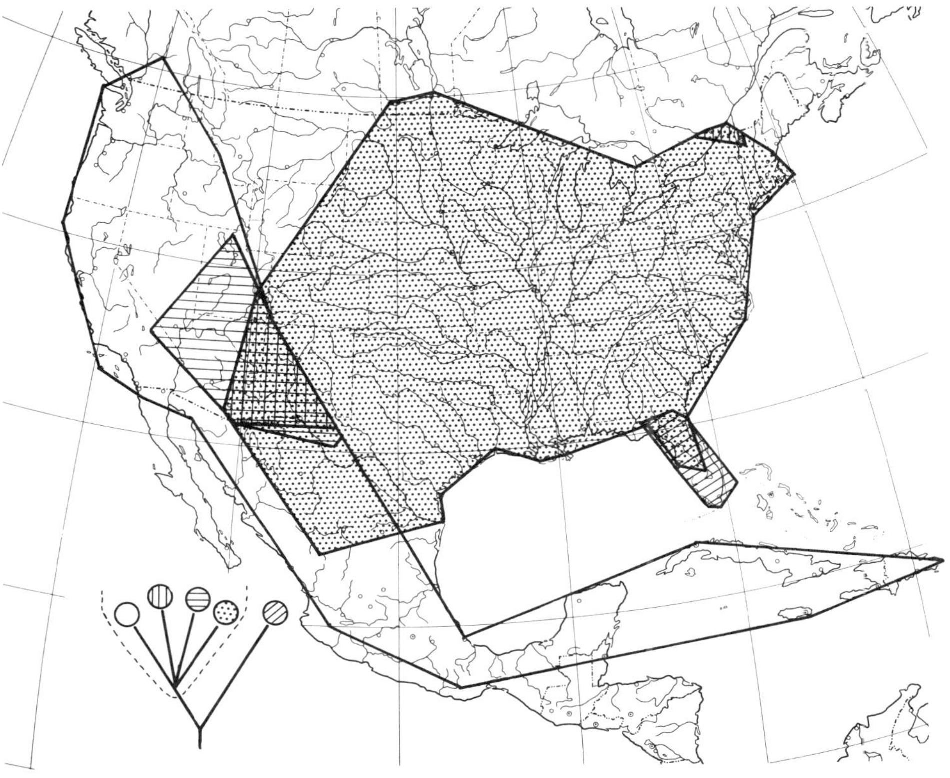

Figure 69. Ranges of forms in the *A. decorum* complex determined by connecting outlying distributional records, accompanied by hypothesis of phylogenetic relationships within the complex. Transversely striped region represents *A. elongatulum*. Dashed line on cladogram encloses forms *A. decorum*: open area represents green-glabrous form; vertical stripes represent green-hirsute form; horizontal stripes represent red-glabrous form; dots represent red-hirsute form.

A. decorum is restricted to lowland lacustrine and riparian marshes. The geographic patterns of the setational and color polymorphisms in *A. decorum* can be associated with climatic changes in the Arizona lowlands. The complementary pattern of the red-hirsute and green-glabrous morphs (fig. 69) may indicate an east-west vicariance in the recent past within this species. Origin of the derived red and hirsute alleles in eastern North America, with subsequent spread, best explains the present day distribution.

As mentioned in Liebherr (1983a), only red-glabrous and green-glabrous morphs occur in the Lake Bonneville drainage in Utah and in several places in eastern California. This could indicate an earlier diffusion of the red allele throughout the western United States, or independent origins of the red allele in several disjunct areas. A reversion to the green state evidently has occurred in the Lake Champlain region, probably since the last glaciation, indicating lability in the state of prothoracic color. Regardless of the history of the polymorphisms within *A. decorum*, present distributions of the alleles governing these polymorphisms indicate very restricted gene flow across the lowlands of Arizona.

The sister to *A. decorum*, *A. elongatulum*, is restricted to peninsular Florida, the Bimini Islands, and a possible founding population along the Atlantic coast (fig. 58). *A. elongatulum* differs biometrically from populations of *A. decorum* on the Greater Antilles. It is most likely that *A. decorum* colonized the Antilles over water. With known Pleistocene sea level changes, relatively small distances separated the Yucatan Peninsula from Cuba (Pregill and Olson 1981; Cronin et al. 1981). The propensity for dispersal in *A. decorum* is shown by a black-light trap capture on an oil rig 18 km off the coast of Louisiana (Erwin 1979a). Overwater dispersal during crepuscular flight is likely, given warm conditions. Colonization of the Antilles has not produced biometric change, as the Antillean populations are similar to those from mainland Mexico (figs. 55, 56). The populations in Florida have diverged, indicating the presence of special conditions in their history.

The special conditions causing divergence of the populations now considered *A. elongatulum* could have been the sea level fluctuations during the Pleistocene. Florida was divided into small islands during periods of sea level rises, and the habitat was rather xeric (Watts 1980). Populations of proto-*A. elongatulum* living there would have been reduced in effective size by habitat restriction. Adaptation to more open, riparian situations may have occurred in these smaller populations, causing divergence. This process might have entailed restriction of founding populations during high sea levels; inter-demic selection at lower sea levels would then have led to adaptation, enabling this species to cope with open riparian habitat. Lande (1980) discusses the theoretical foundation of this mechanism.

Available evidence indicates that the source of the propagules giving rise to *A. elongatulum* is the Greater Antilles. *A. elongatulum* is most similar genetically to *A. decorum* populations in Arizona and California, not populations in eastern North America (fig. 14). Moreover, *A. elongatulum* is biometricaly most similar to *A. decorum* from western North America. It is maximally separated from *A. decorum* populations in the southeastern United States (figs. 20, 21). Finally, *A. elongatulum* does not possess the derived states of hirsute upper body surface (character 11) and red prothorax and mesosternum (character 16) seen in *A. decorum* (fig. 16). This suggests that the source of propagules for *A. elongatulum* should be green-glabrous, at least by the most parsimonious explanation. Colonization of Florida from the south would entail one more occurrence of over water dispersal, in a lineage which colonized the Antilles in such a manner.

The sister species of *A. decorum* and *A. elongatulum* is *A. extensicolle*, which has a large distributional range, principally east of the Rocky Mountains and Sierra Madre Occidental. If we assume that the present range of the green-glabrous form of *A. decorum* best represents the ancestral range of this species, an interesting complementarity of species distributions is seen between *A. decorum* and *A. extensicolle* (fig. 70). One cannot determine the ancestral range of a species with certainty, but for *A. decorum*, an estimate is made in order to maximize the species range complementarity with *A. extensicolle*. This approach assumes these lineages speciated allopatrically with minimal subsequent dispersal.

A. extensicolle occurs in upland situations throughout its range as well as in lowland habitats in eastern North America. The Rocky Mountains form an effective barrier—colonization along riparian corridors accounts for distribution in the American Southwest. The most logical mechanism for the divergence of *A. extensicolle* and the *A. decorum* lineage is allopatric speciation due to east-west vicariance of sets of populations across the filter bridge. Vicariance would be accomplished by xeric conditions in the Arizona lowlands. Subsequent dispersal would have expanded the ranges of *A. extensicolle* to the west of the filter barrier, and *A. decorum* to the east of the filter barrier.

A good complementary fit of present distributions is also seen between *A. texanum* and green-glabrous *A. decorum*. *A. texanum* is found in lowland *Acacia* grassland east of the Cochise filter

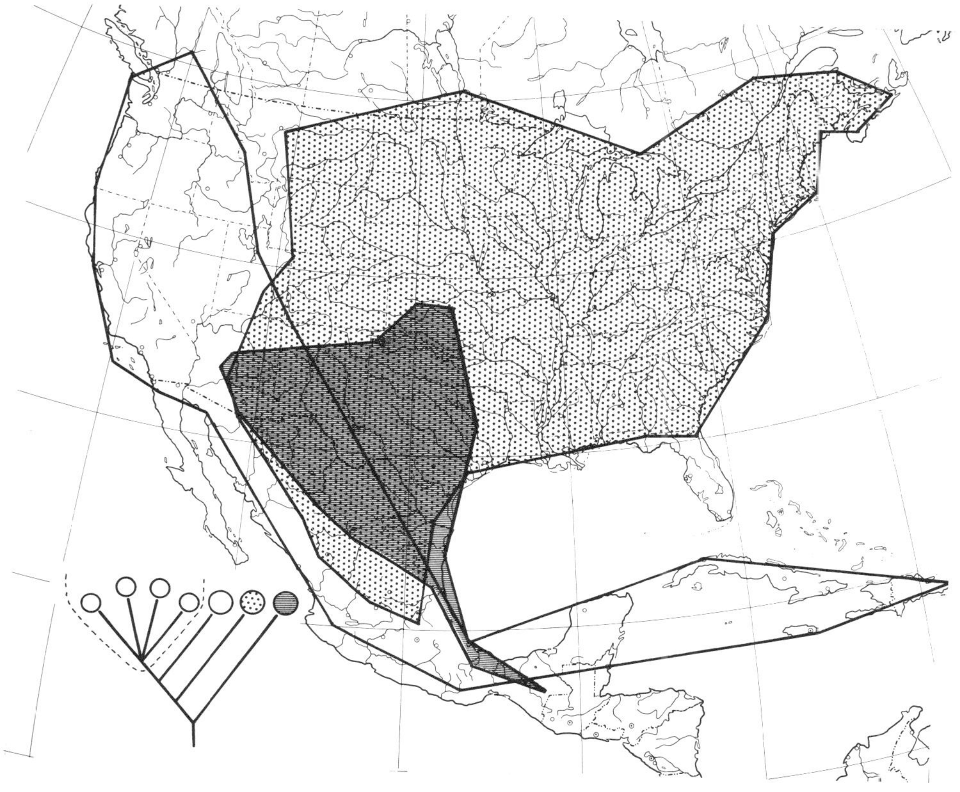

Figure 70. Ranges of *A. extensicolle* (dotted shading) and *A. texanum* (horizontal shading) determined by connecting outlying distributional records, accompanied by hypothesis of phylogenetic relationships between these species and the *A. decorum* complex. Apical unresolved quadricotomy represents *A. decorum,* open large circle represents *A. elongatulum.* Open enclosed area on map is distribution of green-glabrous form of *A. decorum.*

barrier. Its distribution in the east is similar to that of catclaw acacia, *Acacia greggii* A. Gray (Little 1976). Several populations occur west of the filter barrier, but they are divergent biometrically (figs. 44a,b) and probably are recently established. If the common ancestor of the *A. decorum* - *A. elongatulum* - *A. extensicolle* lineage occurred west of the filter barrier, and proto-*A. texanum* populations occurred to the east, the speciation of *A. texanum* can be explained by vicariance across the Arizona lowlands. This vicariance event would predate the origin of *A. extensicolle*, and would involve the development of xeric conditions in the filter barrier, isolating the *Acacia* grassland to the east and marshy *A. decorum*-like habitat to the west.

If we assume that *A. texanum* is not a member of the *A. extensicolle* lineage, but is equally related to the *A. cyanope* and *A. extensicolle* lineages as shown by the relaxed consensus network (fig. 66), the initial vicariance in this group must be explained differently. If a common ancestor were uniformly distributed across California, Nevada, Sonora, and east through the Arizona lowlands into the lowland areas of Chihuahua and Texas, more xeric conditions would separate this distribution into three parts. The Colorado desert would isolate California and Nevada from the Sonoran area, and xeric conditions in the Cochise filter barrier would isolate the Sonoran and Chihuahuan regions. Fragmentation of this original range would produce the three lineages shown in the consensus diagram (fig. 66).

Comparable Biogeographic Patterns in Other Groups. In discussing the biogeography of the *A. extensicolle* group, three patterns are evident. The first involves *A. cyanope* and the ancestor of *A. extimum* and *A. parextimum*. This vicariance event separates an upland species with reduced metathoracic development, *A. cyanope*, and a lowland lineage with well-developed metathoracic phragmata, *A. extimum* and *A. parextimum*. East-west vicariance of sister taxa is the second pattern, with the disjunction of ranges corresponding to the Cochise filter barrier. Divergence events of this type produced the sister taxa *A. extimum* and *A. parextimum*, green-glabrous *A. decorum* and red-hirsute *A. decorum*, *A. decorum* and *A. extensicolle*, and *A. texanum* and the *A. decorum* - *A. extensicolle* lineage. Colonization of the Antilles without speciation from a source in Mexico, with subsequent colonization of Pleistocene Florida from the Antilles with speciation is the third pattern. This pattern is exemplified by the range of *A. decorum*, and divergence of its sister species, *A. elongatulum*. There are similar patterns of geographic distribution in other animals, which suggest events parallel to those theorized for the *A. extensicolle* group.

A. cyanope divergence. The hypothesized divergence of *A. cyanope* from the *A. extimum* - *A. parextimum* stock includes a habitat shift. *A. cyanope* prefers mesic upland habitat, whereas its sister lineage is composed of lowland desert inhabitants. Findley (1969) discussed such a pattern of habitat preference for a number of mammal species groups, calling the phenomenon "the boreal pattern." In bats of the genus *Myotis*, the *lucifugus* and *evotis* species groups both consist of three species, one with an upland forest habitat preference, one found at lower elevations, and a third found in desert to grassland habitats. Climatic fluctuations are hypothesized as the driving force behind these species-habitat relationships. During pluvial periods, mesic-adapted species would be distributed uniformly across the Southwest. During xeric periods, the ranges of such species would be fragmented, and mountain relictual populations formed. These montane populations would be more likely to diverge because of the more restricted habitat and presumably smaller effective population size. Reestablishment of sympatry during the next (or subsequent) pluvial periods would result in altitudinal substitution of closely related species. Findley (1969) cites other examples of this pattern in chipmunks (*Eutamias*), voles (*Microtus*), and rabbits (*Sylvilagus*).

The present day range of *A. cyanope* extends from the Mogollon Rim in Arizona south along the Sierra Madre Occidental to the southern Central Plateau of Mexico (fig. 37). Separation of proto-*A. cyanope* from its lowland sister could have involved any part of this range, and possibly a geographic area not presently occupied by the species. Isolation of marginal populations in mesic canyons north of the Cochise filter barrier is one possible vicariance mechanism. That the *A. cyanope* lineage exhibits a pattern of altitudinal zonation similar to that seen in other groups lends support to the above speciation mechanism.

Cochise filter barrier vicariance. The hypothesized sister group relationships east and west of the Cochise filter barrier in the *A. extensicolle* group are of different ages. In the *A. extensicolle* lineage, three successive east-west vicariance events are hypothesized. If such a process caused the diversity seen in the *A. extensicolle* group, we should find evidence of its effect on other groups restricted to similar habitat.

Among the Carabidae, the genus *Brachinus* contains a number of sister species pairs that conform to the pattern seen in the *A. extensicolle* group (Erwin 1970). These bombardier beetles often are taken with *Agonum* species in riparian and marshy habitats. They are likely candidates for vicariance of sets of populations by the same xeric climatic regimes that would cause disjunctions in the ranges of *Agonum* beetles. Erwin (1970) produced a phylogeny based on cladistic principles for *Brachinus* in North and Middle America. His cladogram includes 29 sister group relationships involving groups of two to four species. These components, representing recently divergent groups, are the ones in which allopatric species pair distributions should be expected. Of the 29 components specified on the cladogram, three involve east-west vicariant species pairs: *B. phaeocerus* Chaudoir and *B. imporcitis* Erwin, *B. kavanaughi* Erwin and *B. mexicanus* Erwin, and *B. aabaaba* Erwin and *B. sonorous* Erwin. *Brachinus favicollis* Erwin is the sister of three other species in its species group. It is found in southern Arizona and the Colorado River Valley. The sister group to *B. favicollis* consists of *B. perplexus* Dejean and *B. imperialensis* Erwin, both predominantly eastern species, and *B. velutinus* Erwin, a species restricted to the Central Valley of California. In this case successive isolation of peripheral isolates would be compatible with the species' distributions. This phenomenon has been termed *centrifugal speciation* (Brown 1957). In at least one divergence event in this species group, east-west vicariance in Arizona corresponds to present day distributions. A fifth vicariance event compatible with action of the Cochise filter barrier involves the *Brachinus texanus* group. *B. elongatulus* Chaudoir is found from west Texas west through the filter bridge and on the Pacific Coast from Oregon south to Oaxaca. Its sister species are *B. rhytiderus* Chaudoir and *B. texanus* Chaudoir. These species are found on the Gulf Coast from Texas to Yucatan and in Central America, and in the eastern United States, respectively. In this case an east-west vicariance event before separation of *B. texanus* and *B. rhytiderus* is compatible with present day distributions.

I have observed *Brachinus mexicanus*, *B. favicollis*, and *B. elongatulus* in sympatry with *A. extensicolle* and *A. decorum* in Arizona. Records (Erwin 1970) for all other *Brachinus* species mentioned above correspond to localities at which one or more species of the *A. extensicolle* group have been taken. Such sympatry connotes similar ecological preference, supporting use of *Brachinus* zoogeographic patterns to corroborate events in the *A. extensicolle* group.

Whitehead (1972) supplied cladograms and distributional maps for the carabid genus *Schizogenius*. There are two species pairs that exhibit east-west vicariance at the Cochise filter barrier: *S. ferrugineus* Putzeys and *S. auripennis* Bates, and *S. scopaeus* Whitehead and *S. pygmaeus* Van Dyke. The western species of both of these pairs, *S. ferrugineus* and *S. pygmaeus*, occur in the same habitats as *A. decorum* or *A. extensicolle*. As in *Brachinus*, similar habitat preferences connote similar vicariating mechanisms.

Other carabid genera that exhibit similar sister species patterns are *Calybe* and *Colliuris*. *Calybe (Ega) sallei* Chevrolat is distributed in the southeastern United States from Texas to Florida, and in Mexico, Guatemala, and Nicaragua (Liebke 1937). *C. (E.) laetula* LeConte, a very similar species, and probably a sister, is found in Arizona, southern California, and Sonora (Liebherr, unpub. data). Liebke (1937) placed *C. pensylvanica* Linné and *C. lengi* Schaeffer in the subgenus *Odacanthella*. These are the only two species in this subgenus found north of Mexico. *C. pensylvanica* is found from east of Arizona through the eastern United States, whereas *C. lengi* occurs in Arizona. If cladistic analysis affirms that these are sister species, another east-west vicariant pair will have been substantiated.

Other groups of animals restricted to habitats that are moist for at least part of the year should exhibit this pattern of vicariance. Morafka (1977) analyzed the affinities of the Chihuahuan, Sonoran, and Mojavean desert herpetofauna, summarizing the biotic diversity of the Chihuahuan desert. Based on his work, that of Stebbins (1966), and the *Catalogue of American Amphibians and Reptiles* (Amer. Soc. of Ichthyologists and Herpetologists 1963 et seq.), two east-west species pairs from at least temporarily moist habitats can be listed: the green toads, *Bufo debilis* Girard and *B. retiformis* Sanders and Smith, and the blind snakes, *Leptotyphlops dulcis* (Baird and Girard) and *L. humilis* (Baird and Girard). Species less restricted to moist habitat but nonetheless occurring as east-west vicariant pairs include the hook-nosed snakes *Ficimia cana* Cope and *F. quadrangularis* Gunther, and the banded geckos, *Coleonyx brevis* Stejneger and *C. variegatus* Cope. Three species of the lyre snake genus *Trimorphodon* occur in parapatric distribution: *T. vandenburghi* Klauber in coastal southern California, *T. lambda* Cope in south and western Arizona, Sonora, Sinaloa, and Baja California Sur, and *T. vilkonsoni* Cope in southern New Mexico and west Texas south into Mexico east of the Sierra Madre Occidental.

Numerous occurrences of subspecific differentiation at the Cochise filter barrier can be cited: the western whiptail lizard, *Cnemidophorus tigris* Baird and Girard (Zweifel 1962); the Great Plains narrow-mouthed toad, *Gastrophryne olivacea* (Hallowell); the kingsnake, *Lampropeltis getulus* Linné; and the coachwhip snake, *Masticophis flagellum* (Shaw) (Stebbins 1966).

Mammalian distributions also reflect disjunctions caused by the Cochise filter barrier. Examples include: the pocket gopher, *Thomomys bottae* (Eydoux and Gervais) (Patton and Yang 1977; Patton and Smith 1981); the pocket mouse, *Perognathus penicillatus* Woodhouse (Patton, Sherwood, and Yang 1981); the harvest mice, *Reithrodontomys montanus* (Baird) and *R. burti* Benson (Hall 1981); the grasshopper mice, *Onychomys arenicola* Mearns and *O. torridus* (Coues) (Hinesley 1979); the antelope squirrels, *Ammospermophilus interpres* (Merriam) and *A. harrisi* Audubon and Bachman (Hall 1981); and the jackrabbits, *Lepus callotis* Wagler and *L. alleni* Mearns (Findley 1969).

The time of divergence of many of these species pairs is unknown. Zweifel (1962) hypothesizes that subspecies of *Cnemidophorus tigris* may have started introgressing in the San Simon Valley of Arizona only in historical time, owing to the habitat alterations made by man. Findley (1969) suggests Pleistocene climatic fluctuations as the cause for species pairs he discusses. Vicariance in the region of the Cochise filter barrier may have been occurring well before the Pleistocene. Cohn (1965) presents a detailed analysis of biogeographic patterns in the long horned grasshopper genus *Neobarretia* (Orthoptera: Tettigoniidae). According to Cohn, diversification of this group into eastern and western lineages occurred in the Miocene. The habitat preference of *Neobarretia* species is specific, and conditions have not been appropriate for east-west exchange since then.

Caribbean relationships. The origin of *A. elongatulum* is hypothesized as a peripheral isolate derived from Antillean green-glabrous *A. decorum*. Pleistocene Florida bore a closer biotic similarity to the Antilles than it did to the rest of southeastern North America. No species pairs could be found that replicated the distributions of *A. elongatulum* and *A. decorum*. A number of distributions indicate relative ease of dispersal from Mexico to the Antilles; these examples parallel the circumscription and distribution of *A. decorum*.

Among carabid beetles, the genus *Loxandrus* includes several species with distributions on the Antilles and in Mexico (Allen 1972; Allen and Ball 1980). *Loxandrus cubanus* Tschitcherine is found on the west coast of Mexico south to Costa Rica, and on the Bahamas, Cuba, Hispaniola, and Puerto Rico. *Loxandrus infimus* Chaudoir is found in Mexico and on Hispaniola. *Loxandrus rectangulus* LeConte is known from the Isthmus of Tehuantepec, Grand Cayman island, and Central Florida, and its sister species *L. nocticolor* Darlington is restricted to Cuba. *Loxandrus floridanus* LeConte is found from east Texas to Florida and on the Bahamas. No meaningful area-taxon statements can be drawn from these distributions, but *L. cubanus* and *L.*

infimus occur in Mexico and the West Indies. Numerous other carabids exhibit species ranges that contain both Mexico and the Antilles (Blackwelder 1944).

The moth genus *Ethmia* (Lepidoptera: Ethmiidae) contains five species groups having members found in southern Mexico and the Antilles (Powell 1973). Of the 48 species in these groups, four are present on keys off the coast of Florida and on at least one Antillean island. The restriction of the Florida records to offshore islands implies that portions of the ranges of these species have been colonized over water. In no case is an Antillean *Ethmia* found in central or northern Florida.

Among mammals, aerial dispersers are often found in Mexico and on the Antilles. Bat species of six genera have colonized the Antilles (Hall 1981). Of these, *Lasiurus intermedius* is represented by ssp. *insularis* (Hall and Jones) on Cuba. This subspecies is more similar to the Mexican subspecies than either is to the Florida subspecies. At least for bats, the straits of Florida appear to be correlated with a number of species boundaries.

Little firm support for the speciation mechanism that created the *A. elongatulum* - *A. decorum* species pair can be drawn from other groups. The choice between deriving the proto-*A. elongatulum* stock from Cuba or from southern North America must be judged on the electrophoretic (fig. 14), biometric (figs. 20, 21) and chorological (fig. 69) data. Both the electrophoretic and biometric data show greater similarity between *A. elongatulum* and Mexican and southwestern *A. decorum* than between southeastern *A. decorum* and *A. elongatulum*. As the chorological data are equivocal, the former two data sets provide the evidence in hypotheses regarding speciation.

Timing of Divergence Events. The repeated involvement of the Cochise filter barrier in speciation events means that speciation has occurred by similar mechanisms at different times. The determination of timing of such events is difficult.

If we assume that structural proteins assayed by electrophoresis change at a relatively constant rate, genetic distances can be used to estimate the time of divergence of taxa. It has been shown that many proteins evolve at a regular rate, with rates of evolution dependent upon the function and attributes of the protein molecule (Wilson, Carlson, and White 1977). Sarich (1977) analyzed data from a number of reptile and mammal species and found a correlation between albumin immunological distance, obtained via microcomplement fixation, and genetic distance, as determined by electrophoresis of proteins. Maxson and Maxson (1979) reported corroborative results using data from plethodontid salamanders. The average rate of change per locus for the electrophoretic data set is dependent on the types of proteins assayed. Secreted proteins, such as those in the variable substrate class (Johnson 1974), acquire substitutions rapidly. Sarich (1977) estimates the average time for substitution of one allele by another for this class of proteins to be 2.4 million years. Other enzymes used in central metabolism would acquire substitutions more slowly, with allelic substitution occurring on the average every 30 million years. In an average study using 25% fast loci and 75% slow loci, 1 unit of Nei's genetic distance would correspond to approximately 20 million years of divergence (Sarich 1977).

Beverley and Wilson (1984) determined an immunological clock for larval hemolymph protein of *Drosophila*. Using a variety of calibration dates, including fossil evidence and geological evidence for the age of the Hawaiian islands, they determined that larval hemolymph protein changes approximately 1.25 unit immunological distance every million years. A phylogeny of representative *Drosophila* species based on larval hemolymph protein immunological distance (Beverley and Wilson 1982) is compatible with that of Throckmorton (1975). The stem lengths of several species, representing species groups, are shown in table 36.

Electrophoretic Nei's genetic distances from three studies (Ayala 1975; Lakovaara, Saura, and Falk 1972; Nei 1971) have been compared with estimated divergence times based on immunological distance to obtain time estimates corresponding to unit Nei's distance. This calibration is a rough estimate, as signified by the wide range of estimates for the three species groups. The actual time period corresponding to unit Nei's distance for *Drosophila* is likely to be shorter, since the genetic distances used for calibration are the range of distances between the most divergent lineages within the species group. The genetic distance from the maximal within group divergence back to the splitting of the species group from its sister lineage is not included in the electrophoretic distance, but is accounted for by the immunological distance. The similarity of the electrophoretic clock based on reptiles, mammals and salamanders (Sarich 1977; Maxson and Maxson 1979) to the estimates made here implies a uniformity in substitution of electrophoretic alleles through time. This uniformity is the basis for rejecting the much shorter evolutionary period for electrophoretic alleles predicted by the model of Nei (1971).

Table 36. Immunological distance to nearest shared ancestor and time of divergence based on *Drosophila* phylogeny of Beverley and Wilson (1982, 1984), range of genetic distance for earliest divergence within species group, and estimate of time per unit Nei D.

Species group	Immunological distance	Time of divergence (m.y.)	Electrophoretic Nei's D	m.y. / unit Nei's D
D. virilis[a]	24	19.2	.65–1.27	15.1–29.5
D. pseudoobscura[b]	30	24	1.30–1.91	12.6–18.5
D. willistoni[c]	35	28	.43–1.00	28–65

[a] Nei (1971)
[b] Lakovaara, Saura, and Falk (1972)
[c] Ayala (1975)

The estimate used for million years per unit Nei's distance is dependent on the percentage of fast and slow loci (Sarich 1977). Rather than fine-tune an electrophoretic clock using limited data, an estimate of 20-25 million years per unit Nei's distance is used to date divergence times in this study. Based on distances in the exemplar electrophoretic data set, the times of divergence for some of the sets of taxa can be determined. The average distance between *A. extimum*, *A. texanum* and the other four taxa off the root of the electrophoretic network (table 30, relationships same as in fig. 59) is .315 (n = 9). This corresponds to a maximal divergence time among these five species of 6.3-7.9 million years. The divergence of *A. extensicolle* from *A. decorum* and *A. elongatulum* occurred at 3.90-4.8 million years ago, based on an average Nei's distance of .193. The next divergence event for which information is available is the separation of *A. elongatulum* and *A. decorum*. The Nei's distance of .097 corresponds to a time of divergence of 1.9-2.4 million years based on the assumed rate.

The large data set used for species circumscription can be used as a check on the above divergence times. The range of Nei's distance values between the *A. extimum* sample and the other 31 samples is .322-.567 (Appendix 4), corresponding to a divergence time between 6.4 and 14.2 million years. This is roughly similar to the 6.3-7.9 million year estimate derived from the exemplar data set. The range of genetic distances between the eight *A. extensicolle* samples and the samples of *A. decorum* and *A. elongatulum* is .180 to .351, corresponding to a divergence time of 3.6-8.8 million years for this speciation event. This broad estimate encompasses the estimate of 3.9-4.8 m.y. based on the exemplar data set.

The separation of *A. decorum* and *A. elongatulum* may have occurred as recently as 160,000 years ago, based on the Nei's distance of .008 between the Canelo and Santa Fe River samples. The maximum distance between these species in the large data set is .079, corresponding to a maximum divergence time of 2.0 million years. This time corresponds well to that estimated from the exemplar data set.

Scenario for Diversification of *A. extensicolle* Group. Based on the estimates of Cenozoic conditions of Cohn (1965) and Morafka (1977), the following scenario can be presented to summarize diversification within the *A. extensicolle* group. The Miocene was warmer and moister than later periods. Such conditions from 8-15 million years ago would have resulted in a continuous mesic region from California through the present day Mojave, Sonoran, and Chihuahuan deserts, as well as the *Acacia* grassland areas of Texas. During this time, the *A. extensicolle* group stock may have been distributed more or less continuously over this area. During early to mid-Pliocene times, 6-12 million years ago, drier climatic conditions caused constriction of the mesic habitats and disjunction of the original stock into three areas, California containing the *A. extensicolle* lineage, the Sonoran desert thorn scrub association containing the *A. cyanope* lineage, and the *Acacia* grassland of coastal Texas containing ancestral *A. texanum*. This event may have occurred in one step or in two. If the latter, *A. texanum* is assumed to be a later vicariant.

Climatic oscillations caused vicariance of *A. cyanope* in mesic areas north of the Cochise filter barrier, whereas the ancestor of *A. extimum* and *A. parextimum* remained in the Sonoran desert region. The timing of this event is unknown, but could have occurred as early as 6 million years ago, when the cool, high-elevation Tertiary forest was isolated by a warmer drier climate (Morafka 1977). This drying was accentuated by the opening of the Gulf of California.

The stock that produced *A. extimum* and *A. parextimum* remained a Sonoran isolate until the late Pliocene or Pleistocene. The relative distinctiveness of the Sonoran biota supports the contention that it did not interact with the Chihuahuan biota until recently. Movement of proto-*A. extimum* - *A. parextimum* stock east into the Chihuahuan desert during Pleistocene climatic fluctuations is hypothesized as the source of *A. extimum*. Whether movement was west to east in all species pairs is unknown, but it has been seen three times in the other lineage. Vicariance at the Cochise filter barrier in the Pleistocene, at most one to two million years ago, is hypothesized as the speciation event producing *A. extimum* and *A. parextimum*.

The late Pliocene saw renewed mountain building activity, and uplift in the Cochise filter barrier. This separated the Madro-Tertiary geoflora into Mojavean and Chihuahuan entities. Subsequent to isolation of *A. texanum* to the east, proto-*A. extensicolle* would have dispersed into the Arizona mountains. A gradual increase in aridity eliminated mesic habitats in the Cochise filter barrier, isolating proto-*A. extensicolle* either to the east, or at higher elevations in the Arizona-New Mexico mountains. Constriction of *A. extensicolle* to upland riparian habitats would produce the needed reduction of population size to facilitate divergence. The electrophoretic clock sets this time from 3.6-8.8 million years ago.

The origin of *A. elongatulum* as a peripheral isolate is dated at less than two million years b.p. If *A. elongatulum* colonized Pleistocene Florida at that time, it must have survived numerous sea level fluctuations due to Pleistocene glaciations (Cronin et al. 1981). A mechanism of population flushes and crashes (Carson 1968; Templeton 1980) is hypothesized as the means of divergence for this species.

The final action of the Cochise filter barrier on the *A. extensicolle* group involves the polymorphisms seen in *A. decorum*. The coincidence of the green-glabrous to red-hirsute transition with the Cochise filter barrier lends support to vicariance within this species as responsible for the distribution of the polymorphisms. Subsequent conjunction of the range and introgression has produced the melange of forms in the southwest. The contraction and expansion of *A. decorum*'s range is an effect of late Pleistocene climatic fluctuations.

APPENDIXES 1-7

APPENDIX 1

Taxon	1	2	3	4	5	6	7	8	9	10	11	12	13	14	15	16	17	18	19	20	21	22	23	24	25	26	27
	Characters																										
Synuchus impunctatus Say	1	0	0	0	0	1	0	0	0	0	0	0	0	0	1	1	0	1	0	1	0	1	0	1	0	1	0
Pristonychus complanatus Dej.	0	0	0	0	1	1	0	0	0	0	0	0	0	0	1	1	0	1	0	0	0	0	1	1	1	1	0
Calathus ruficollis Dej.	0	0	0	0	0	1	0	0	0	0	0	0	0	0	1	1	0	1	0	0	0	0	0	1	1	1	0
Olisthopus parmatus Say	1	0	0	0	0	0	0	0	0	0	0	0	0	1	0	0	0	0	0	0	0	0	0	0	0	1	0
Elliptoleus sp.	0	0	1	0	0	0	1	0	0	0	0	0	0	0	0	0	0	0	0	0	0	0	0	0	0	0	0
Sericoda obsoleta Say	0	0	1	0	0	0	1	0	0	0	0	0	0	0	0	0	0	1	0	0	0	0	0	0	0	0	0
Anchomenus dorsalis Pont.	0	0	1	0	0	0	1	0	0	0	0	0	0	0	0	0	0	0	0	0	0	0	0	0	0	1	0
Anchomenus picticornis Newm.	0	0	1	0	0	0	1	0	0	0	0	0	0	0	0	0	0	0	0	0	0	0	0	0	0	1	0
Pseudanchus funebris LeC.	0	0	1	0	0	0	1	0	0	0	0	0	0	0	0	0	0	0	0	0	0	0	0	0	0	1	0
Pseudanchus quadratus LeC.	0	0	1	0	0	0	1	0	0	0	0	0	0	0	0	0	0	0	0	0	0	0	0	0	0	1	0
Atranus pubescens Dej.	0	0	0	0	0	0	0	1	0	1	0	0	0	0	1	0	1	-	0	0	0	0	0	0	0	1	0
Rhadine caudata LeC.	0	0	0	1	0	0	0	1	0	0	0	0	0	0	1	0	1	-	0	0	0	0	0	0	0	1	1
Rhadine nivalis Horn	0	0	0	1	0	0	0	1	0	0	0	0	0	0	1	0	0	1	0	0	0	0	0	0	0	1	1
Tanystoma striata Dej.	0	0	0	1	0	0	0	1	0	0	0	0	0	0	0	0	0	0	0	0	0	0	0	0	0	1	1
Tanystoma maculicolle Dej.	0	0	0	1	0	0	0	1	0	0	0	0	1	0	0	0	0	0	0	0	0	0	0	0	0	1	1
Anchus pusillus LeC.	0	0	0	1	0	0	0	1	0	0	0	0	0	0	0	0	0	1	0	0	0	0	0	0	0	1	0
Paranchus albipes Fabr.	0	0	0	1	0	0	0	1	0	0	0	0	0	0	0	0	0	0	0	0	0	0	0	0	0	1	0
Agonum collare Say	0	0	0	1	0	0	0	1	0	0	0	0	0	0	0	0	0	0	0	0	0	0	0	0	0	1	1
Agonum mutatum Gemm. & Har.	0	0	0	1	0	0	0	1	0	0	0	0	0	0	0	0	0	0	0	0	0	0	0	0	0	1	1
Agonum 4-maculatum Horn	0	0	0	0	1	0	1	0	0	0	0	0	0	0	0	0	0	1	1	0	0	0	0	0	0	1	0
Agonum extensicolle Say	0	0	0	0	1	0	1	0	0	0	0	0	0	0	0	0	0	0	0	0	0	0	0	0	0	1	0
Agonum cyanope Bates	0	0	0	0	1	0	1	0	0	0	0	1	1	0	0	0	0	0	0	0	0	0	0	0	0	1	0
Agonum thoreyi Dej.	0	0	0	0	1	0	1	0	0	0	0	0	0	0	0	0	0	0	0	0	0	0	0	0	0	1	0
Agonum anchomenoides Rand.	0	0	0	0	1	0	1	0	0	0	0	0	0	0	0	0	0	0	0	0	0	0	0	0	0	1	0
Agonum variolatum LeC.	0	0	0	0	1	0	1	0	0	0	0	0	0	0	0	0	0	0	1	0	0	0	0	0	0	1	0
Agonum nutans Say	0	0	0	0	1	0	1	0	0	0	0	0	0	0	0	0	0	0	0	0	0	0	0	0	0	1	0
Agonum punctiforme Say	0	0	0	0	1	0	1	0	0	0	0	0	0	0	0	0	0	0	0	0	0	0	0	0	0	1	0
Agonum excavatum Dej.	0	0	0	0	1	0	1	0	0	0	0	0	0	0	0	0	0	0	0	0	0	0	0	0	0	1	0
Agonum pacificum Casey	0	0	0	0	1	0	1	0	0	0	0	0	0	0	0	0	0	0	1	0	0	0	0	0	0	1	0
Agonum sexpunctatum L.	0	0	0	0	1	0	1	0	0	0	0	0	0	0	0	0	0	0	0	0	0	0	0	0	0	1	0
Agonum harrisi LeC.	0	0	0	0	1	0	1	0	0	0	0	0	0	0	0	0	0	0	0	0	0	0	0	0	0	1	0
Agonum placidum Say	0	0	0	0	1	0	1	0	0	0	0	0	0	0	0	0	0	1	0	0	0	0	0	0	0	1	0
Agonum albicrus Dej.	0	0	0	0	1	0	1	0	0	0	0	0	0	0	0	0	0	0	0	0	0	0	0	0	0	1	0
Agonum languidum Horn	0	0	0	0	1	0	0	1	0	0	0	0	0	0	0	0	1	-	1	0	0	0	0	0	0	1	0
Agonum nigriceps LeC.	0	0	0	0	1	0	0	0	0	0	0	0	0	0	0	0	0	0	1	0	0	0	0	0	0	1	0
Agonum sulcipenne Horn	0	1	0	0	1	0	0	0	0	0	0	0	1	0	0	0	0	0	0	0	0	0	0	0	0	1	0
Agonum ferreum Hald.	0	0	0	0	1	0	0	0	0	0	0	0	1	0	0	0	0	0	0	0	0	0	0	0	0	1	0
Agonum errans Say	0	0	0	0	1	0	0	0	0	0	0	0	1	0	0	0	0	0	0	0	0	0	0	0	0	1	0
Platynus decentis Say	0	0	0	0	1	0	0	1	1	0	0	1	1	0	0	0	0	0	0	1	0	0	0	0	0	1	0
Platynus ovipennis Mann.	0	0	0	0	1	0	0	1	0	0	0	1	1	0	0	0	0	0	1	0	0	0	0	0	0	1	0
Platynus brunneomarginatus Mann.	0	0	0	0	1	0	0	0	0	0	0	0	0	0	0	0	0	0	0	0	0	0	0	0	0	1	0
Platynus hypolithos Say	0	0	0	0	1	0	0	0	0	0	1	0	0	0	0	0	0	0	1	0	0	0	0	0	0	1	0
Onypterygia tricolor Dej.	0	0	0	0	1	0	0	0	0	0	0	0	1	0	0	0	0	0	1	0	1	0	0	0	0	1	1
Patrobus longicornis Say	0	0	0	0	1	0	0	0	0	0	0	1	0	0	1	0	1	-	0	1	0	0	0	-	-	1	1
Diplous californicus Mots.	0	0	0	0	1	0	0	0	0	0	0	1	0	0	1	1	1	-	0	1	0	0	0	-	-	0	1
Pterostichus menetriesi Mots.	0	0	0	0	0	1	1	1	0	0	0	0	0	0	0	0	0	1	0	0	0	0	0	0	0	1	1
Pterostichus angustatus Dej.	0	0	0	0	0	1	1	1	0	0	0	0	0	0	1	0	0	0	0	0	0	0	0	0	0	1	0
Pterostichus lucublandus Say	0	0	0	1	0	0	0	1	0	0	0	0	0	0	0	0	0	0	0	0	0	0	1	0	0	1	0
Pterostichus haldemani LeC.	0	0	0	0	0	1	0	0	0	1	0	0	0	0	0	0	0	1	0	0	0	0	0	0	0	1	1
Evarthrus sodalis LeC.	0	0	0	0	0	1	1	0	0	0	0	1	0	0	1	0	1	-	0	1	0	1	0	0	0	1	1
Abacidus permundus Say	0	0	0	0	0	0	0	0	0	0	0	0	0	0	1	0	0	1	0	0	0	0	0	1	1	1	0
Myas coracinus Dej.	0	0	0	0	0	1	0	0	0	1	0	0	0	0	1	1	1	-	0	1	0	1	0	1	0	1	1
Loxandrus celeris Dej.	0	0	0	0	0	1	0	0	0	0	0	1	0	0	1	1	1	-	1	0	0	0	0	0	0	1	1
Harpalus caliginosus Fabr.	0	0	0	0	1	0	0	0	0	0	0	0	0	0	1	1	1	-	0	0	0	0	0	0	0	1	1
Lebia grandis Hentz	0	0	0	0	1	1	1	0	0	0	0	1	0	0	1	1	1	-	0	1	0	1	1	0	0	1	1

Taxon	Characters																												
	28	29	30	31	32	33	34	35	36	37	38	39	40	41	42	43	44	45	46	47	48	49	50	51	52	53	54	55	56
S. impunctatus	0	0	0	0	1	0	0	0	1	1	0	0	0	0	1	0	0	0	1	0	0	1	0	C	0	0	0	1	0
P. complanatus	0	0	0	0	1	1	0	0	1	0	0	0	0	0	1	0	0	0	1	0	0	0	0	C	0	0	0	1	0
C. ruficollis	0	0	0	0	1	1	0	0	0	0	0	0	0	0	0	0	0	0	1	0	0	1	1	C	0	0	0	1	0
O. parmatus	0	0	0	0	1	1	0	1	0	0	0	0	0	0	1	0	0	0	0	0	0	1	1	C	0	0	0	0	0
Elliptoleus sp.	0	0	0	0	0	0	0	0	0	0	0	0	0	0	0	0	0	0	0	0	0	1	1	C	0	0	0	0	0
S. obsoleta	0	0	0	0	0	0	0	0	0	0	0	0	0	0	0	0	0	0	0	0	0	1	0	C	0	0	0	0	0
A. dorsalis	0	0	0	1	1	1	0	0	0	0	0	0	0	1	0	0	0	0	0	0	0	1	1	C	0	0	0	0	0
A. picticornis	0	0	0	0	0	0	1	0	0	1	0	0	0	1	0	1	0	1	0	0	0	1	1	C	0	1	0	0	0
P. funebris	0	0	0	0	0	0	0	0	0	0	0	0	0	0	0	0	0	0	0	1	0	1	1	1	0	1	0	0	0
P. quadratus	0	0	0	0	0	0	0	0	0	0	0	0	0	0	0	0	0	0	0	1	0	1	1	1	0	1	0	0	0
A. pubescens	0	0	1	1	1	1	0	0	1	0	0	1	0	0	1	0	0	1	0	1	1	0	0	0	0	0	0	0	0
R. caudata	0	0	0	0	0	0	0	0	0	0	0	1	0	1	0	1	1	0	0	0	0	1	1	1	0	0	0	0	0
R. nivalis	0	0	0	0	0	0	0	0	1	0	1	1	0	1	0	0	0	0	0	0	0	1	1	C	0	1	1	0	0
T. striata	0	0	0	0	0	1	0	0	0	0	0	0	0	0	0	1	0	0	0	0	0	1	0	0	0	0	0	0	0
T. maculicolle	0	0	0	0	0	1	0	0	0	0	0	0	0	0	0	1	0	0	0	0	0	1	1	C	0	0	0	0	0
A. pusillus	0	0	0	1	1	1	0	0	1	0	0	1	0	0	1	0	0	1	0	1	1	1	1	C	0	0	0	0	0
P. albipes	0	0	0	0	0	0	0	0	0	0	1	1	0	0	1	0	0	0	0	0	0	1	1	1	0	0	0	0	0
A. collare	0	0	0	0	0	0	0	0	0	0	0	0	0	0	0	0	0	0	0	1	0	1	1	C	0	0	0	0	0
A. mutatum	1	0	0	0	0	0	0	0	0	0	0	0	0	0	0	0	0	0	0	1	0	1	1	C	0	0	0	0	0
A. 4-maculatum	0	0	0	1	1	1	0	0	0	0	0	0	0	1	0	0	0	1	0	1	1	1	1	C	0	0	0	0	0
A. extensicolle	1	0	0	0	1	1	0	0	0	0	0	0	0	0	0	1	0	0	0	1	1	1	1	C	0	0	0	0	0
A. cyanope	1	0	0	1	1	1	0	0	0	0	0	0	0	0	0	1	0	0	0	1	1	1	1	C	0	0	0	0	0
A. thoreyi	1	0	0	1	0	1	0	0	0	0	0	0	0	0	0	1	0	0	0	0	0	1	1	1	0	0	0	0	0
A. anchomenoides	0	0	0	1	0	1	0	0	0	0	0	0	0	0	0	1	0	0	0	0	0	1	1	C	0	0	0	0	0
A. variolatum	1	0	0	0	0	1	0	0	0	0	0	0	0	0	0	1	0	0	1	1	1	0	0	C	0	0	0	0	0
A. nutans	1	0	0	0	0	1	0	0	0	0	0	0	0	0	0	1	0	0	0	1	0	1	1	C	0	0	0	0	0
A. punctiforme	1	0	0	0	0	0	0	0	0	0	0	0	0	0	0	0	0	0	0	0	0	1	1	C	0	0	0	0	0
A. excavatum	1	0	0	0	0	0	0	0	0	0	0	0	0	0	0	0	0	0	0	1	1	1	1	C	0	1	0	0	0
A. pacificum	1	1	0	0	0	0	0	0	0	0	0	0	0	0	0	1	0	0	0	1	1	1	1	C	0	0	0	0	0
A. sexpunctatum	1	1	0	0	0	0	0	0	0	0	0	0	0	0	0	1	0	0	0	1	1	1	1	C	0	0	0	0	0
A. harrisi	1	1	0	0	0	0	0	0	0	0	0	0	0	0	0	0	0	0	0	0	0	1	1	C	0	0	0	0	0
A. placidum	0	0	0	0	0	0	0	0	0	0	0	0	0	0	0	1	0	0	0	0	0	1	1	C	0	0	0	0	0
A. albicrus	0	0	0	0	0	0	1	0	0	0	0	0	0	0	0	0	0	0	0	1	1	1	1	C	0	0	0	0	0
A. languidum	0	0	0	0	0	0	0	0	0	0	0	0	0	0	0	0	0	0	0	0	0	1	1	C	0	0	0	0	0
A. nigriceps	0	0	0	0	0	0	0	0	0	0	0	0	0	0	0	0	0	0	0	0	0	1	1	C	0	0	0	0	0
A. sulcipenne	0	0	0	0	0	0	0	0	0	0	0	0	0	0	0	0	0	0	0	0	0	1	1	C	0	1	0	0	0
A. ferreum	0	0	0	0	0	0	0	0	0	0	0	0	0	0	0	0	0	0	0	0	0	1	0	C	0	1	0	0	0
A. errans	1	0	0	0	0	0	0	0	0	0	0	0	0	0	0	0	0	0	0	0	0	1	1	C	0	1	0	0	0
P. decentis	0	0	0	0	0	0	0	0	0	1	0	1	0	0	0	0	0	0	0	0	0	1	1	C	0	1	1	0	0
P. ovipennis	1	0	0	0	0	0	0	0	1	0	0	1	0	0	0	0	0	0	0	0	0	1	1	C	0	0	0	0	0
P. brunneomarginatus	0	0	0	0	0	0	0	0	0	0	0	1	0	0	0	0	0	0	0	0	0	1	1	C	0	1	1	0	0
P. hypolithos	0	0	0	0	0	0	0	0	0	1	1	1	0	0	0	1	1	0	0	0	0	1	1	C	0	0	0	0	0
O. tricolor	0	0	0	0	0	0	0	0	0	0	1	1	1	0	1	0	0	0	0	0	0	1	1	C	1	0	0	1	1
P. longicornis	0	0	0	1	1	1	0	0	1	0	1	1	0	0	0	0	0	0	0	0	0	1	1	C	0	1	1	0	0
D. californicus	0	0	0	1	0	0	0	0	1	0	1	1	0	0	0	0	0	0	0	0	0	1	0	C	0	1	1	0	0
P. menetriesi	0	1	0	0	0	0	0	0	1	0	1	0	0	0	1	0	0	0	0	0	0	1	0	C	0	0	0	0	0
P. angustatus	0	0	0	0	0	0	0	0	0	1	0	1	0	0	1	0	0	0	1	0	0	1	0	C	0	1	1	0	0
P. lucublandus	0	0	0	0	0	0	0	0	1	1	0	0	0	0	1	0	0	0	1	0	0	1	1	C	0	0	0	0	0
P. haldemani	0	1	0	0	0	0	0	0	1	0	0	0	0	0	0	0	0	0	1	0	0	1	0	C	0	1	1	0	0
E. sodalis	0	0	0	0	0	0	0	0	1	-	0	0	0	0	1	0	0	0	1	0	0	1	0	C	0	0	0	0	0
A. permundus	0	0	0	0	0	0	0	0	1	0	1	0	0	0	0	0	0	0	1	0	0	1	0	C	0	0	0	0	0
M. coracinus	0	0	0	0	0	0	0	0	0	0	1	0	0	0	1	0	0	0	1	0	0	0	0	C	0	0	0	0	0
L. celeris	0	0	0	0	0	0	0	0	0	0	0	0	0	0	1	0	0	0	0	0	0	1	1	0	0	0	0	0	0
H. caliginosus	0	0	0	1	0	1	0	1	0	0	0	0	0	1	1	0	0	0	0	1	1	0	0	0	0	0	0	0	0
L. grandis	0	0	0	0	0	0	0	0	0	1	0	1	0	0	1	0	0	0	0	1	1	0	0	0	1	0	0	1	1

APPENDIX 2

Classification of North American Platynini

Listing the Taxa Studied for the Cladistic Analysis

Tribe Platynini
- unplaced genus
  - Olisthopus Dejean (6 spp.)
    - parmatus Say*
- Subtribe Sphodri Laporte
  - Pristonychus Dejean (2 spp., introduced)
    - complanatus Dejean *
  - Calathus Bonelli (22 spp.)
    - gregarius Say*
    - peropacus Casey*
    - ruficollis Dejean*
  - **Synuchus Gyllenhal (2 spp.)
    - dubius LeConte*
    - impunctatus Say*
- Subtribe Platyni Bonelli
  - Elliptoleus Bates (6 spp.)
    - sp. 1*
  - **Sericoda Kirby (4 spp.)
    - bembidioides Kirby (type species)
    - bogemanni Gyllenhal*
    - obsoleta Say*
    - quadripunctata DeGeer
  - **Anchomenus Bonelli (1 sp. in North America)
    - dorsalis Pontipiddan (type species, European)*
    - picticornis Newman*
  - **Pseudanchus Casey (3 spp.)
    - aeneolus LeConte*
    - funebris LeConte (type species)*
    - quadratus LeConte*
  - **Atranus LeConte (1 sp.)
    - pubescens Dejean (type species)*
  - Rhadine LeConte (approx. 60 spp., Barr 1974)
    - caudata LeConte*
    - nivalis Horn*
    - jejuna group sp. 1*
  - **Tanystoma Motschulsky (4 spp.)
    - cuyama Liebherr*
    - maculicolle Dejean*
    - striata Dejean (type species)*
    - sulcata Dejean*
  - **Paranchus Lindroth (1 sp.)
    - albipes Fabricius (type species, European, introduced)*
  - **Anchus LeConte (1 sp.)
    - pusillus LeConte (type species)*
  - **Agonum Bonelli (71 spp. excluding exclusively Mexican spp., A. marginatum L. type species, European)
    - A. collare species group
      - collare Say (type species, hereby designated)*
      - mutatum Gemminger and Harold*
    - A. ferreum species group (Stereagonum Casey)
      - errans Say*
      - ferreum Haldeman (type species)*
      - sulcipenne Horn*
    - A. nigriceps species group (Platynomicrus Casey)
      - bicolor Dejean
      - ferruginosum Dejean*
      - languidum Horn*
      - nigriceps LeConte (type species)*
      - piceolum LeConte*
    - A. marginatum species group (Agonum s.s., European A. marginatum L. type species)
      - affine Kirby
      - albicrus Dejean*
      - arizonense Horn*
      - belleri Hatch*
      - brevicolle Dejean*
      - corvus LeConte*
      - cupreum Dejean*
      - cupripenne Say*
      - deceptivum LeConte
      - excavatum Dejean*
      - fallianum Leng*
      - fidele Casey*
      - fossigerum Dejean*
      - harrisi LeConte*

A. marginatum species group (cont.)
- melanarium Dejean*
- metallescens LeConte*
- moerens Dejean*
- muelleri Herbst (European, introduced)*
- muiri Liebherr*
- octopunctatum Fabricius*
- pacificum Casey*
- placidum Say*
- propinquum Gemminger and Harold*
- quinquepunctatum Motschulsky*
- subsericeum LeConte*
- tenue LeConte*
- trigeminum Lindroth*

A. punctiforme species group (Circinalia Casey)
- crenistriatum LeConte*
- pallipes Fabricius*
- punctiforme Say (type species)*
- rigidulum Casey
- rufipes Dejean

A. nutans species group (Micragonum Casey)
- aeruginosum Dejean*
- basale LeConte*
- crenulatum LeConte
- nutans Say (type species)*
- striatopunctatum Dejean*
- variolatum LeConte*

A. sordens species group (Europhilus Chaudoir, European A. micans Nic. type species)
- anchomenoides Randall*
- canadense Goulet*
- consimile Gyllenhal*
- darlingtoni Lindroth
- exaratum Mannerheim*
- galvestonicum Casey
- gratiosum Mannerheim*
- lutulentum LeConte*
- palustre Goulet*
- picicornoides Lindroth
- retractum LeConte*
- simile Kirby
- sordens Kirby*
- superioris Lindroth
- thoreyi Dejean*

A. extensicolle species group (Stictanchus Casey)
- cyanope Bates*
- decorum Say*
- elongatulum Dejean*
- extensicolle Say (type species)*
- extimum Liebherr*
- parextimum Liebherr*
- texanum LeConte*

A. quadrimaculatum species group (Deratanchus Casey)
- quadrimaculatum Horn (type species)*

Platynus Bonelli (many spp.)
- angustatus Dejean*
- brunneomarginatus Mannerheim*
- cincticollis Say*
- decentis Say*
- hypolithos Say*
- ovipennis Mannerheim*
- tenuicollis LeConte*

Onypterygia Dejean (19 spp.)
- fulgens Dejean*
- tricolor Dejean*

Colpodes MacLeay
- buchanani Hope (introduced)*

Glyptolenus Bates (9 spp.)
- sp. 1*

* Taxon was studied using internal dissection.

** Listing under generic name includes all valid species for America north of Mexico.

APPENDIX 3

Allelic frequencies for variable loci. Alleles labeled by mobility relative to most common allele (1.00). n = number of individuals scored.

Locus	Allele	1 NICA	2 DIXN	3 BONT	4 LAG1	5 LAG2	6 FISH	7 DIAZ	8 LITL	9 DSVY	10 CARL	11 HIKO	12 PAHR	13 SASP	14 NOGA	15 CANO	16 SART	17 NOXB	18 BURL	19 STFE	20 OKEE	21 ARCH	22 MYAK	23 OKEE	24 ITHA	25 VERM	26 BAMA	27 LESL	28 CAVE	29 PEBL	30 RUCK	31 TURK	32 COAH
Est-d	1.14	0.	0.	0.	0.	0.	0.	0.	0.	0.	.055	0.	0.	0.	0.	0.	0.	0.	0.	0.	0.	0.	0.	0.	0.	0.	0.	0.	0.	0.	0.	0.	0.
	1.10	0.	.050	.050	.024	.083	0.	0.	0.	.024	0.	0.	0.	0.	0.	0.	0.	0.	.029	.147	.083	.233	.278	.083	0.	0.	.055	0.	0.	0.	0.	0.	0.
	1.00	.800	.500	.400	.429	.472	.875	.722	.667	.619	.389	.667	.778	.833	.652	.648	.833	.833	.706	.588	.584	.500	.500	.694	1.0	.945	0.	.278	.055	0.	0.	0.	0.
	.87	.200	.400	.550	.500	.445	.125	.278	.222	.333	0.	0.	0.	0.	.303	.315	0.	.167	.265	.265	.330	.267	.194	.167	0.	.055	.945	.722	.890	.945	.813	1.0	0.
	.78	0.	.050	0.	.047	0.	0.	0.	.111	.024	.556	.333	.222	.167	.045	.037	.167	0.	0.	0.	0.	0.	.028	.056	0.	0.	0.	0.	.055	.055	.187	0.	0.
	.64	0.	0.	0.	0.	0.	0.	0.	0.	0.	0.	0.	0.	0.	0.	0.	0.	0.	0.	0.	0.	0.	0.	0.	0.	0.	0.	0.	0.	0.	0.	0.	.055
	.57	0.	0.	0.	0.	0.	0.	0.	0.	0.	0.	0.	0.	0.	0.	0.	0.	0.	0.	0.	0.	0.	0.	0.	0.	0.	0.	0.	0.	0.	0.	0.	.945
	n	10	10	10	22	18	9	9	9	21	9	9	9	12	31	28	12	12	18	18	18	18	18	18	11	9	9	9	9	9	8	9	9
Phpr	1.17	.050	0.	0.	0.	0.	0.	0.	0.	0.	0.	0.	.042	0.	0.	0.	0.	0.	.028	.036	.154	0.	0.	0.	0.	0.	0.	0.	0.	0.	0.	0.	0.
	1.12	.500	.250	.300	.119	.278	.111	.167	.167	.088	0.	.111	.042	0.	0.	.071	.333	.250	.028	0.	.038	.042	.214	.118	0.	0.	0.	0.	0.	0.	0.	0.	0.
	1.00	.300	.600	.650	.548	.528	.722	.611	.555	.529	.611	.444	.666	.450	.559	.286	.417	.625	.611	.571	.538	.500	.536	.441	.045	0.	0.	0.	0.	.187	.125	0.	.055
	.87	0.	0.	0.	.333	.111	.055	.055	.055	.206	.333	.222	.042	.300	.382	.500	.062	.111	.143	.192	.250	.071	.176	.176	.227	.222	.250	0.	0.	.063	0.	0.	.222
	.81	.100	.100	.050	0.	.055	.111	.167	.167	.176	.055	.222	.166	.200	.059	.143	0.	0.	.194	.214	0.	.208	.071	.147	.409	.444	.250	.750	.714	.750	.666	1.0	0.
	.70	0.	.050	0.	0.	0.	0.	0.	.055	0.	0.	0.	.042	.050	0.	0.	0.	.062	.028	0.	0.	0.	.071	.059	.045	0.	0.	0.	0.	0.	.042	0.	.500
	.60	.050	0.	0.	0.	.028	0.	0.	0.	0.	0.	0.	0.	0.	0.	0.	0.	0.	0.	.036	.077	0.	.036	.059	.273	.333	.500	.083	.143	0.	.042	0.	.222
	.52	0.	0.	0.	0.	0.	0.	0.	0.	0.	0.	0.	0.	0.	0.	0.	0.	0.	0.	0.	0.	0.	0.	0.	0.	0.	0.	.167	.143	0.	.125	0.	0.
	n	10	10	10	21	18	9	9	9	17	9	0	12	10	17	7	6	8	18	14	13	12	14	17	11	9	2	6	7	8	12	4	9
Acph	2.05	0.	0.	0.	0.	0.	0.	0.	0.	0.	0.	0.	0.	0.	0.	0.	.055	0.	0.	0.	0.	0.	0.	0.	0.	0.	0.	0.	0.	0.	0.	0.	0.
	1.72	.100	.100	0.	0.	0.	.125	0.	0.	0.	0.	0.	0.	.042	.047	.019	.055	0.	.083	0.	.083	.083	0.	0.	0.	0.	0.	0.	0.	0.	0.	0.	0.
	1.00	.900	.900	1.0	1.0	1.0	.875	1.0	.889	1.0	1.0	1.0	1.0	.958	.937	.962	.890	.962	.917	.971	.862	.806	1.0	1.0	0.	.062	.062	0.	0.	0.	0.	0.	0.
	.50	0.	0.	0.	0.	0.	0.	0.	.111	0.	0.	0.	0.	0.	.016	.019	0.	.038	0.	.029	.055	.111	0.	0.	1.0	.938	.938	1.0	1.0	1.0	1.0	1.0	1.0
	n	10	10	10	21	18	8	9	9	20	8	9	9	12	32	26	9	13	18	17	18	18	18	18	11	8	8	8	9	7	9	9	9
Pgi	1.72	0.	0.	0.	0.	0.	0.	0.	0.	0.	0.	0.	0.	0.	0.	0.	0.	0.	0.	0.	0.	.042	0.	0.	.045	0.	0.	0.	0.	0.	0.	0.	0.
	1.46	0.	.055	0.	0.	0.	0.	0.	0.	0.	0.	0.	.055	0.	0.	0.	0.	0.	0.	0.	0.	0.	.038	.042	0.	0.	0.	.083	0.	0.	0.	0.	0.
	1.37	.050	0.	0.	.024	.111	0.	0.	.062	0.	.111	0.	0.	0.	.055	0.	0.	0.	.031	.077	.056	.083	0.	.042	.136	0.	.250	0.	0.	.250	.375	0.	0.
	1.30	0.	0.	0.	0.	0.	0.	0.	.062	.036	0.	0.	0.	.250	0.	0.	0.	0.	0.	0.	0.	0.	.038	.042	0.	0.	0.	0.	0.	0.	0.	0.	0.
	1.21	.050	0.	0.	0.	0.	.111	.214	.062	0.	0.	0.	0.	0.	.055	0.	0.	.071	0.	0.	0.	0.	0.	0.	.045	0.	0.	0.	0.	0.	0.	0.	0.
	1.00	.900	.890	.814	.857	.695	.889	.786	.752	.964	.695	.929	.723	.688	.890	.929	.800	.786	.844	.808	.555	.750	.731	.707	.638	.889	.750	.917	1.0	.750	.625	1.0	.938
	.70	0.	0.	0.	0.	0.	0.	0.	0.	0.	0.	0.	0.	0.	0.	.071	0.	0.	0.	0.	.056	0.	0.	0.	0.	0.	0.	0.	0.	0.	0.	0.	0.
	.62	0.	.055	.062	.095	.194	0.	0.	.062	0.	.194	.071	.222	0.	0.	0.	.200	.143	.125	.077	.333	.125	.154	.167	0.	.111	0.	0.	0.	0.	0.	0.	0.
	.37	0.	0.	0.	0.	0.	0.	0.	0.	0.	0.	0.	0.	0.	0.	0.	0.	0.	0.	.038	0.	0.	0.	0.	.091	0.	0.	0.	0.	0.	0.	0.	.062
	.30	0.	0.	.062	0.	0.	0.	0.	0.	0.	0.	0.	0.	0.	0.	0.	0.	0.	0.	0.	0.	0.	.038	0.	0.	0.	0.	0.	0.	0.	0.	0.	0.
	.20	0.	0.	0.	.024	0.	0.	0.	0.	0.	0.	0.	0.	.062	0.	0.	0.	0.	0.	0.	0.	0.	0.	0.	.045	0.	0.	0.	0.	0.	0.	0.	0.
	0.00	0.	0.	.062	0.	0.	0.	0.	0.	0.	0.	0.	0.	0.	0.	0.	0.	0.	0.	0.	0.	0.	0.	0.	0.	0.	0.	0.	0.	0.	0.	0.	0.
	n	10	9	8	21	18	9	7	8	14	9	7	9	8	9	7	5	7	16	13	9	12	13	12	11	9	2	6	5	6	4	6	8
Pgm-1	1.18	0.	0.	0.	0.	0.	0.	0.	0.	0.	0.	0.	0.	0.	0.	0.	0.	0.	0.	0.	0.	0.	0.	0.	0.	0.	0.	1.0	.556	1.0	.917	.833	0.
	1.14	0.	0.	0.	0.	0.	0.	0.	0.	0.	0.	0.	0.	0.	0.	0.	0.	0.	0.	0.	0.	0.	0.	.045	0.	0.	0.	0.	.444	0.	.083	.167	0.
	1.10	0.	0.	0.	.075	0.	0.	0.	.062	0.	0.	.214	.100	0.	.167	0.	0.	0.	0.	.269	.150	.333	.083	.045	0.	0.	0.	0.	0.	0.	0.	0.	0.
	1.05	.200	.125	.111	.150	.222	.375	.357	.125	.125	0.	.500	.450	.062	.250	.800	.167	.083	.156	.693	.700	.542	.750	.773	0.	0.	0.	0.	0.	0.	0.	0.	0.
	1.00	.200	.500	.833	.200	.278	.500	.571	.625	.625	.750	.214	.450	.562	.583	.200	.833	.750	.813	.038	.150	.083	.167	.136	0.	.167	0.	0.	0.	0.	0.	0.	.071
	.94	.550	.313	.055	.275	.250	.062	0.	.062	.250	.187	.072	0.	.375	0.	0.	0.	0.	.031	0.	0.	.042	0.	0.	.833	.611	0.	0.	0.	0.	0.	0.	.929
	.89	.050	0.	0.	.300	.222	.062	.071	.125	0.	.062	0.	0.	0.	0.	0.	0.	.167	0.	0.	0.	0.	0.	0.	.167	.222	1.0	0.	0.	0.	0.	0.	0.
	.85	0.	.062	0.	0.	.028	0.	0.	0.	0.	0.	0.	0.	0.	0.	0.	0.	0.	0.	0.	0.	0.	0.	0.	0.	0.	0.	0.	0.	0.	0.	0.	0.
	n	10	8	9	20	18	8	7	8	4	8	7	10	8	6	5	6	6	16	13	10	12	12	11	9	9	2	6	9	7	6	3	7
Pgm-2	1.45	0.	0.	0.	0.	0.	0.	0.	0.	0.	0.	0.	0.	0.	.024	0.	0.	0.	0.	0.	0.	0.	0.	0.	.111	.222	.250	.071	.167	0.	.133	.125	0.
	1.29	0.	0.	0.	.048	0.	0.	0.	0.	0.	0.	0.	0.	0.	.024	.029	0.	0.	0.	0.	0.	.033	0.	0.	.778	.778	.750	.929	.833	1.0	.867	.875	.389
	1.16	0.	.100	.111	0.	.031	.111	.055	.062	0.	0.	0.	.038	.182	.262	.324	.143	.150	.222	.154	.179	.067	.167	.062	.111	0.	0.	0.	0.	0.	0.	0.	.556
	1.00	.350	.750	.222	.476	.438	.167	.667	.313	.579	.667	.722	.615	.409	.571	.529	.643	.550	.555	.614	.607	.700	.600	.750	0.	0.	0.	0.	0.	0.	0.	0.	.055
	.85	.650	.150	.667	.452	.531	.556	.222	.563	.395	.333	.278	.346	.364	.071	.118	.214	.250	.167	.231	.214	.133	.200	.188	0.	0.	0.	0.	0.	0.	0.	0.	0.
	.75	0.	0.	0.	.024	0.	.111	0.	0.	0.	0.	0.	0.	.045	.024	0.	0.	.050	0.	0.	0.	.033	0.	0.	0.	0.	0.	0.	0.	0.	0.	0.	0.
	.67	0.	0.	0.	0.	0.	.055	.055	.062	.026	0.	0.	0.	0.	.024	0.	0.	0.	.056	0.	0.	.033	.033	0.	0.	0.	0.	0.	0.	0.	0.	0.	0.
	n	10	10	9	21	16	9	9	8	19	9	9	13	11	21	17	7	10	18	13	14	15	15	16	9	9	2	7	15	8	15	4	9

Locus	Allele	1 NICA	2 DIXN	3 BONT	4 LAG1	5 LAG2	6 FISH	7 DIAZ	8 LITL	9 DSVY	10 CARL	11 HIKO	12 PAHR	13 SASP	14 NOGA	15 CANO	16 SART
Mpi	1.09	.050	0.	0.	.026	.088	.187	.071	.071	0.	0.	.214	.167	.100	0.	0.	0.
	1.04	0.	.125	.111	.026	.088	.187	0.	.215	0.	.375	0.	0.	.200	.125	0.	0.
	1.00	.800	.625	.833	.842	.647	.375	.572	.572	.300	.625	.429	.333	.450	.750	.700	.583
	.94	.150	.187	.055	0.	.059	0.	.071	.071	.100	0.	0.	0.	0.	0.	.100	.250
	.90	0.	0.	0.	.026	.088	.062	.215	0.	.500	0.	.286	.222	.200	0.	.200	.083
	.82	0.	.062	0.	.079	.030	.187	.071	.071	.100	0.	.071	.278	.050	.125	0.	.083
	.71	0.	0.	0.	0.	0.	0.	0.	0.	0.	0.	0.	0.	0.	0.	0.	0.
	n	10	8	9	19	17	8	7	7	5	8	7	9	10	4	10	6
Aldh	1.20	0.	0.	0.	0.	0.	0.	0.	0.	0.	0.	0.	0.	0.	0.	0.	0.
	1.00	1.0	1.0	1.0	1.0	1.0	1.0	1.0	1.0	1.0	1.0	1.0	1.0	1.0	.984	1.0	1.0
	.73	0.	0.	0.	0.	0.	0.	0.	0.	0.	0.	0.	0.	0.	.016	0.	0.
	n	10	10	10	22	18	9	9	9	21	9	9	9	12	31	28	9
Gapd	1.17	0.	0.	0.	0.	0.	0.	0.	0.	0.	0.	0.	0.	0.	0.	0.	0.
	1.00	1.0	1.0	1.0	1.0	1.0	1.0	1.0	1.0	1.0	1.0	1.0	1.0	1.0	1.0	1.0	1.0
	.79	0.	0.	0.	0.	0.	0.	0.	0.	0.	0.	0.	0.	0.	0.	0.	0.
	n	9	9	9	24	16	5	5	3	5	5	2	5	9	5	8	9
Ada	1.00	1.0	1.0	1.0	1.0	1.0	1.0	1.0	1.0	1.0	1.0	1.0	1.0	1.0	1.0	1.0	1.0
	.80	0.	0.	0.	0.	0.	0.	0.	0.	0.	0.	0.	0.	0.	0.	0.	0.
	n	10	10	10	22	18	9	9	8	6	9	9	7	8	29	13	9
Eno	1.18	0.	0.	0.	.023	0.	0.	0.	0.	0.	0.	0.	0.	0.	0.	.029	0.
	1.00	.945	1.0	.950	.977	1.0	.938	.945	.945	1.0	.945	1.0	1.0	1.0	.967	.971	1.0
	.65	.055	0.	.050	0.	0.	.062	.055	.055	0.	.055	0.	0.	0.	.033	0.	0.
	n	9	10	10	22	18	8	9	9	23	9	9	14	15	15	17	9
6-pgd	1.08	.150	.055	.143	.075	.125	0.	0.	0.	0.	0.	0.	0.	0.	.500	0.	0.
	1.00	.300	.555	.214	.575	.531	.688	.857	.688	.875	.625	.667	.636	.900	.500	.800	.600
	.84	.550	.222	.214	.325	.156	.312	.143	.187	.125	0.	.333	.364	.100	0.	.200	0.
	.78	0.	.168	.429	.025	.031	0.	0.	.125	0.	0.	0.	0.	0.	0.	0.	.400
	.67	0.	0.	0.	0.	.156	0.	0.	0.	0.	.187	0.	0.	0.	0.	0.	0.
	.58	0.	0.	0.	0.	0.	0.	0.	0.	0.	.187	0.	0.	0.	0.	0.	0.
	n	10	9	7	20	16	8	7	8	4	8	6	11	5	2	5	5
Sod	1.30	0.	0.	0.	0.	.094	.062	0.	.071	.146	0.	0.	.054	.062	.019	.050	0.
	1.00	1.0	1.0	1.0	1.0	.906	.938	1.0	.929	.854	1.0	1.0	.946	.938	.981	.950	1.0
	n	9	6	6	19	16	8	6	7	24	7	7	13	16	26	20	9
Acon-1	1.10	0.	0.	0.	0.	0.	0.	0.	0.	0.	0.	.062	0.	.050	0.	0.	.071
	1.00	.850	.800	.833	.842	.869	.889	.833	.889	.818	.900	.563	.875	.800	.767	.792	.929
	.89	.150	.200	.167	.132	.105	.111	.167	.111	.182	.100	.375	.125	.150	.233	.208	0.
	-.78	0.	0.	0.	.026	.026	0.	0.	0.	0.	0.	0.	0.	0.	0.	0.	0.
	n	10	10	9	19	19	9	9	9	11	10	8	8	10	15	12	7
Acon-2	1.00	1.0	1.0	1.0	1.0	1.0	1.0	1.0	1.0	1.0	1.0	1.0	1.0	1.0	1.0	1.0	1.0
	.84	0.	0.	0.	0.	0.	0.	0.	0.	0.	0.	0.	0.	0.	0.	0.	0.
	-.78	0.	0.	0.	0.	0.	0.	0.	0.	0.	0.	0.	0.	0.	0.	0.	0.
	n	10	9	8	19	18	8	7	9	7	10	8	7	6	11	9	7
Idh	1.46	0.	0.	0.	0.	0.	0.	0.	0.	0.	0.	0.	0.	0.	0.	0.	0.
	1.29	0.	0.	0.	0.	0.	0.	0.	0.	0.	0.	0.	0.	0.	0.	0.	0.
	1.00	.900	1.0	1.0	1.0	1.0	1.0	.930	1.0	1.0	1.0	1.0	1.0	1.0	1.0	1.0	1.0
	.64	0.	0.	0.	0.	0.	0.	.062	0.	0.	0.	0.	0.	0.	0.	0.	0.
	.37	.100	0.	0.	0.	0.	0.	0.	0.	0.	0.	0.	0.	0.	0.	0.	0.
	n	10	10	9	21	17	9	8	9	15	7	8	7	9	16	15	7
Mdh-2	1.33	0.	0.	0.	0.	0.	0.	0.	0.	0.	0.	0.	0.	0.	0.	0.	0.
	1.00	1.0	1.0	1.0	1.0	1.0	1.0	1.0	1.0	1.0	1.0	1.0	1.0	1.0	1.0	1.0	1.0
	.33	0.	0.	0.	0.	0.	0.	0.	0.	0.	0.	0.	0.	0.	0.	0.	0.
	n	10	10	10	17	18	9	8	9	19	8	10	10	13	20	28	8

Locus	Allele	17 NOXB	18 BURL	19 STFE	20 OKEE	21 ARCH	22 MYAK	23 OKEE	24 ITHA	25 VERM	26 BAMA	27 LESL	28 CAVE	29 PEBL	30 RUCK	31 TURK	32 COAH
Mpi	1.09	0.	.031	.042	0.	0.	0.	0.	.062	0.	0.	0.	0.	0.	0.	0.	0.
	1.04	.167	.031	.083	.100	0.	0.	0.	0.	0.	0.	0.	0.	0.	0.	0.	0.
	1.00	.583	.625	.792	.550	.682	.818	.682	.813	.687	.500	.625	.500	.750	.643	.833	.071
	.94	.250	.156	0.	.050	.045	0.	.091	0.	.313	0.	.250	.333	0.	.143	0.	0.
	.90	0.	.156	.083	.250	.227	.136	.227	0.	0.	.500	.125	.167	0.	.214	0.	.786
	.82	0.	0.	0.	.050	.045	.045	0.	.125	0.	0.	0.	0.	.250	0.	.167	0.
	.71	0.	0.	0.	0.	0.	0.	0.	0.	0.	0.	0.	0.	0.	0.	0.	.143
	n	6	16	12	10	11	11	11	8	8	1	4	6	2	7	3	7
Aldh	1.20	0.	.083	0.	0.	0.	0.	0.	0.	0.	0.	0.	0.	0.	0.	0.	0.
	1.00	1.0	.917	1.0	1.0	1.0	1.0	1.0	1.0	1.0	1.0	1.0	1.0	1.0	1.0	1.0	1.0
	.73	0.	0.	0.	0.	0.	0.	0.	0.	0.	0.	0.	0.	0.	0.	0.	0.
	n	12	18	18	18	18	18	18	11	9	9	8	9	9	9	9	9
Gapd	1.17	0.	0.	0.	.042	0.	0.	0.	0.	0.	0.	0.	0.	0.	0.	0.	0.
	1.00	.600	1.0	1.0	.958	1.0	1.0	1.0	1.0	1.0	1.0	.938	1.0	1.0	1.0	1.0	1.0
	.79	.400	0.	0.	0.	0.	0.	0.	0.	0.	0.	.062	0.	0.	0.	0.	0.
	n	5	18	15	12	15	15	15	11	9	4	8	6	6	6	4	9
Ada	1.00	1.0	1.0	1.0	.941	.971	.969	.929	1.0	1.0	1.0	1.0	1.0	1.0	1.0	1.0	1.0
	.80	0.	0.	0.	.059	.029	.031	.071	0.	0.	0.	0.	0.	0.	0.	0.	0.
	n	11	18	18	17	17	16	14	11	7	3	5	9	7	8	9	9
Eno	1.18	.038	.028	0.	0.	0.	0.	0.	0.	0.	.100	0.	0.	0.	0.	0.	0.
	1.00	.924	.944	1.0	1.0	1.0	1.0	.969	1.0	1.0	.900	1.0	1.0	1.0	1.0	1.0	1.0
	.65	.038	.028	0.	0.	0.	0.	.031	0.	0.	0.	0.	0.	0.	0.	0.	0.
	n	13	18	18	17	18	18	16	11	9	5	7	9	9	9	9	9
6-pgd	1.08	0.	.125	0.	.167	0.	0.	.167	0.	0.	0.	0.	0.	0.	0.	.250	0.
	1.00	.687	.562	.846	.722	.682	.731	.722	.136	.055	0.	0.	0.	.200	.100	0.	.071
	.84	.250	.281	.115	0.	.318	.192	0.	.091	.444	0.	.300	.400	.400	.400	.125	0.
	.78	.062	0.	0.	.111	0.	.038	.111	.545	.500	1.0	.700	.600	.350	.500	.625	.929
	.67	0.	.031	0.	0.	0.	.038	0.	.227	0.	0.	0.	0.	.050	0.	0.	0.
	.58	0.	0.	.038	0.	0.	0.	0.	0.	0.	0.	0.	0.	0.	0.	0.	0.
	n	8	16	13	9	11	13	9	11	9	3	5	5	10	5	4	7
Sod	1.30	.038	0.	.194	.111	.118	.056	.111	.136	.111	0.	.143	.055	.143	.111	.250	0.
	1.00	.962	1.0	.806	.889	.882	.944	.889	.864	.889	1.0	.857	.945	.857	.889	.759	1.0
	n	13	18	18	18	17	18	18	11	9	5	7	9	7	9	6	9
Acon-1	1.10	0.	0.	.071	.038	0.	0.	.038	0.	.167	0.	0.	0.	0.	0.	0.	0.
	1.00	.875	.833	.822	.885	.885	.750	.885	.864	.778	1.0	1.0	.929	1.0	1.0	1.0	0.
	.89	.125	.167	.071	.077	.038	.125	.077	0.	.055	0.	0.	.071	0.	0.	0.	1.0
	-.78	0.	0.	.036	0.	.077	.125	0.	.136	0.	0.	0.	0.	0.	0.	0.	0.
	n	8	18	14	13	13	16	13	11	9	2	7	7	4	7	3	9
Acon-2	1.00	1.0	1.0	1.0	1.0	1.0	1.0	1.0	1.0	1.0	1.0	1.0	1.0	1.0	1.0	1.0	.835
	.84	0.	0.	0.	0.	0.	0.	0.	0.	0.	0.	0.	0.	0.	0.	0.	.110
	-.78	0.	0.	0.	0.	0.	0.	0.	0.	0.	0.	0.	0.	0.	0.	0.	.055
	n	6	17	13	10	13	14	10	10	9	2	6	2	3	6	1	9
Idh	1.46	0.	0.	.036	0.	0.	0.	0.	0.	0.	0.	0.	0.	0.	0.	0.	0.
	1.29	0.	0.	0.	.100	.077	.033	.100	0.	0.	0.	.071	0.	0.	0.	0.	0.
	1.00	1.0	1.0	.964	.900	.885	.967	.900	1.0	1.0	1.0	.929	1.0	1.0	.938	1.0	1.0
	.64	0.	0.	0.	0.	.038	0.	0.	0.	0.	0.	0.	0.	0.	0.	0.	0.
	.37	0.	0.	0.	0.	0.	0.	0.	0.	0.	0.	0.	0.	0.	.062	0.	0.
	n	9	17	14	15	13	15	15	8	9	2	7	7	5	8	2	9
Mdh-2	1.33	.042	0.	0.	0.	0.	0.	0.	0.	0.	0.	0.	0.	0.	0.	0.	0.
	1.00	.958	1.0	1.0	.971	1.0	1.0	.971	1.0	1.0	1.0	1.0	1.0	1.0	1.0	1.0	0.
	.33	0.	0.	0.	.029	0.	0.	.029	0.	0.	0.	0.	0.	0.	0.	0.	1.0
	n	12	18	15	17	15	16	17	10	9	8	7	9	5	9	5	6

APPENDIX 4

Rogers' genetic distance (upper right) and Nei's genetic distance corrected for sample size (lower left).

	Population	1 NICA	2 DIXN	3 BONT	4 LAG1	5 LAG2	6 FISH	7 DIAZ	8 LITL	9 DSVY	10 CARL	11 HIKO	12 PAHR	13 SASP	14 NOGA	15 CANO	16 SART	17 NOXB	18 BURL	19 STFE	20 OKEE	21 ARCH	22 MYAK	23 PAHO	24 ITHA	25 VERM	26 BAMA	27 LESL	28 CAVE	29 PEBL	30 RUCK	31 TURK	32 COAH
1	NICA	--	.121	.129	.113	.117	.124	.139	.121	.157	.178	.160	.157	.165	.149	.156	.175	.157	.133	.164	.184	.165	.158	.160	.274	.249	.328	.300	.288	.304	.299	.321	.424
2	DIXN	.029	--	.129	.091	.089	.119	.091	.095	.099	.121	.117	.116	.136	.009	.118	.096	.111	.081	.118	.132	.118	.108	.123	.268	.240	.310	.282	.267	.286	.288	.304	.408
3	BONT	.054	.024	--	.103	.103	.133	.130	.102	.143	.139	.178	.155	.171	.130	.168	.111	.139	.111	.164	.184	.181	.148	.176	.264	.249	.295	.279	.268	.280	.281	.289	.428
4	LAG1	.027	.014	.034	--	.064	.125	.113	.073	.112	.122	.119	.117	.132	.108	.115	.123	.136	.106	.113	.142	.119	.108	.125	.247	.235	.287	.288	.273	.275	.285	.293	.423
5	LAG2	.021	.009	.023	.002	--	.115	.108	.083	.106	.128	.132	.107	.122	.125	.128	.118	.128	.112	.110	.115	.115	.100	.110	.247	.242	.293	.285	.274	.274	.272	.301	.430
6	FISH	.035	.028	.039	.034	.022	--	.094	.073	.117	.144	.127	.084	.099	.129	.134	.162	.114	.105	.138	.149	.139	.136	.133	.307	.291	.332	.317	.297	.311	.316	.334	.441
7	DIAZ	.048	.004	.043	.025	.016	.008	--	.087	.082	.121	.105	.088	.098	.102	.100	.123	.106	.076	.100	.121	.114	.098	.091	.308	.297	.323	.317	.305	.318	.309	.342	.447
8	LITL	.028	.007	.012	.015	.003	0.	0.	--	.101	.108	.133	.099	.103	.112	.130	.125	.093	.090	.120	.135	.128	.119	.118	.274	.263	.306	.294	.280	.284	.286	.312	.434
9	DSVY	.056	.013	.052	.031	.021	.018	.001	.008	--	.123	.109	.102	.102	.120	.110	.134	.134	.102	.109	.135	.128	.128	.117	.298	.278	.321	.306	.291	.312	.312	.327	.411
10	CARL	.081	.025	.054	.042	.039	.046	.031	.016	.034	--	.135	.126	.132	.123	.159	.135	.137	.121	.149	.168	.162	.151	.148	.304	.299	.445	.338	.319	.325	.325	.349	.445
11	HIKO	.050	.024	.081	.035	.034	.026	.010	.027	.019	.042	--	.077	.100	.133	.111	.161	.156	.123	.116	.144	.108	.110	.105	.318	.291	.336	.328	.305	.326	.323	.347	.405
12	PAHR	.053	.023	.062	.040	.027	.005	.004	.009	.015	.038	.001	--	.085	.132	.127	.145	.119	.101	.114	.125	.115	.103	.098	.314	.299	.338	.326	.305	.313	.314	.346	.442
13	SASP	.064	.036	.075	.043	.033	.007	.007	.009	.017	.036	.007	.006	--	.123	.096	.159	.139	.120	.102	.118	.124	.108	.103	.306	.298	.333	.327	.310	.321	.316	.348	.434
14	NOGA	.058	.010	.036	.023	.025	.030	.010	.016	.027	.027	.028	.029	.023	--	.097	.131	.135	.085	.121	.134	.135	.128	.136	.286	.278	.321	.311	.297	.302	.312	.311	.426
15	CANO	.066	.036	.084	.035	.038	.040	.017	.040	.034	.068	.015	.033	.010	.020	--	.147	.146	.106	.090	.112	.108	.085	.079	.289	.272	.315	.299	.284	.305	.300	.319	.422
16	SART	.077	.007	.018	.034	.025	.057	.021	.020	.027	.039	.063	.052	.059	.025	.049	--	.132	.102	.149	.141	.144	.136	.147	.261	.239	.277	.269	.256	.277	.271	.293	.425
17	NOXB	.052	.013	.039	.040	.029	.019	.007	.003	.025	.034	.044	.024	.032	.023	.052	.030	--	.096	.147	.158	.158	.134	.136	.332	.305	.352	.323	.323	.341	.339	.366	.468
18	BURL	.050	.005	.030	.032	.026	.018	.002	.005	.013	.030	.030	.016	.026	.004	.035	.021	.008	--	.115	.131	.120	.116	.116	.298	.271	.319	.302	.287	.304	.298	.327	.432
19	STFE	.067	.033	.081	.032	.028	.037	.017	.032	.040	.065	.019	.026	.015	.028	.008	.060	.054	.042	--	.089	.073	.068	.077	.278	.278	.330	.300	.302	.292	.297	.308	.454
20	OKEE	.075	.031	.073	.037	.023	.037	.016	.032	.036	.064	.027	.020	.015	.022	.012	.039	.048	.037	.003	--	.088	.091	.081	.295	.293	.321	.308	.308	.312	.296	.337	.448
21	ARCH	.058	.024	.079	.025	.026	.041	.016	.033	.033	.062	.015	.021	.023	.026	.011	.048	.052	.034	0.	.004	--	.080	.074	.265	.263	.306	.280	.281	.275	.268	.306	.440
22	MYAK	.056	.024	.064	.030	.023	.037	.013	.031	.042	.062	.017	.022	.019	.029	.009	.046	.046	.035	0.	.004	.003	--	.062	.286	.286	.331	.314	.305	.305	.304	.328	.441
23	PAHO	.058	.030	.085	.037	.029	.037	.009	.033	.033	.062	.012	.014	.010	.032	.004	.054	.044	.035	.001	0.	.001	0.	--	.300	.291	.328	.308	.305	.308	.295	.336	.456
24	ITHA	.230	.215	.227	.200	.209	.290	.284	.235	.270	.288	.306	.319	.307	.248	.258	.212	.309	.270	.264	.257	.228	.273	.292	--	.087	.153	.160	.142	.130	.142	.135	.326
25	VERM	.199	.187	.197	.180	.197	.260	.256	.208	.240	.269	.265	.284	.286	.228	.233	.183	.267	.230	.252	.248	.209	.257	.267	.009	--	.148	.142	.103	.140	.133	.146	.335
26	BAMA	.284	.248	.229	.215	.224	.300	.286	.242	.277	.312	.311	.321	.314	.265	.269	.203	.306	.282	.288	.255	.244	.291	.299	.043	.030	--	.190	.162	.184	.171	.188	.344
27	LESL	.266	.250	.251	.259	.264	.293	.296	.250	.293	.335	.308	.312	.315	.278	.278	.245	.308	.276	.287	.287	.246	.299	.301	.080	.060	.076	--	.081	.085	.075	.090	.381
28	CAVE	.253	.230	.233	.238	.246	.283	.282	.240	.272	.319	.280	.301	.308	.265	.262	.220	.303	.260	.278	.276	.233	.287	.290	.063	.032	.054	.005	--	.110	.086	.092	.357
29	PEBL	.274	.247	.251	.236	.248	.289	.288	.245	.291	.319	.309	.307	.314	.267	.274	.240	.318	.274	.271	.279	.232	.286	.297	.069	.064	.093	.006	.019	--	.064	.081	.409
30	RUCK	.266	.244	.245	.242	.245	.292	.287	.242	.286	.312	.299	.301	.308	.273	.274	.234	.314	.270	.275	.271	.230	.285	.290	.069	.057	.068	0.	.006	0.	--	.098	.390
31	TURK	.293	.265	.254	.263	.271	.324	.318	.270	.315	.351	.338	.347	.346	.276	.296	.251	.350	.297	.294	.304	.261	.314	.325	.067	.059	.081	.004	.009	.004	.011	--	.398
32	COAH	.499	.459	.503	.504	.510	.535	.537	.506	.469	.542	.461	.540	.521	.497	.494	.498	.567	.514	.561	.518	.518	.532	.558	.322	.322	.327	.405	.370	.459	.414	.443	--

APPENDIX 5

Means (standard deviation) of 13 mensural characters for males.

Population Code Number	hw	pl	apw	mpw	bpw	huw	mew	el	4al	pfl	pfw	3sl	pd
01	.83(.03)	1.58(.05)	1.30(.04)	1.88(.06)	1.40(.09)	1.09(.04)	1.49(.03)	4.95(.13)	.53(.02)	1.66(.09)	.59(.02)	.57(.02)	1.68(.07)
02	.84(.04)	1.58(.11)	1.32(.10)	1.91(.13)	1.41(.11)	1.10(.06)	1.50(.07)	5.05(.29)	.54(.04)	1.69(.13)	.57(.03)	.64(.09)	1.74(.11)
03	.86(.03)	1.50(.08)	1.30(.05)	1.86(.08)	1.33(.08)	1.14(.06)	1.53(.04)	5.10(.22)	.54(.03)	1.67(.10)	.54(.04)	.62(.06)	1.68(.09)
04	.83(.03)	1.49(.08)	1.29(.05)	1.76(.08)	1.36(.08)	1.15(.06)	1.49(.04)	5.13(.21)	.53(.03)	1.68(.10)	.53(.03)	.64(.06)	1.68(.09)
05	.86(.04)	1.51(.13)	1.34(.07)	1.89(.17)	1.44(.10)	1.26(.06)	1.61(.09)	5.25(.30)	.55(.03)	1.72(.12)	.56(.04)	.68(.08)	1.73(.11)
06	.84(.03)	1.42(.07)	1.28(.08)	1.81(.08)	1.40(.05)	1.17(.06)	1.53(.05)	4.86(.16)	.48(.03)	1.55(.10)	.50(.03)	.63(.02)	1.62(.05)
07	.88(.03)	1.52(.07)	1.37(.08)	1.94(.11)	1.56(.12)	1.28(.06)	1.59(.08)	5.16(.22)	.52(.02)	1.71(.09)	.54(.02)	.61(.04)	1.74(.07)
08	.82(.04)	1.38(.11)	1.27(.07)	1.75(.14)	1.37(.10)	1.18(.10)	1.44(.13)	4.68(.31)	.46(.04)	1.50(.10)	.49(.03)	.56(.04)	1.58(.09)
09	.77(.05)	1.43(.11)	1.29(.08)	1.77(.11)	1.44(.10)	1.12(.06)	1.41(.11)	4.61(.31)	.47(.05)	1.58(.10)	.51(.04)	.57(.05)	1.55(.11)
10	.74(.05)	1.36(.08)	1.28(.09)	1.73(.10)	1.36(.11)	1.08(.08)	1.34(.09)	4.34(.20)	.46(.04)	1.47(.04)	.48(.02)	.54(.05)	1.48(.09)
11	.82(.02)	1.36(.06)	1.30(.08)	1.82(.09)	1.40(.12)	1.16(.07)	1.47(.07)	4.51(.22)	.43(.03)	1.49(.05)	.50(.02)	.57(.03)	1.57(.05)
12	.71(.05)	1.47(.04)	1.20(.06)	1.71(.08)	1.27(.07)	1.07(.05)	1.44(.10)	4.74(.20)	.54(.03)	1.61(.06)	.55(.03)	.59(.03)	1.56(.08)
13	.74(.04)	1.52(.10)	1.24(.06)	1.81(.08)	1.36(.08)	1.12(.07)	1.49(.08)	4.90(.23)	.55(.04)	1.64(.08)	.56(.03)	.62(.06)	1.61(.08)
14	.80(.02)	1.50(.07)	1.22(.05)	1.74(.08)	1.33(.08)	1.13(.05)	1.49(.06)	4.77(.17)	.53(.03)	1.64(.05)	.51(.02)	.57(.06)	1.62(.09)
15	.74(.06)	1.47(.10)	1.28(.09)	1.81(.13)	1.42(.13)	1.21(.12)	1.54(.13)	4.88(.31)	.49(.04)	1.64(.11)	.51(.03)	.58(.03)	1.63(.12)
16	.73(.04)	1.58(.07)	1.29(.06)	1.89(.08)	1.39(.10)	1.25(.06)	1.60(.04)	5.02(.17)	.53(.02)	1.76(.06)	.55(.03)	.63(.05)	1.69(.10)
17	.84(.04)	1.61(.06)	1.38(.06)	1.94(.08)	1.51(.06)	1.27(.06)	1.60(.07)	5.20(.20)	.57(.03)	1.82(.06)	.55(.03)	.65(.04)	1.76(.05)
18	.81(.02)	1.58(.06)	1.37(.07)	1.94(.09)	1.49(.09)	1.22(.07)	1.56(.06)	5.04(.14)	.54(.03)	1.79(.07)	.57(.04)	.62(.06)	1.73(.08)
19	.79(.10)	1.51(.10)	1.33(.09)	1.80(.15)	1.42(.13)	1.21(.11)	1.50(.11)	5.15(.32)	.48(.05)	1.71(.10)	.50(.05)	.62(.06)	1.68(.10)
20	.80(.04)	1.50(.08)	1.33(.05)	1.85(.08)	1.43(.07)	1.26(.09)	1.62(.08)	5.21(.25)	.54(.04)	1.78(.08)	.50(.02)	.63(.04)	1.66(.06)
21	.84(.04)	1.57(.09)	1.38(.08)	1.98(.12)	1.52(.10)	1.30(.08)	1.64(.07)	5.38(.22)	.55(.05)	1.77(.09)	.52(.04)	.71(.06)	1.72(.10)
22	.87(.03)	1.56(.06)	1.39(.09)	1.94(.07)	1.49(.04)	1.28(.07)	1.60(.08)	5.43(.19)	.53(.04)	1.76(.09)	.54(.05)	.62(.05)	1.72(.07)
23	.83(.04)	1.57(.08)	1.39(.05)	1.94(.09)	1.48(.05)	1.31(.06)	1.60(.07)	5.37(.22)	.53(.02)	1.71(.09)	.52(.03)	.64(.05)	1.73(.07)
24	.88(.03)	1.68(.05)	1.47(.07)	2.04(.06)	1.62(.09)	1.36(.07)	1.77(.07)	5.71(.14)	.54(.02)	1.80(.06)	.53(.03)	.73(.04)	1.83(.04)
25	.84(.05)	1.51(.09)	1.36(.09)	1.88(.13)	1.38(.10)	1.25(.08)	1.59(.07)	5.27(.24)	.53(.03)	1.73(.10)	.50(.02)	.64(.05)	1.68(.11)
26	.82(.03)	1.58(.12)	1.30(.07)	1.78(.10)	1.34(.07)	1.21(.09)	1.63(.10)	5.15(.34)	.61(.09)	1.88(.08)	.53(.04)	.72(.08)	1.77(.08)
27	.80(.05)	1.52(.14)	1.30(.11)	1.72(.18)	1.27(.12)	1.09(.09)	1.55(.12)	4.85(.44)	.59(.07)	1.83(.15)	.54(.06)	.65(.06)	1.69(.13)
28	.83(.02)	1.55(.09)	1.35(.09)	1.85(.08)	1.38(.04)	1.22(.06)	1.68(.09)	5.21(.22)	.56(.03)	1.87(.10)	.56(.03)	.82(.03)	1.73(.09)
29	.82(.02)	1.61(.08)	1.29(.08)	1.80(.08)	1.34(.12)	1.10(.08)	1.53(.08)	4.83(.29)	.56(.06)	1.86(.11)	.55(.05)	.69(.06)	1.72(.10)
30	.83(.02)	1.62(.06)	1.34(.06)	1.85(.05)	1.40(.05)	1.15(.05)	1.62(.06)	5.01(.14)	.59(.01)	1.88(.05)	.58(.04)	.69(.04)	1.74(.05)
31	.85(.04)	1.68(.11)	1.37(.08)	1.85(.11)	1.40(.10)	1.19(.06)	1.64(.06)	5.14(.20)	.61(.06)	1.86(.10)	.59(.04)	.72(.03)	1.76(.09)
32	.82(.04)	1.67(.14)	1.32(.10)	1.87(.11)	1.38(.08)	1.23(.12)	1.67(.13)	5.22(.22)	.61(.05)	1.86(.08)	.59(.04)	.72(.03)	1.75(.09)
33	.81(.01)	1.66(.09)	1.35(.06)	1.91(.08)	1.40(.06)	1.29(.09)	1.70(.08)	5.46(.39)	.63(.06)	1.89(.08)	.59(.02)	.73(.02)	1.77(.07)
34	.79(.04)	1.67(.11)	1.33(.06)	1.90(.13)	1.38(.10)	1.29(.12)	1.72(.09)	5.42(.31)	.68(.04)	1.94(.09)	.58(.04)	.74(.05)	1.80(.11)
35	.82(.02)	1.74(.11)	1.42(.06)	1.97(.10)	1.46(.06)	1.37(.03)	1.75(.07)	5.70(.22)	.69(.03)	2.01(.10)	.61(.04)	.78(.03)	1.85(.04)
36	.81(.04)	1.65(.09)	1.37(.08)	1.88(.09)	1.36(.09)	1.26(.08)	1.68(.10)	5.23(.30)	.65(.05)	1.90(.12)	.56(.03)	.72(.04)	1.76(.10)
37	.84(.03)	1.65(.10)	1.37(.07)	1.85(.10)	1.41(.07)	1.29(.10)	1.68(.08)	5.33(.25)	.66(.04)	1.87(.11)	.57(.05)	.70(.04)	1.77(.07)
38	.82(.03)	1.66(.12)	1.41(.07)	1.90(.15)	1.42(.10)	1.30(.08)	1.68(.07)	5.22(.30)	.65(.07)	1.85(.11)	.57(.05)	.71(.05)	1.76(.10)
39	.82(.03)	1.60(.09)	1.34(.07)	1.88(.09)	1.44(.10)	1.32(.11)	1.68(.07)	5.17(.19)	.59(.06)	1.79(.10)	.58(.03)	.69(.04)	1.76(.08)
40	.82(.05)	1.77(.12)	1.44(.08)	2.02(.09)	1.52(.13)	1.40(.10)	1.82(.11)	5.88(.28)	.69(.06)	1.98(.12)	.61(.06)	.75(.03)	1.90(.11)
41	.85(.02)	1.78(.08)	1.40(.06)	1.96(.10)	1.49(.08)	1.36(.04)	1.81(.08)	5.74(.30)	.69(.04)	1.94(.10)	.58(.05)	.71(.03)	1.89(.09)
42	.82(.02)	1.69(.09)	1.40(.07)	1.93(.08)	1.45(.05)	1.40(.06)	1.79(.07)	5.78(.25)	.63(.05)	1.96(.08)	.58(.03)	.72(.03)	1.84(.06)
44	.80(.07)	1.72(.18)	1.43(.16)	1.95(.19)	1.51(.14)	1.35(.12)	1.78(.16)	5.76(.45)	.62(.08)	1.89(.13)	.58(.05)	.67(.12)	1.81(.19)
45	.81(.03)	1.72(.11)	1.44(.05)	2.01(.06)	1.50(.03)	1.40(.04)	1.79(.08)	5.71(.24)	.64(.04)	1.90(.08)	.59(.04)	.74(.02)	1.87(.07)
46	.71(.05)	1.51(.13)	1.32(.14)	1.89(.13)	1.47(.16)	1.25(.13)	1.57(.11)	4.82(.38)	.47(.09)	1.56(.14)	.50(.04)	.58(.07)	1.64(.11)
47	.78(.05)	1.71(.08)	1.46(.04)	2.13(.09)	1.73(.08)	1.42(.09)	1.68(.13)	5.26(.36)	.54(.03)	1.82(.06)	.55(.03)	.69(.04)	1.87(.10)
48	.78(.05)	1.72(.09)	1.44(.09)	2.09(.08)	1.73(.09)	1.43(.09)	1.80(.07)	5.49(.21)	.58(.06)	1.86(.05)	.54(.04)	.70(.07)	1.87(.08)
49	.87(.04)	1.56(.13)	1.41(.11)	2.04(.12)	1.58(.11)	1.37(.09)	1.72(.11)	5.29(.33)	.54(.03)	1.69(.11)	.55(.04)	.70(.08)	1.75(.10)
50	.93(.02)	1.77(.10)	1.54(.08)	2.25(.11)	1.81(.08)	1.49(.05)	1.88(.07)	5.93(.28)	.58(.02)	1.88(.07)	.59(.04)	.80(.04)	1.92(.08)
51	.83(.02)	1.75(.10)	1.52(.06)	2.28(.09)	1.82(.08)	1.49(.06)	1.84(.07)	5.82(.24)	.55(.02)	1.86(.07)	.58(.05)	.73(.04)	1.92(.06)
52	.91(.04)	1.74(.12)	1.52(.07)	2.16(.12)	1.72(.13)	1.45(.07)	1.86(.06)	5.73(.23)	.55(.04)	1.81(.06)	.57(.03)	.77(.03)	1.85(.12)
53	.90(.06)	1.70(.12)	1.48(.07)	2.17(.15)	1.70(.12)	1.44(.05)	1.83(.07)	5.76(.31)	.54(.04)	1.81(.10)	.56(.05)	.74(.06)	1.83(.12)
55	.89(.04)	1.71(.12)	1.49(.09)	2.14(.13)	1.67(.14)	1.44(.07)	1.82(.08)	5.78(.23)	.54(.02)	1.83(.05)	.54(.04)	.75(.05)	1.87(.09)
56	.92(.04)	1.78(.07)	1.55(.11)	2.25(.10)	1.79(.04)	1.50(.04)	1.86(.03)	5.99(.15)	.57(.02)	1.84(.07)	.59(.02)	.98(.04)	1.94(.05)
57	.84(.02)	1.66(.09)	1.44(.06)	2.12(.06)	1.68(.09)	1.43(.04)	1.79(.07)	5.68(.19)	.52(.03)	1.80(.10)	.56(.03)	.74(.05)	1.81(.06)

APPENDIX 6

Means (standard deviation) of 13 mensural characters for females.

Population Code Number	hw	pl	apw	mpw	bpw	huw	mew	el	4al	pfl	pfw	3sl	pd
01	.88(.02)	1.69(.09)	1.41(.07)	2.02(.08)	1.51(.08)	1.18(.04)	1.61(.04)	5.30(.20)	.56(.04)	1.77(.06)	.54(.02)	.67(.03)	1.77(.06)
02	.85(.03)	1.62(.04)	1.33(.08)	1.93(.13)	1.41(.10)	1.15(.07)	1.60(.06)	5.15(.16)	.55(.05)	1.75(.07)	.51(.02)	.67(.03)	1.73(.09)
03	.87(.04)	1.62(.09)	1.38(.07)	1.94(.11)	1.46(.10)	1.28(.10)	1.66(.10)	5.37(.25)	.55(.03)	1.74(.11)	.52(.03)	.72(.03)	1.76(.11)
04	.87(.05)	1.58(.11)	1.35(.09)	1.89(.08)	1.43(.11)	1.27(.11)	1.68(.10)	5.46(.33)	.55(.05)	1.71(.10)	.53(.04)	.71(.07)	1.74(.10)
05	.89(.04)	1.57(.08)	1.42(.05)	1.98(.07)	1.56(.07)	1.36(.06)	1.70(.06)	5.47(.16)	.53(.03)	1.75(.07)	.52(.02)	.70(.03)	1.80(.05)
06	.87(.03)	1.52(.09)	1.40(.06)	1.90(.16)	1.53(.10)	1.32(.11)	1.63(.06)	5.10(.16)	.50(.03)	1.64(.07)	.49(.02)	.69(.03)	1.68(.05)
07	.92(.03)	1.61(.05)	1.48(.06)	2.09(.07)	1.67(.11)	1.43(.02)	1.74(.04)	5.45(.08)	.51(.02)	1.80(.05)	.53(.03)	.70(.00)	1.85(.03)
08	.84(.04)	1.45(.10)	1.37(.08)	1.86(.11)	1.47(.09)	1.26(.10)	1.53(.10)	4.90(.35)	.45(.05)	1.54(.13)	.47(.02)	.63(.07)	1.63(.12)
09	.83(.04)	1.49(.06)	1.38(.06)	1.88(.09)	1.51(.08)	1.20(.06)	1.50(.07)	4.81(.23)	.48(.02)	1.58(.08)	.51(.02)	.60(.03)	1.61(.07)
10	.84(.03)	1.49(.05)	1.40(.09)	1.89(.09)	1.55(.09)	1.19(.04)	1.49(.07)	4.78(.20)	.48(.02)	1.56(.04)	.51(.02)	.61(.03)	1.61(.08)
11	.87(.04)	1.46(.07)	1.37(.08)	1.87(.10)	1.45(.07)	1.25(.08)	1.54(.10)	4.82(.30)	.41(.04)	1.50(.07)	.48(.03)	.63(.05)	1.69(.08)
12	.82(.05)	1.56(.08)	1.32(.07)	1.86(.07)	1.45(.08)	1.16(.03)	1.56(.02)	5.03(.22)	.56(.02)	1.73(.70)	.53(.04)	.69(.07)	1.68(.08)
13	.82(.02)	1.64(.09)	1.36(.04)	1.91(.06)	1.46(.04)	1.17(.10)	1.61(.13)	5.20(.28)	.56(.03)	1.75(.11)	.55(.04)	.67(.07)	1.71(.12)
14	.79(.05)	1.50(.10)	1.26(.09)	1.75(.12)	1.35(.10)	1.20(.14)	1.57(.14)	4.89(.33)	.52(.02)	1.67(.11)	.50(.03)	.59(.05)	1.66(.11)
15	.81(.02)	1.58(.09)	1.40(.04)	1.90(.06)	1.49(.04)	1.35(.10)	1.68(.08)	5.17(.31)	.52(.02)	1.70(.13)	.51(.02)	.66(.07)	1.75(.08)
16	.85(.05)	1.70(.11)	1.43(.08)	2.02(.11)	1.55(.12)	1.31(.14)	1.70(.11)	5.24(.18)	.55(.04)	1.80(.09)	.57(.04)	.70(.05)	1.80(.12)
17	.82(.02)	1.56(.07)	1.32(.06)	1.86(.08)	1.45(.11)	1.16(.10)	1.74(.12)	5.38(.20)	.56(.03)	1.86(.05)	.52(.03)	.71(.03)	1.84(.05)
18	.89(.03)	1.68(.09)	1.49(.06)	2.09(.08)	1.67(.09)	1.35(.11)	1.70(.10)	5.22(.18)	.57(.03)	1.86(.06)	.57(.02)	.71(.02)	1.83(.07)
19	.86(.05)	1.63(.12)	1.43(.10)	2.00(.12)	1.57(.15)	1.35(.10)	1.73(.14)	5.58(.33)	.52(.03)	1.80(.13)	.53(.06)	.70(.06)	1.78(.10)
20	.86(.05)	1.59(.14)	1.44(.11)	1.95(.15)	1.51(.11)	1.36(.10)	1.71(.14)	5.60(.35)	.53(.05)	1.77(.13)	.50(.04)	.73(.08)	1.75(.12)
21	.87(.03)	1.62(.12)	1.46(.07)	2.01(.13)	1.55(.12)	1.39(.09)	1.81(.08)	5.71(.30)	.55(.02)	1.80(.12)	.52(.04)	.75(.06)	1.84(.11)
22	.88(.04)	1.63(.11)	1.46(.11)	1.95(.16)	1.51(.12)	1.33(.09)	1.73(.11)	5.58(.35)	.53(.02)	1.76(.14)	.52(.04)	.69(.10)	1.80(.12)
23	.89(.03)	1.59(.10)	1.50(.06)	2.02(.07)	1.53(.07)	1.39(.05)	1.78(.09)	5.63(.25)	.54(.04)	1.81(.08)	.53(.03)	.74(.04)	1.84(.08)
24	.84(.05)	1.57(.08)	1.40(.08)	1.93(.08)	1.49(.10)	1.32(.10)	1.70(.11)	5.39(.22)	.54(.03)	1.74(.09)	.54(.02)	.70(.05)	1.77(.08)
25	.88(.04)	1.63(.13)	1.49(.07)	2.03(.15)	1.58(.11)	1.39(.11)	1.80(.11)	5.57(.31)	.54(.04)	1.81(.09)	.52(.03)	.73(.07)	1.84(.09)
26	.85(.04)	1.65(.14)	1.39(.07)	1.90(.12)	1.44(.12)	1.36(.06)	1.77(.12)	5.40(.41)	.59(.05)	1.95(.14)	.53(.04)	.78(.03)	1.82(.14)
27	.87(.03)	1.64(.12)	1.39(.08)	1.88(.13)	1.36(.06)	1.17(.11)	1.62(.10)	5.16(.32)	.62(.06)	1.95(.05)	.57(.05)	.72(.05)	1.82(.13)
28	.86(.03)	1.66(.11)	1.46(.13)	1.95(.14)	1.48(.17)	1.31(.11)	1.74(.11)	5.57(.34)	.58(.04)	1.96(.11)	.58(.05)	.76(.07)	1.79(.11)
29	.82(.09)	1.61(.15)	1.32(.15)	1.77(.18)	1.35(.17)	1.14(.13)	1.57(.18)	4.82(.38)	.54(.06)	1.82(.18)	.53(.06)	.72(.03)	1.73(.15)
30	.89(.03)	1.72(.09)	1.44(.06)	1.95(.08)	1.46(.05)	1.20(.04)	1.76(.06)	5.22(.19)	.59(.04)	1.87(.05)	.57(.03)	.77(.05)	1.86(.04)
31	.92(.04)	1.76(.08)	1.46(.08)	1.98(.11)	1.54(.15)	1.28(.12)	1.81(.11)	5.40(.20)	.62(.06)	1.92(.09)	.60(.05)	.78(.03)	1.88(.09)
32	.88(.04)	1.73(.08)	1.44(.08)	1.98(.12)	1.52(.10)	1.29(.08)	1.82(.08)	5.57(.24)	.65(.05)	1.92(.08)	.57(.03)	.79(.04)	1.85(.09)
33	.86(.04)	1.78(.12)	1.46(.09)	2.01(.13)	1.54(.15)	1.41(.12)	1.79(.12)	5.71(.36)	.66(.07)	1.97(.10)	.58(.04)	.79(.03)	1.86(.12)
34	.85(.03)	1.84(.05)	1.49(.12)	2.02(.12)	1.55(.13)	1.42(.09)	1.85(.08)	5.82(.29)	.70(.04)	1.99(.09)	.57(.04)	.82(.04)	1.90(.11)
35	.87(.03)	1.85(.10)	1.52(.12)	2.08(.12)	1.67(.17)	1.36(.11)	1.82(.08)	6.02(.28)	.67(.07)	2.05(.11)	.61(.04)	.82(.04)	1.92(.10)
36	.85(.02)	1.72(.11)	1.46(.11)	1.95(.09)	1.48(.09)	1.33(.07)	1.75(.09)	5.37(.27)	.64(.08)	1.91(.08)	.52(.04)	.76(.03)	1.82(.11)
37	.86(.03)	1.69(.15)	1.46(.13)	1.99(.15)	1.49(.11)	1.40(.10)	1.78(.08)	5.49(.36)	.66(.04)	1.88(.11)	.50(.04)	.76(.05)	1.80(.13)
38	.87(.02)	1.78(.09)	1.50(.07)	2.05(.09)	1.52(.06)	1.43(.06)	1.85(.07)	5.66(.22)	.62(.06)	1.94(.07)	.56(.04)	.79(.04)	1.88(.06)
39	.88(.02)	1.75(.10)	1.44(.08)	2.01(.11)	1.56(.08)	1.41(.05)	1.77(.09)	5.53(.25)	.59(.04)	1.86(.03)	.54(.03)	.76(.04)	1.82(.04)
40	.85(.03)	1.77(.06)	1.47(.06)	2.01(.07)	1.52(.09)	1.43(.07)	1.88(.09)	5.99(.26)	.69(.03)	1.98(.08)	.56(.03)	.75(.04)	1.91(.09)
41	.84(.03)	1.78(.14)	1.43(.11)	2.02(.20)	1.52(.16)	1.38(.12)	1.90(.14)	5.99(.40)	.66(.08)	2.02(.13)	.57(.06)	.76(.07)	1.95(.16)
42	.86(.03)	1.82(.09)	1.45(.07)	2.02(.11)	1.54(.12)	1.46(.07)	1.96(.08)	6.06(.26)	.63(.04)	2.01(.12)	.58(.05)	.79(.04)	1.94(.11)
43	.81(.03)	1.76(.12)	1.48(.07)	2.03(.10)	1.56(.10)	1.40(.06)	1.87(.10)	6.02(.26)	.59(.03)	1.92(.07)	.57(.04)	.76(.04)	1.89(.09)
44	.85(.01)	1.83(.05)	1.49(.05)	2.09(.06)	1.57(.08)	1.47(.01)	1.93(.05)	6.11(.09)	.67(.03)	1.98(.04)	.57(.02)	.77(.01)	1.94(.04)
45	.84(.02)	1.81(.09)	1.46(.10)	2.01(.10)	1.55(.13)	1.44(.04)	1.90(.06)	6.04(.27)	.67(.03)	1.98(.08)	.57(.03)	.77(.03)	1.94(.10)
46	.72(.03)	1.52(.06)	1.37(.08)	1.92(.09)	1.56(.06)	1.31(.09)	1.61(.08)	4.98(.25)	.46(.03)	1.51(.08)	.49(.03)	.62(.05)	1.68(.08)
47	.83(.02)	1.81(.08)	1.54(.08)	2.24(.11)	1.84(.09)	1.51(.07)	1.87(.10)	5.74(.31)	.58(.03)	1.85(.10)	.56(.03)	.75(.04)	1.98(.11)
48	.82(.04)	1.82(.11)	1.55(.08)	2.23(.15)	1.79(.16)	1.47(.11)	1.88(.10)	5.81(.31)	.58(.04)	1.86(.09)	.55(.04)	.76(.05)	1.98(.13)
49	.82(.06)	1.54(.18)	1.39(.17)	2.03(.24)	1.58(.24)	1.32(.14)	1.72(.18)	5.32(.58)	.50(.06)	1.60(.18)	.48(.05)	.67(.11)	1.71(.12)
50	.90(.03)	1.86(.12)	1.64(.14)	2.34(.19)	1.88(.13)	1.56(.09)	1.95(.14)	6.28(.34)	.58(.03)	1.90(.14)	.57(.05)	.83(.07)	1.99(.15)
52	.88(.05)	1.83(.14)	1.62(.12)	2.32(.15)	1.87(.16)	1.57(.13)	1.93(.13)	6.14(.41)	.57(.05)	1.87(.12)	.56(.06)	.81(.06)	1.97(.14)
53	.84(.03)	1.77(.11)	1.55(.08)	2.22(.13)	1.83(.12)	1.48(.11)	1.90(.12)	5.99(.29)	.54(.04)	1.83(.09)	.54(.04)	.79(.05)	1.92(.10)
54	.87(.04)	1.87(.11)	1.68(.13)	2.40(.14)	1.94(.10)	1.58(.11)	2.00(.11)	6.32(.42)	.58(.05)	1.91(.12)	.60(.06)	.82(.05)	2.03(.11)
55	.85(.04)	1.78(.12)	1.58(.14)	2.26(.16)	1.78(.14)	1.51(.11)	1.94(.11)	6.03(.45)	.56(.07)	1.86(.10)	.53(.05)	.78(.04)	1.96(.12)
56	.87(.06)	1.84(.03)	1.62(.11)	2.35(.06)	1.92(.05)	1.53(.06)	1.95(.05)	5.93(.27)	.58(.03)	1.87(.08)	.58(.04)	.81(.02)	1.97(.04)
57	.86(.04)	1.70(.11)	1.53(.09)	2.22(.15)	1.79(.10)	1.48(.07)	1.90(.09)	5.98(.33)	.51(.04)	1.85(.07)	.54(.02)	.80(.05)	1.92(.10)

APPENDIX 7

Allelic frequencies for exemplar data set. Alleles labeled by relative mobility to most common allele (1.00). n = number of individuals scored.

Locus	Allele	EXTM	EXTN	TEXA	ELON	DECg	DECr
Est-d	1.10	0.	0.	0.	0.	.08	.25
	1.00	0.	0.	0.	0.	.59	.50
	.87	0.	.79	.20	.42	.33	.25
	.78	0.	.14	.60	.58	0.	0.
	.64	.07	.07	.20	0.	0.	0.
	.57	.93	0.	0.	0.	0.	0.
	n	7	7	5	6	6	6
Phpr	1.64	0.	0.	.10	0.	0.	0.
	1.59	0.	0.	.10	0.	0.	0.
	1.41	0.	0.	0.	.12	.17	0.
	1.34	0.	0.	0.	0.	.33	0.
	1.23	.07	.07	.10	.12	.50	.33
	1.09	.22	.07	.10	.25	0.	.33
	1.05	.07	.07	.30	0.	0.	.17
	1.00	.43	.65	0.	.50	0.	.17
	.73	0.	0.	.10	0.	0.	0.
	.68	0.	.14	.20	0.	0.	0.
	.64	.21	0.	0.	0.	0.	0.
	n	7	7	5	4	6	6
Acph	2.05	0.	0.	.75	0.	0.	0.
	1.72	0.	0.	.25	0.	.08	0.
	1.00	0.	.08	0.	1.0	.92	.92
	.50	1.0	.92	0.	0.	0.	.08
	n	7	6	4	4	6	6
Pgi	1.37	0.	.07	0.	0.	0.	0.
	1.00	1.0	.86	1.0	.75	.83	.92
	.58	0.	.07	0.	.12	.17	.08
	.47	0.	0.	0.	.12	0.	0.
	n	6	7	4	4	6	6
Pgm-1	1.10	0.	0.	0.	0.	0.	.08
	1.05	0.	0.	0.	.75	0.	.08
	1.00	.60	.22	0.	.25	.40	.84
	.94	.40	.71	0.	0.	.30	0.
	.89	0.	.07	0.	0.	.30	0.
	.78	0.	0.	.90	0.	0.	0.
	.74	0.	0.	.10	0.	0.	0.
	n	5	7	5	2	5	6
Pgm-2	1.45	.07	.14	0.	0.	0.	0.
	1.26	.50	.86	.50	.40	0.	.25
	1.00	.43	0.	.30	.40	.43	.42
	.88	0.	0.	.20	.10	0.	.08
	.78	0.	0.	0.	.10	.57	.08
	.66	0.	0.	0.	0.	0.	.17
	n	7	7	5	5	6	6
Mpi	1.09	0.	0.	.25	0.	0.	.17
	1.04	.08	.58	0.	0.	0.	.08
	1.00	0.	0.	0.	.50	.60	.25
	.94	.50	.17	.75	0.	.30	.42
	.90	.42	.08	0.	.50	.10	0.
	.82	0.	.17	0.	0.	0.	0.
	.71	0.	0.	0.	0.	0.	.08
	n	6	6	4	2	5	6
6-pgd	1.17	.30	.07	0.	0.	0.	0.
	1.11	.20	.07	0.	0.	0.	0.
	1.00	0.	0.	.90	.17	.10	0.
	.83	.40	.15	.10	.08	.40	.25
	.75	.10	.64	0.	0.	.10	.67
	.69	0.	.07	0.	.08	.10	0.
	.61	0.	0.	0.	.67	.10	.08
	.48	0.	0.	0.	0.	.20	0.
	n	5	7	5	6	5	6
Sod	1.30	0.	0.	0.	.08	.08	.08
	1.00	.93	1.0	1.0	.92	.92	.92
	.63	.07	0.	0.	0.	0.	0.
	n	7	7	5	6	6	6
Acon-1	1.10	0.	.14	0.	0.	0.	0.
	1.00	0.	.86	0.	1.0	.67	.75
	.90	1.0	0.	1.0	0.	.25	.25
	.81	0.	0.	0.	0.	.08	0.
	n	7	7	5	4	6	6
Acon-2	1.00	.90	1.0	1.0	1.0	1.0	1.0
	.84	.10	0.	0.	0.	0.	0.
	n	5	7	5	2	5	6
Idh	1.00	1.0	.93	1.0	1.0	1.0	1.0
	.34	0.	.07	0.	0.	0.	0.
	n	7	7	5	5	6	6

REFERENCES CITED

Allen, R. T.

1972 A revision of the genus *Loxandrus* LeConte (Coleoptera: Carabidae) in North America. *Entomol. Amer.* 46(1):1-184.

Allen, R. T., and G. E. Ball.

1980 Synopsis of Mexican taxa of the *Loxandrus* series (Coleoptera: Carabidae: Pterostichini). *Trans. Amer. Entomol. Soc.* 105:481-576.

American Society of Ichthyologists and Herpetologists.

1963 et seq. *Catalogue of American Amphibians and Reptiles.*

Andrewes, H. E.

1919 Note on Bonelli's "Tableau Synoptique." *Trans. Entomol. Soc. London* 1919:89-92.

Audouin, J. V., and A. Brullé.

1834 *Histoire Naturelle des Insectes, traitant de leur Organization et de leurs moeurs en général*, et comprenant leur classification et la description des espèces. 4. Coléoptères. Paris: I. F. D. Pillot.

Avise, J. C., and R. K. Selander.

1972 Evolutionary genetics of cave-dwelling fishes of the genus *Astyanax*. *Evolution* 26:1-19.

Ayala, F. J.

1975 Genetic differentiation during the speciation process. *Evol. Biol.* 8:1-78.

1982 Genetic variation in natural populations: problem of electrophoretically cryptic alleles. *Proc. Nat'l. Acad. Sci.* 79:550-54.

Ayala, F. J., J. R. Powell, M. L. Tracey, C. A. Mourao, and S. Perez-Salas.

1972 Enzyme variability in the *Drosophila willistoni* group. IV. Genetic variation in natural populations of *Drosophila willistoni*. *Genetics* 70:113-59.

Barr, T. C., Jr.

1974 Revision of *Rhadine* LeConte (Coleoptera, Carabidae). I. The *subterranea* group. *Amer. Mus. Novit.* no. 2539.

Bates, H. W.

1882 *Biologia Centrali-Americana Insecta Coleoptera Carabidae* 1(1):41-152.

1884 *Biologia Centrali-Americana Insecta Coleoptera Carabidae* 1(4):256-99.

1891 Additions to the carabideous fauna of Mexico, with remarks on some of the species previously recorded. *Trans. Roy. Ent. Soc. London.* 1891:223-78.

Becker, E. C.
1956 Revision of the Neartic Species of *Agriotes* (Coleoptera: Elateridae) *Can. Entomol.* 88(suppl. 1): 101.

Berlocher, S. H., and G. L. Bush.
1982 An electrophoretic analysis of *Rhagoletis* (Diptera: Tephritidae) phylogeny. *Syst. Zool.* 31:136-55.

Beverley, S. M., and A. C. Wilson.
1982 Molecular evolution in *Drosophila* and higher Diptera. I. Micro-complement fixation studies of a larval hemolymph protein. *J. Mol. Evol.* 18:251-64.
1984 Molecular evolution in *Drosophila* and the higher Diptera. II. A time scale for fly evolution. *J. Mol. Evol.* 21:1-13.

Blackwelder, R. E.
1944 *Checklist of the coleopterous insects of Mexico, Central America, the West Indies, and South America*, pt. 1. Smithsonian Institution, U. S. National Museum Bulletin 185. Washington, D.C.

Blatchley, W. S.
1910 The Coleoptera of Indiana. *Bull. Ind. Dept. Geol. and Nat. Res.* 1:1-1385.

Boer, P. J. Den.
1970 On the significance of dispersal power for populations of carabid-beetles. *Oecologia* (Berlin) 4:1-28.

Boer, P. J. Den, T. H. P. Van Huizen, W. Den Boer-Daanje, and B. Aukema.
1980 Wing polymorphism and dimorphism in ground beetles as stages in an evolutionary process. *Entomol. Generalis* 6:107-34.

Bonelli, F. A.
1810 Observationes entomologiques, Premiere Partie, including *Tabula Synoptica exhiben genera Carabicorum in sectiones et stirpes disposita. Mém. Acad. imp. Sci. Litt. et Beaux-Arts de Turin*, 18 (*Mém. pres.* vol. 4):21-78 plus table.

Brown, W. L., Jr.
1957 Centrifugal speciation. *Quart. Rev. Biol.* 32:247-77.

Bush, G. L.
1966 The taxonomy, cytology and evolution of the genus *Rhagoletis* in North America (Diptera: Tephritidae). *Bull. Mus. Comp. Zool.* 134(11):431-562.

Cain, A. J., and G. A. Harrison.
1958 An analysis of the taxonomist's judgement of affinity. *Proc. Zool. Soc. London* 131:85-98.

Campbell, N. A., and W. R. Atchley.
1981 The geometry of canonical variate analysis. *Syst. Zool.* 30:268-80.

Carson, H. L.
1968 The population flush and its genetic consequences. In *Population Biology and Evolution*, ed. R. C. Lewontin, 123-37. Syracuse: Syracuse Univ. Press.

Carter, A.
1976 Wing polymorphism in the insect species *Agonum retractum* LeConte. *Can. J. Zool.* 54:1375-82.

Casey, T. L.
1920 *Memoirs on the Coleoptera*, vol. 9. Lancaster, Pa.: New Era Printing Co.
1924 *Memoirs on the Coleoptera* vol. 11. Lancaster, Pa.: New Era Printing Co.

Chaudoir, M. de.
1843 Carabiques nouveaux. *Bull. Soc. Imp. Nat. Moscou* 16:671-791.
Cohn, T. J.
1965 *The arid-land katydids of the North American genus Neobarretia (Orthoptera: Tettigoniidae): their systematics and a reconstruction of their history. Misc. Publ. Mus. Zool. Univ. Mich.* no. 126.
Cronin, T. M., B. J. Szabo, T. A. Ager, J. E. Hazel, and J. P. Owens.
1981 Quaternary climates and sea levels of the U. S. Atlantic coastal plain. *Science* 211:233-40.
Csiki, E.
1931 Carabidae: Harpalinae 5. *Coleopterorum Catalogus* 115:739-1022. W. Junk, Berlin.
Darlington, P. J., Jr.
1934 New West Indian Carabidae, with a list of Cuban species. *Psyche* 41:66-130.
1936 Variation and atrophy of flying wings of some carabid beetles. *Ann. Entomol. Soc. Amer.* 24:136-79.
1943 Carabidae of mountains and islands: data on the evolution of isolated faunas, and on atrophy of wings. *Ecol. Monog.* 13:37-61.
1953 West Indian Carabidae (Coleoptera): the Bahama species. *Amer. Mus. Novit.* no. 1650.
Dejean, P. F. M. A.
1828 *Spécies général des Coléoptères* 3:1-520.
Dover, G.
1982 Molecular drive: a cohesive mode of species evolution. *Nature* 299:111-17.
Doyen, J. T., and C. N. Slobodchikoff.
1974 An operational approach to species classification. *Syst. Zool.* 23:239-47.
Doyen, J. T., and W. R. Tschinkel.
1982 Phenetic and cladistic relationships among tenebrionid beetles. *Syst. Entomol.* 7:127-83.
Ehrlich, P. R., and P. H. Raven.
1969 Differentiation of populations. *Science* 165:1228-32.
Ehrman, L.
1965 Direct observation of sexual isolation between allopatric and between sympatric strains of the different *Drosophila paulistorum* races. *Evolution* 19:459-64.
Erwin, T. L.
1970 A reclassification of bombardier beetles and a taxonomic revision of the North and Middle American species (Carabidae: Brachinida). *Quaest. Entomol.* 6:4-215.
1979a The American connection, past and present, as a model blending dispersal and vicariance in the study of biogeography. In *Carabid Beetles: Their Evolution, Natural History, and Classification,* ed. T. L. Erwin, G. E. Ball, D. R. Whitehead, and A. L. Halpern, 355-67. The Hague: W. Junk Publ.
1979b Thoughts on the evolutionary history of ground beetles: hypotheses generated from comparative faunal analyses of lowland forest sites in temperate and tropical regions. In *Carabid Beetles*, 539-92. *See* Erwin 1979a.

Erwin, T. L., and D. H. Kavanaugh.
1981 Systematics and zoogeography of *Bembidion* Latreille. I. the *carlhi* and *erasum* groups of western North America (Coleoptera: Carabidae: Bembidiini). *Entomol. Scand. Suppl.* 15:33-72.
Estabrook, G. F., J. G. Strauch, Jr., and K. L. Fiala.
1977 An application of compatibility analysis to the Blackiths' data on orthopteroid insects. *Syst. Zool.* 26:269-76.
Farris, J. S.
1970 Methods for computing Wagner trees. *Syst. Zool.* 19:83-92.
1981 Distance data in phylogenetic analysis. In *Advances in Cladistics*, ed. V. A. Funk and D. R. Brooks, 3-23. New York: New York Botanical Garden.
Farris, J. S., A. G. Kluge, and M. J. Eckhardt.
1970 A numerical approach to phylogenetic systematics. *Syst. Zool.* 19:172-91.
Findley, J. S.
1969 Biogeography of southwestern boreal and desert mammals. In *Contributions in Mammalogy*, 113-28. *Univ. of Kansas Mus. Nat. Hist. Misc. Pub.* no. 51. Lawrence.
Forsyth, D. J.
1972 The structure of the pygidial defense glands of Carabidae (Coleoptera). *Trans. Zool. Soc. London* 32:249-309.
Freitag, R.
1979 Reclassification, phylogeny, and zoogeography of the Australian species of *Cicindela. Aust. J. Zool. Suppl. Ser.* no. 66:1-99.
Giuseffi, S., T. C. Kane, and W. F. Duggleby.
1978 Genetic variability in the Kentucky cave beetle *Neaphaenops tellkampfii* (Coleoptera: Carabidae). *Evolution* 32:679-81.
Goldschmidt, R.
1940 *The Material Basis of Evolution.* New Haven: Yale University Press.
Gorman, G. C., and J. Renzi, Jr.
1979 Genetic distance and heterozygosity estimates in electrophoretic studies: effects of sample size. *Copeia* 1979:242-49.
Grant, P. R.
1972 Convergent and divergent character displacement. *Biol. J. Linn. Soc.* 4:39-68.
Gray, B., and M. H. Hatch.
1941 The Coleoptera of Washington, Carabidae: Agonini. *Univ. Wash. Publ. Biol.* 10:7-27.
Habu, A.
1973 Notes on the generic name *Agonum* (Coleoptera: Carabidae). *Entomol. Rev. Japan* 25:65-70.
1978 Carabidae: Platynini (Insecta: Coleoptera). In *Fauna Japonica.* Tokyo: Keigaku Publ. Co.
Hall, E. R.
1981 *The Mammals of North America.* 2 vols. New York: John Wiley and Sons.
Harris, H., and D. A. Hopkinson.
1976 *Handbook of Enzyme Electrophoresis in Human Genetics.* Amsterdam, North-Holland Publ. Co.

Harris, T. W.
1828 Contributions to entomology IV. *New England Farmer* 7(17):132.

Hatch, M. H.
1926 *Anchomenus decorus* ab. syracusensis nov. J. New York Entomol. Soc. 34:247-48.

Hausen, J. T.
1890 A new Canadian *Platynus*. *Can. Rec. Sci.* 4:235-36.
1891 Notes entomologiques. *Natur. Can.* 20:155-60.

Hennig, W.
1966 *Phylogenetic Systematics*. Urbana: University of Illinois Press.

Hinesley, L. L.
1979 Systematics and distribution of two chromosome forms in the southern grasshopper mouse, genus *Onychomys*. *J. Mamm.* 60:117-28.

Iuga, V. G.
1966 Morphologie du sommet abdominal des Caraboïdes, comparé à celui des Coléoptères avec ovipositeur. *Trav. Mus. Hist. Nat. "Grigore Antipa"* (Bucharest). 6:171-226.

Jeannel, R.
1942 Coléoptères Carabiques II. *Faune de France* 40: 573-1173.

Johnson, C. G.
1969 *Migration and Dispersal of Insects by Flight*. London: Methuen & Co.

Johnson, G. B.
1974 Enzyme polymorphism and metabolism. *Science* 184:28-37.

Kanehisa, K., and M. Murase.
1977 Comparative study of the pygidial defensive systems of carabid beetles. *Appl. Entomol. Zool.* 12:225-35.

Kim, J. O.
1975 Factor analysis. In *Statistical Package for the Social Sciences, Second Edition*. ed. N. H. Nie, C. H. Hull, J. G. Jenkins, K. Steinbrenner, and D. H. Bent, 468-514. McGraw-Hill Book Co.

Klecka, W. R.
1975 Discriminant analysis. In *Statistical Package for the Social Sciences, Second Edition*, 434-67. *See* Kim 1975.
1980 Discriminant Analysis. Beverly Hills: Sage Publications.

Kryzhanovsky, O. L.
1976 Revised classification of the family Carabidae. *Entomol. Rev.* 1:80-91.

Laing, C., G. R. Carmody, and S. B. Peck.
1976 Population genetics and evolutionary biology of *Ptomophagus hirtus*. *Evolution* 30:484-99.

Lakovaara, S., A. Saura, and C.T. Falk.
1972 Genetic distance and evolutionary relationships in the *Drosophila obscura* group. *Evolution* 26:177-84.

Lande, R.
1980 Genetic variation and phenotypic evolution during allopatric speciation. Amer. Natur. 116:463-79.

LeConte, J. L.
1848 A descriptive catalogue of the geodephagous Coleoptera inhabiting the United States east of the Rocky Mountains. *Ann. Lyc. Nat. Hist. New York.* 4:173-374.
1854 Synopsis of the species of *Platynus* and allied genera, inhabiting the United States. *Proc. Acad. Nat. Sci. Phila.* 7:35-59.
1861 Classification of the Coleoptera of North America. Pt I. *Smithson. Misc. Coll.* 3:1-286.
1878 Additional descriptions of new species. *Proc. Amer. Philos. Soc. Phila.* 17:373-472.
LeConte, J. L., and G. H. Horn.
1883 Classification of the Coleoptera of North America. *Smithson. Misc. Coll.* 507.
Levene, H.
1949 On a matching problem arising in genetics. *Ann. Math. Stat.* 20:91-94.
Lewontin, R. C., and J. L. Hubby.
1966 A molecular approach to the study of genic heterozygosity. II. Amount of variation and degree of heterozygosity in natural populations of *Drosophila pseudoobscura. Genetics* 54:595-609.
Liebherr, J. K.
1982 Lectotype designations for *Agonum cyanope* (Bates) and *Agonum megillum* (Bates). *Coleopt. Bull.* 36:153-54.
1983a The genetic basis for polymorphism in the ground beetle, *Agonum decorum* (Say) (Coleoptera: Carabidae). *Ann. Entomol. Soc. Amer.* 76:349-58.
1983b Larval description of *Calybe* (*Ega*) *sallei* (Chevrolat) with a preliminary assessment of lachnophorine affinities (Carabidae: Lachnophorini). *Coleopt. Bull.* 37:254-60.
1984 *Agonum muiri* n. sp. from California, and *Agonum pacificum* Casey valid species. (Coleoptera: Carabidae: Platynini). *Coleopt. Bull.* 38:374-83.
Liebke, M.
1937 Denkschrift über die Carabiden-tribus Colliurini. *Fest. zum 60. Geburstage von Prof. Dr. Embrik Strand* 4:37-141.
Lindroth, C. H.
1946 Inheritance of wing dimorphism in *Pterostichus anthracinus* I11. *Hereditas* 32:37-40.
1949 Die Fennoskandischen Carabidae. Eine tiergeographische Studie III. *Allgemeiner Teil. Götesborgs Kungl. Veterskaps-och Vitterhets-Samhälles Handlingar*, Ser. B 4(3):1-911.
1955 Dejean's types of North American Carabidae. *Opusc. Entomol.* 20:10-34.
1956 A revision of the genus *Synuchus* Gyllenhal in the widest sense, with notes on *Pristosia* Motschulsky (*Eucalathus* Bates) and *Calathus* Bonelli. *Trans. Roy. Entomol. Soc. London* 108:485-574.
1961 The ground beetles of Canada and Alaska. Part 2. *Opusc. Entomol. Suppl.* 20:1-200.
1966 The ground beetles of Canada and Alaska. Part 4. *Opusc. Entomol. Suppl.* 29:409-648.
1974a Coleoptera: Carabidae. In *Handbooks for the Identification of British Insects* 4(2):1-148.
1974b On the elytral microsculpture of carabid beetles. *Entomol. Scand.* 5:251-64.

Lindroth, C. H., and R. Freitag.
1969 North American ground beetles described by Thomas Say: designation of lectotypes and neotypes. *Psyche* 76:326-61.

Little, E. L., Jr.
1976 *Atlas of United States Trees. Vol. 3, Minor Western Hardwoods.* U.S.D.A. Misc. Publ. no. 1314.

Madge, R. G.
1978 Comment on the proposed addition of Bonelli's Tabula Synoptica to the official list, Z.N.(S.)2135. *Bull. Zool. Nom.* 35:9-12.

Mannerheim, C. G.
1843 Beitrage zur Käfer-Fauna der Aleutischen Inseln, der Insel Sithka und Neu-Californiens. *Bull. Soc. Imp. Nat. Moscou* 16:175-314.

Maslin, T. P.
1952 Morphological criteria of phyletic relationships. *Syst. Zool.* 1:49-70.

Maxson, L. R., and R. D. Maxson.
1979 Comparative albumin and biochemical evolution in plethodontid salamanders. *Evolution* 33:1057-62.

Mayr, E.
1940 Speciation phenomena in birds. *Amer. Natur.* 74:249-78.
1963 *Animal Species and Evolution.* Cambridge: Harvard University Press, Belknap Press.
1969 *Principles of Systematic Zoology.* New York: McGraw-Hill Book Co.

Melville, R. V.
1982 Opinion 1226, Bonelli's *Tabula Synoptica* ruled to be an available work and to have been published in 1810. *Bull Zool. Nom.* 39:192-93.

Michener, C. D.
1977 Discordant evolution and the classification of allodapine bees. *Syst. Zool.* 26:32-56.

Mickevich, M. F., and M. S. Johnson.
1976 Congruence between morphological and allozyme data in evolutionary inference and character evolution. *Syst. Zool.* 25:260-70.

Mickevich, M. F., and C. Mitter.
1981 Treating polymorphic characters in systematics: a phylogenetic treatment of electrophoretic data. In *Advances in Cladistics.*, 45-58. *See* Farris 1981.

Moore, B. P.
1979 Chemical defense in carabids and its bearing on phylogeny. In *Carabid Beetles,* 193-203. *See* Erwin 1979a.

Moore, B. P., and B. E. Wallbank.
1968 Chemical composition of the defensive secretion in carabid beetles and its importance as a taxonomic character. *Proc. Roy. Entomol. Soc. London*(B) 37:62-72.

Morafka, D. J.
1977 *A Biogeographic Analysis of the Chihuahuan Desert through its Herpetofauna.* The Hague: W. Junk Publ.

Motschulsky, V. de.
1859 Coléoptères nouveaux de la Californie. *Bull. Soc. Imp. Nat. Moscou* 32:122-268.

Mroczkowski, M.
1977 Bonelli, F. A., "Tabula Synoptica 1811": proposed addition to the official list. *Bull. Zool. Nom.* 34:61-62.

Nei, M.
1971 Interspecific gene differences and evolutionary time estimated from electrophoretic data on protein identity. *Amer. Natur.* 105:385-98.
1978 Estimation of average heterozygosity and genetic distance from a small number of individuals. *Genetics* 89:583-90.

Nelson, G. H.
1979 Cladistic analysis and synthesis: principles and definitions, with a historical note on Adanson's *Familles des Plantes* (1763-1764). *Syst. Zool.* 28:1-39.

Nevo, E.
1978 Genetic variation in natural populations: patterns and theory. *Theor. Pop. Biol.* 13:121-77.

Palmén, E.
1944 Die anemohydrochore Ausbreitung der Insekten als zoogeographischer Faktor. *Ann. Soc. Zool. Bot. Fenn. "Vanamo"* 10:1-262.

Patton, J. L., S. W. Sherwood, and S. Y. Yang.
1981 Biochemical systematics of chaetopodine pocket mice, genus *Perognathus*. *J. Mamm.* 62:477-92.

Patton, J. L., and M. F. Smith.
1981 Molecular evolution in *Thomomys*: phyletic systematics, paraphyly, and rates of evolution. *J. Mamm.* 62:493-500.

Patton, J. L., and S. Y. Yang.
1977 Genetic variation in *Thomomys bottae* pocket gophers: macrogeographic patterns. *Evolution* 31:697-720.

Platnick, N. I.
1981 Widespread taxa and biogeographic congruence. In *Advances in Cladistics*, 223-27. *See* Farris 1981.

Powell, J. A.
1973 *A systematic monograph of New World ethmiid moths (Lepidoptera: Gelechioidea).* Smithsonian Contributions to Zoology no. 120.

Powell, J. R.
1975 Protein variation in natural populations of animals. *Evol. Biol.* 8:79-119.

Pregill, G. K., and S. L. Olson.
1981 Zoogeography of West Indian invertebrates in relation to Pleistocene climatic cycles. *Ann. Rev. Ecol. Syst.* 12:75-98.

Raven, P. H., B. Berlin, and D. E. Breedlove.
1971 The origins of taxonomy. *Science* 174:1210-13.

Reichardt, H.
1977 A synopsis of the genera of Neotropical Carabidae. *Quaest. Entomol.* 13:346-493.

Rogers, J. S.
1972 Measures of genetic similarity and genetic distance. *Studies in Genetics* 7:145-53.

Sarich, V. M.
1977 Rates, sample sizes, and the neutrality hypothesis for electrophoresis in evolutionary studies. *Nature* 265:24-28.

Say, T.
1823 Descriptions of insects of the families of Carabici and Hydrocanthari of Latreille, inhabiting North America. *Trans. Amer. Philos. Soc.*, n.s. 2, 1-109. Philadelphia.
1834 Descriptions of new North American insects, and observations on some already described. *Trans. Amer. Philos. Soc.*, n.s. 4, 409-70. Philadelphia.

Schildknecht, H., U. Maschwitz, and H. Winkler.
1968 Zur Evolution der Carabiden-Wehrdrüsensekrete. *Naturwissenshaften* 55:112-17.

Schuler, L.
1960 Les spermathèques dans la tribu des Bembidiini Jeannel (Coléoptères: Trechidae). *Rev. Franc. Entomol.* 27:24-47.
1962 La spermathèque des *Chlaenius* de France (Callistimorphi Jeannel). *L'Entomologiste* 18:77-86.
1963 Les organes genitaux femelles chez les Pterostichidae de France. Les tribus Anchomenini et Sphodrini (suite). Le cas des Patrobidae (Coléoptères: Carabiques). *Bull. Soc. Entomol. France* 68:13-26.

Selander, R. K., M. H. Smith, S. Y. Yang, W. E. Johnson, and J. B. Gentry.
1971 Biochemical polymorphism and systematics in the genus *Peromyscus*. I. Variation in the old-field mouse (*Peromyscus polionotus*). *Studies in Genetics* 6 (Univ. of Texas Publ. no. 7103): 49-90.

Selvin, S.
1980 *Multivariate Statistics in the Health Sciences.* Berkeley: School of Biological and Environmental Health Sciences, Univ. California, Berkeley.

Sharp, D., and F. Muir.
1912 The comparative anatomy of the male genital tube in Coleoptera. *Trans. Entomol. Soc. London* 1912:477-642.

Simpson, G. G.
1961 *Principles of Animal Taxonomy.* New York: Columbia University Press.

Smith, D. S.
1964 The structure and development of flightless Coleoptera: a light and electron microscopic study of the wings, thoracic exoskeleton and rudimentary flight musculature. *J. Morph.* 114:107-84.

Smithe, F. B.
1974 *Naturalist's Color Guide and Supplement.* New York: American Museum of Natural History.

Sneath, P. H. A., and R. R. Sokal.
1973 *Numerical Taxonomy.* San Francisco: W. H. Freeman and Co.

Soltz, D. L., and R. J. Naiman.
1978 The natural history of native fishes in the Death Valley system. *Los Angeles Co. Mus. Sci. Ser.* 30:1-76.

Stebbins, R. C.
1966 *A Field Guide to Western Reptiles and Amphibians.* Boston: Houghton Mifflin Co.

Suomalainen, E., and A. Saura.
1973 Genetic polymorphism and evolution in parthenogenetic animals. I.: polyploid Curculionidae. *Genetics* 74:489-508.

Templeton, A. R.
1980 The theory of speciation via the founder principle. *Genetics* 94:1011-38.

Throckmorton, L. H.
1975 The phylogeny, ecology, and geography of *Drosophila*. In *Handbook of Genetics, vol. 3*, ed. R. C. King, 421-69. New York: Plenum Publ. Corp.

Turanchik, E. J., and T. C. Kane.
1979 Ecological genetics of the cave beetle *Neaphaenops tellkampfii*. *Oecologia* (Berlin) 44:63-67.

Turner, B. J.
1974 Genetic divergence of Death Valley pupfish species: biochemical versus morphological evidence. *Evolution* 28:281-94.

Van Huizen, T. H. P.
1979 Individual and environmental factors determining flight in carabid beetles. In *On the Evolution of Behaviour in Carabid Beetles*. ed. P. J. Den Boer, H. V. Thiele, and F. Weber, 199-211. Wageningen: Agric. Univ. Wageningen Misc. Paper No. 18.

Wagner, W. H., Jr.
1980 Origin and philosophy of the groundplan-divergence method of cladistics. *Syst. Bot.* 5:173-93.

Ward, P. S.
1980 Genetic variation and population differentiation in the *Rhytidoponera impressa* group, a species complex of ponerine ants. *Evolution* 34:1060-76.

Ward, R. D.
1979 Metathoracic wing structures as phylogenetic indicators in the Adephaga. In *Carabid Beetles*, 181-91. *See* Erwin 1979a.

Watrous, L. E., and Q. D. Wheeler.
1981 The out-group comparison method of character analysis. *Syst. Zool.* 30:1-11.

Watts, W. A.
1980 The late Quaternary vegetation history of the southeastern United States. *Ann. Rev. Ecol. Syst.* 11:387-408.

Whitehead, D. R.
1972 Classification, phylogeny, and zoogeography of *Schizogenius* Putzeys. (Coleoptera: Carabidae: Scaritini). *Quaest. Entomol.* 8:131-348.

Wilson, A. C., S. S. Carlson, and T. J. White.
1977 Biochemical evolution. *Ann. Rev. Biochem.* 46:573-639.

Wilson, E. O.
1961 The nature of the taxon cycle in the Melanesian ant fauna. *Amer. Natur.* 95:169-93.

Workman, P. L., and J. D. Niswander.
1970 Population studies on southwestern indian tribes. II. Local genetic differentiation in the Papago. *Amer. J. Human. Genet.* 22:24-49.

Wright, S.
1965 The interpretation of population structure by F-statistics with special regard to systems of mating. *Evolution* 19:395-420.
1969 *Evolution and the Genetics of Populations, vol. 2. The Theory of Gene Frequencies*. Chicago: University of Chicago Press.
1978 *Evolution and the Genetics of Populations, vol. 4. Variability within and among Natural Populations.* Chicago: University of Chicago Press.

Zweifel, R. G.
1962 Analysis of hybridization between two subspecies of the desert whiptail lizard, *Cnemidophorus tigris*. *Copeia* 1962:749-66.